Estimating Groundwater Recharge

Understanding groundwater recharge is essential for the successful management of water resources and modeling fluid and contaminant transport within the subsurface. This book provides a critical evaluation of the theory and assumptions that underlie methods for estimating rates of groundwater recharge. Detailed explanations of the methods are provided – allowing readers to apply many of the techniques themselves without needing to consult additional references. Numerous practical examples highlight the benefits and limitations of each method and provide guidance on the selection and application of methods under both ideal and less-than-ideal conditions. More than 800 references allow advanced practitioners to pursue additional information on any method.

For the first time, theoretical and practical considerations for selecting and applying methods for estimating groundwater recharge are covered in a single volume with uniform presentation. Hydrogeologists, water-resource specialists, civil and agricultural engineers, earth and environmental scientists, and agronomists will benefit from this informative and practical book, which is also a useful adjunct text for advanced courses in groundwater or hydrogeology.

For more than 30 years, Rick Healy has been conducting research for the US Geological Survey on groundwater recharge, water budgets of natural and human-impacted hydrologic systems, and fluid and contaminant transport through soils. He has taught numerous short courses on unsaturated zone flow and transport, and groundwater flow modeling. He first presented a short course on methods for estimating recharge in 1994, and over the intervening 15 years the course has been presented to several hundred professionals and students. The material in that course has been expanded and refined over the years and forms the basis of *Estimating Groundwater Recharge*. Rick has authored more than 60 scientific publications and developed the VS2DI suite of models for simulating water, solute, and heat transport through variably saturated porous media. He is a member of the Soil Science Society of America, the American Geophysical Union, and the Geological Society of America.

Estimating Groundwater Recharge

Richard W. Healy
US Geological Survey
Lakewood, Colorado

With contributions by
Bridget R. Scanlon
Bureau of Economic Geology
Jackson School of Geosciences
University of Texas, Austin

CAMBRIDGE
UNIVERSITY PRESS

CAMBRIDGE
UNIVERSITY PRESS

University Printing House, Cambridge CB2 8BS, United Kingdom

One Liberty Plaza, 20th Floor, New York, NY 10006, USA

477 Williamstown Road, Port Melbourne, VIC 3207, Australia

314-321, 3rd Floor, Plot 3, Splendor Forum, Jasola District Centre, New Delhi-110025, India

79 Anson Road, #06-04/06, Singapore 079906

Cambridge University Press is part of the University of Cambridge.

It furthers the University's mission by disseminating knowledge in the pursuit of education, learning and research at the highest international levels of excellence.

www.cambridge.org
Information on this title: www.cambridge.org/9781108446945

First published 2010
4th printing 2016
First paperback edition 2017

A catalogue record for this publication is available from the British Library

Library of Congress Cataloging in Publication data
Healy, R. W.
   Estimating groundwater recharge / Richard W. Healy ; with contributions
   by Bridget R. Scanlon.
     p. ; cm.
   Includes index.
   ISBN 978-0-521-86396-4 (hardback)
   1. Groundwater recharge–Mathematical models.   I. Scanlon, Bridget R.   II. Title.
   GB1197.77.H43 2010
   551.49–dc22   2010027384

ISBN 978-0-521-86396-4 Hardback
ISBN 978-1-108-44694-5 Paperback

Contents

Preface

Groundwater is an integral part of natural hydrologic systems. Humans have used groundwater for thousands of years. Its use has increased greatly over time, but only in the last few decades has our appreciation of the limitations of its supply and its vulnerability to contamination grown to the point where steps are being taken to protect this valuable resource. One of the most important components in any assessment of groundwater supply or aquifer vulnerability is the rate at which water in the system is replenished – the rate of recharge.

A number of textbooks are devoted to hydrogeology, groundwater flow, and contaminant transport (e.g. Freeze and Cherry, 1979; Domenico and Schwartz, 1998; Todd and Mays, 2005). The importance of recharge is cited in all of these textbooks, but only limited information is provided on the description and analysis of techniques for estimating recharge. Similarly, undergraduate and graduate courses on hydrogeology, groundwater flow, and contaminant transport are offered at many universities, but we know of no university level courses specifically devoted to groundwater recharge. This book attempts to fill these gaps by providing a systematic and comprehensive analysis of methods for estimating recharge.

The book is aimed at practicing hydrogeologists who are actively involved in groundwater studies. The material contained in the text should also be useful to water-resource specialists, civil and agricultural engineers, geologists, geochemists, environmental scientists, soil physicists, agriculturalists, irrigators, and scientists from other fields that have an elemental understanding of hydrologic processes. The book can be used as an adjunct text or reference in an advanced undergraduate or graduate groundwater or hydrogeology course; it can also serve as a primary text in courses on groundwater recharge. Theoretical as well as practical considerations for selecting and applying techniques are discussed. Theoretical analysis of the methods allows the evaluation of assumptions inherent in each method. Practical examples of applications provide guidance for readers in applying methods in their own studies.

Over the years, hydrology has become a diverse field with the development of many new topic areas. Few hydrologists can claim expertise in all areas of hydrology; specialization in groundwater, surface water, unsaturated-zone flow and transport, geochemistry, or other subfields has become more the norm. We anticipate that most readers will have a background in groundwater hydrology. However, application of many of the methods described herein (e.g. streamflow hydrograph separation, the zero-flux plane method, and watershed modeling) requires knowledge of areas outside of groundwater hydrology. A challenge in writing this text was to bring together a number of methods that are drawn from fields outside of groundwater hydrology, fields such as surface-water hydrology, flow and transport through the unsaturated zone, geophysics, remote sensing, and water chemistry. Unsaturated-zone processes, in particular, are described in some detail. Many methods for estimating recharge require assumptions about the mechanisms by which water moves through the unsaturated zone; insight into unsaturated-zone processes provides a basis for evaluating the validity of those assumptions.

Acknowledgments

This text was largely derived from lecture notes for short courses on groundwater recharge that the authors have presented over the last 15 years. Many thanks are due to the following individuals who reviewed one or more parts of the book; unless otherwise noted, these individuals are with the US Geological Survey: Kyle Blasch, Jim Bartolino, J. K. Böhlke, Alissa Coes, John Czarnecki, Geoff Delin, Keith Halford, Randy Hanson, Bill Herkelrath, Randy Hunt, Eve Kuniansky, Steve Loheide (University of Wisconsin), Andy Manning, Dennis Risser, Don Rosenberry, Marios Sophocleous (Kansas Geological Survey), Dave Stannard, Katie Walton-Day, and Tom Winter. Special thanks are owed to Stan Leake and Ed Weeks who were kind enough to provide reviews of the entire text. Finally, we would like to express our gratitude to the US Geological Survey and the Bureau of Economic Geology, University of Texas, Austin, for allowing us to invest time in this endeavor.

Groundwater recharge

1.1 | Introduction

Groundwater is a critical source of fresh water throughout the world. Comprehensive statistics on groundwater abstraction and use are not available, but it is estimated that more than 1.5 billion people worldwide rely on groundwater for potable water (Clarke *et al.*, 1996). Other than water stored in icecaps and glaciers, groundwater accounts for approximately 97% of fresh water on Earth (Nace, 1967; Shiklomanov and Rodda, 2003). As the world population continues to grow, more people will come to rely on groundwater sources, particularly in arid and semiarid areas (Simmers, 1990). Long-term availability of groundwater supplies for burgeoning populations can be ensured only if effective management schemes are developed and put into practice. Quantification of natural rates of groundwater recharge (i.e. the rates at which aquifer waters are replenished) is imperative for efficient groundwater management (Simmers, 1990). Although it is one of the most important components in groundwater studies, recharge is also one of the least understood, largely because recharge rates vary widely in space and time, and rates are difficult to directly measure.

The rate, timing, and location of recharge are important issues in areas of groundwater contamination as well as groundwater supply. In general, the likelihood for contaminant movement to the water table increases

as the rate of recharge increases. Areas of high recharge are often equated with areas of high aquifer vulnerability to contamination (ASTM, 2008; US National Research Council, 1993). Locations for subsurface waste-disposal facilities often are selected on the basis of relative rates of recharge, with ideal locations being those with low aquifer vulnerability so as to minimize the amount of moving water coming into contact with waste (e.g. US Nuclear Regulatory Commission, 1993). A high profile example of the importance of susceptibility to contamination is the study for the proposed high-level radioactive-waste repository at Yucca Mountain, Nevada. Tens of millions of dollars were invested over the course of two decades in efforts to determine recharge rates at the site (Flint *et al.*, 2001a).

Computer models of groundwater-flow are perhaps the most useful tools available for groundwater-resource management. Models are applied in both water-supply and aquifer-vulnerability studies. We expect that many readers of this book will be modelers seeking recharge estimates for use in groundwater-flow models or for evaluating model results.

The primary objective of this text is to provide a critical evaluation of the theory and assumptions that underlie methods for estimating rates of groundwater recharge. A complete understanding of theory and assumptions is fundamental to proper application of any method. Good practice dictates that recharge estimation techniques be matched to conceptual models of recharge processes at individual sites to ensure that assumptions underlying the techniques are consistent with conceptual models. As such, the text should serve as a resource to which hydrologists can refer for making informed decisions on the selection and application of methods. A thorough understanding of methods also provides a framework for the analysis of implications of modifying methods or applying them under less-than-ideal conditions.

A conceptual model of recharge processes attempts to answer the questions of where, when, and why recharge occurs. The model will thus identify the prominent recharge mechanisms, perhaps provide initial estimates of recharge rates, and serve as a guide for the selection of methods and for deciding on locations and time frames for data collection. The importance of matching methods for estimating recharge with conceptual models cannot be overemphasized. Development of a sound conceptual model is imperative for selecting proper methods and obtaining meaningful recharge estimates, but this process can be difficult, complicated by both natural and anthropogenic factors. A conceptual model often evolves over time as data are collected and interpreted; there may be a dynamic feedback effect – recharge estimates may support revision of the conceptual model or suggest the application of alternative methods.

Nature is complex, and each study site is unique. Although conceptual models of recharge processes are important, the development of a conceptual model is not the main focus of this book. Because of the great complexity and limitless variability in hydrologic systems, it is beyond the scope of this text to provide more than general guidelines for developing a conceptual model of recharge processes. It is simply not practical to describe or examine every scenario under which a method will be applied. Section 1.4 provides a general review of critical components of a conceptual model. For illustrative purposes, typical recharge processes in groundwater regions of the United States are briefly discussed in the final chapter.

This text is not intended as a cookbook that provides a recipe for estimating recharge for any and all situations; application of any method requires some hydrologic analysis. However, many of the methods described are simple enough that all the details required for their application are contained herein. Other methods, such as the use of complex models, require training that is beyond the scope of this text. Information is provided on these methods to assist the reader in deciding whether the cost of such training will be balanced by the benefits gained from applying the methods. Applications are illustrated with examples to highlight benefits and limitations. Many references are provided to allow the interested reader to pursue more details on any of the methods discussed.

Most of the discussion in this text is directed toward quantifying rates of natural recharge; however, many methods can and have been used to estimate recharge from artificial recharge operations, irrigated areas, and human-made drainage features, such as canals and urban water-delivery systems. In addition, many of the methods can be used to provide qualitative information on recharge rates (i.e. identifying areas of high and low relative recharge rates) for purposes of determining aquifer vulnerability to contamination from surface sources.

Numerous journal articles and reports describe the theory and details of the various techniques for estimating recharge. Applications of methods are discussed in many other papers. Given the importance of the subject matter, the paucity of textbooks devoted to this topic is surprising. Lerner *et al.* (1990) is the most thorough of these publications in terms of method descriptions. That text provides generic descriptions of physical controls that influence recharge in different hydrogeological provinces and discussion of techniques based on source of recharge water (i.e. precipitation, rivers, irrigation, and urbanization). Wilson (1980), Simmers (1997), and Kinzelbach *et al.* (2002) provide informative discussions on recharge processes in arid and semiarid regions and the techniques that are applicable in those regions. Simmers (1988) is a compendium of papers associated with a conference devoted to groundwater recharge. Hogan *et al.* (2004) and Stonestrom *et al.* (2007) each comprise a series

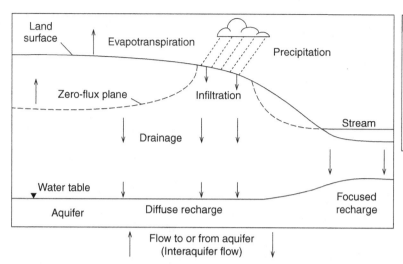

Figure 1.1 Vertical cross section showing infiltration at land surface, drainage through the unsaturated zone, diffuse and focused recharge to an unconfined aquifer, flow between the unconfined aquifer and an underlying confined aquifer (interaquifer flow), and the zero-flux plane.

of papers on recharge processes and case studies of recharge in arid and semiarid regions of the southwestern United States.

1.2 | Terminology

Recharge is defined, herein, as the downward flow of water reaching the water table, adding to groundwater storage. This definition is similar to those given by Meinzer (1923), Freeze and Cherry (1979), and Lerner *et al.* (1990). Strictly speaking, this definition does not include water flow to an aquifer from an adjoining groundwater system (such as water movement from an unconfined aquifer across a confining bed to an underlying aquifer); we refer to this flow as *interaquifer flow*. Others include this flow in their definition of recharge. Interaquifer flow has also been referred to as groundwater underflow. Regardless of terminology, methods for estimating interaquifer flow are included in this text. Recharge is usually expressed as a volumetric flow, in terms of volume per unit time (L^3/T), such as m^3/d, or as a flux, in terms of volume per unit surface area per unit time (L/T), such as mm/yr.

Recharge occurs through diffuse and focused mechanisms (Figure 1.1). *Diffuse* recharge is recharge that is distributed over large areas in response to precipitation infiltrating the soil surface and percolating through

the unsaturated zone to the water table; diffuse recharge is sometimes referred to as *local* recharge (Allison, 1987) or *direct* recharge (Simmers, 1997). *Focused* recharge is the movement of water from surface-water bodies, such as streams, canals, or lakes, to an underlying aquifer. Focused recharge generally varies more in space than diffuse recharge. A distinction between different types of focused recharge has been proposed by Lerner *et al.* (1990), with *localized* recharge defined as concentrated recharge from small depressions, joints, or cracks, and *indirect* recharge defined as recharge from mappable features such as rivers, canals, and lakes. Groundwater systems receive both diffuse and focused recharge, but the importance of each mechanism varies from region to region and even from site to site within a region. Generally, diffuse recharge dominates in humid settings; as the degree of aridity increases, the importance of focused recharge in terms of total aquifer replenishment also tends to increase (Lerner *et al.*, 1990). Some methods addressed in this book are designed to estimate diffuse recharge; others are specific to focused recharge.

Infiltration is the entry of water into the subsurface. Infiltrating water can be viewed as *potential recharge*; it may become recharge, but it may instead be returned to the atmosphere by evapotranspiration, or it may simply remain in storage in the unsaturated zone for some period of time. The *zero-flux plane* (ZFP) is the horizontal plane at some depth within the unsaturated zone that separates upward and downward moving water; the ZFP is sometimes equated with the bottom of the root zone (Figure 1.1). Water above the ZFP moves upward in response to evapotranspiration demand; water beneath the ZFP drains downward, eventually arriving at the water table. The depth of the ZFP changes in response to infiltration and evapotranspiration, ranging from land surface (for the case of downward water movement throughout the unsaturated zone) to some depth beneath the water table (for the case of groundwater evapotranspiration). Water draining beneath the ZFP in the unsaturated zone is referred to as *drainage, percolation,* or *net infiltration*; it becomes actual recharge when it arrives at the water table. Some techniques described in this book provide estimates of potential recharge; others provide estimates of drainage; and some methods provide estimates of actual recharge.

For clarity, we use the term *groundwater* to refer to water beneath the water table (within the saturated zone) and the term *pore water* to refer to water above the water table (within the unsaturated zone). A *point* estimate pertains to recharge at a specific point in space or time, whereas an *integrated* estimate refers to a value of recharge that is averaged over some larger space or time scale.

Different climatic regions are referred to throughout the text. Climatic regions are classified on the basis of annual precipitation. An *arid* climate is one with annual precipitation of less than 250 mm; a *semiarid* region has precipitation rates between 250 and 500 mm/yr; a *subhumid* climate refers to precipitation rates between 500 and 1000 mm/yr; and *humid* climates have annual precipitation rates that exceed 1000 mm.

1.3 | Overview of the text

This text is organized by methods, which are grouped on the basis of types of required or available data (e.g. methods based on water budgets, or on data obtained from the unsaturated zone, or on streamflow data). Our approach differs from that of Lerner *et al.* (1990) and Wilson (1980), who chose to organize methods on the basis of source of recharge (precipitation, rivers, etc.). While there is perhaps no ideal format for this presentation, the format used in this text has proved workable within the classroom over the decade and a half that we have taught this material. Examples are given to show how methods can be applied for different sources of recharge water.

This first chapter provides an introduction to the book, emphasizing the importance of developing a conceptual model of recharge processes for the area of interest. Chapters 2 through 8 are the heart of the book. They provide in-depth analysis of methods for estimating recharge. The format for each presentation is similar: discussion of theory and assumptions, advantages and limitations of the methods, and description of example case studies. Each chapter is devoted to a particular family of methods. Water-budget methods (Chapter 2) are presented first to emphasize the importance of water budgets in all studies.

Water-budget methods are widely used; indeed, most methods for estimating recharge could be classified as water-budget methods. To avoid making Chapter 2 too long, its content is limited to the use of the residual water-budget method, whereby a water-budget equation is derived for a control volume, such as a watershed or an aquifer. All components within that equation, except for recharge, are measured or estimated; recharge is then set equal to the residual in the equation. Other methods that can be categorized as water-budget methods (e.g. the water-table fluctuation method, the zero-flux plane method, and modeling methods) are described in other chapters. Remote-sensing tools are described in Chapter 2, although they may be useful in other methods as well.

Discussion in Chapter 3 is devoted to the use of models for estimating recharge. A general approach to modeling, applicable to all models, is presented first; a brief description of inverse techniques is included. Unsaturated zone water-budget models, watershed models, groundwater flow models, and integrated surface- and subsurface-flow models are then discussed. Because of the complexities of some models, detailed model descriptions are avoided. Instead, examples are used to highlight capabilities of complex models, resources required for model application, and benefits and limitations of using models to generate estimates of recharge. Empirical equations, which are widely used for predicting recharge, are also described, as are regression and geostatistical techniques for upscaling point estimates of recharge to obtain average values for an aquifer or watershed.

Chapter 4 addresses physical methods that are based on surface-water data. Included are stream water-budget methods, seepage meters, streamflow-duration curves, streamflow hydrograph analysis (hydrograph separation), and chemical or isotopic hydrograph separation.

Chapter 5 describes physical methods that can be applied on the basis of data collected in the unsaturated zone. These methods include the zero-flux plane, the Darcy method, and the use of lysimeters. Physical methods based on data collected in the saturated zone form the basis of Chapter 6. The primary method in this group is the water-table fluctuation method. The Darcy method and methods based on time series of measured groundwater levels are also discussed.

Chapters 7 and 8 are devoted to the use of tracers for estimating recharge. Chemical and isotopic tracer methods are described in Chapter 7. Tracers can be naturally occurring (e.g. chloride and isotopes of carbon and hydrogen), can occur as an indirect outcome of anthropogenic activity (e.g. tritium, chlorine-36, and chlorofluorocarbon gases), or can be intentionally applied to the surface or subsurface for experimental purposes (e.g. bromide, fluorescent dyes). Tracers can be used to study water from any source. Use of heat as a tracer for estimating recharge is described in Chapter 8.

The final chapter, Chapter 9, attempts to link conceptual models of recharge processes with estimation methods. It begins with a discussion of considerations important in selecting methods. Figures and tables are presented to compare methods in terms of spatial and temporal scales of applicability. Typical recharge processes and methods that have been used to study these processes are described for groundwater regions of the United States. This discussion is not an attempt at a comprehensive summary of recharge processes and studies; such an attempt is neither practical nor feasible. Rather, the idea is to illustrate how conceptual models of recharge processes can be formed and used to select appropriate methods. The closing section presents some final thoughts on good practices for any recharge study.

1.4 | Developing a conceptual model of recharge processes

The development of a conceptual model of recharge processes (Figure 1.2) is an important step in any recharge study. The conceptual model should be developed at the beginning of a study; it can be revised and adjusted as additional data and analyses provide new insights to the hydrologic system (Zheng and Bennett, 2002; Bredehoeft, 2005). Although this book is focused on methods, the reader should bear in mind the importance of a conceptual model when reviewing various methods. This section provides some discussion on factors that can influence a conceptual model – climate, geology, topography, hydrology, vegetation, and land use. The contents of this section are by no means comprehensive; the intent is to illustrate some of the factors that can help to shape a conceptual model.

Water budgets are fundamental components of any conceptual model of a hydrologic system, providing a link between recharge processes and other processes in the hydrologic cycle. Water-budget equations can be derived for one or more control volumes, such as an aquifer, a watershed, a stream, or even a column of soil (Healy et al., 2007). A water-budget equation allows consideration of the entire hydrology of the system under study, providing information not only on recharge, but also on interrelationships among recharge, discharge, and change in storage. Preliminary water budgets can be readily constructed with existing data and refined as various measurements and

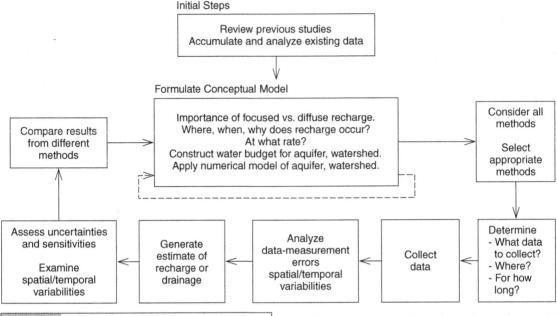

Initial Steps

Review previous studies
Accumulate and analyze existing data

Formulate Conceptual Model

Compare results
from different
methods

Importance of focused vs. diffuse recharge.
Where, when, why does recharge occur?
At what rate?
Construct water budget for aquifer, watershed.
Apply numerical model of aquifer, watershed.

Consider all
methods

Select
appropriate
methods

Assess uncertainties
and sensitivities

Examine
spatial/temporal
variabilities

Generate
estimate of
recharge or
drainage

Analyze
data-measurement
errors
spatial/temporal
variabilities

Collect
data

Determine
- What data
to collect?
- Where?
- For how
long?

Figure 1.2 Schematic showing iterative process for developing a conceptual model of recharge processes.

recharge estimates are obtained. As noted by Lerner *et al.* (1990), a good method for estimating recharge provides not only an estimate of how much water becomes recharge, but also explains the fate of the remaining water that does not become recharge. Water-budget analyses serve that purpose. In addition, a water-budget equation provides a convenient context for the analysis of assumptions inherent in various estimation techniques.

Although recharge is important in water-supply studies, recharge rates are sometimes incorrectly equated with the sustainable yield of an aquifer (Meinzer, 1923; Bredehoeft *et al.*, 1982; Bredehoeft, 2002; Alley and Leake, 2004). The term *sustainable yield* or *safe yield* refers to the rate at which water can be withdrawn from an aquifer without causing adverse impacts. Those impacts could be in the form of decreased discharge to streams and wetlands, land subsidence, or induced contamination of groundwater, for example, by seawater intrusion. The notion that recharge is equivalent to sustainable yield is based on an incomplete or incorrect conceptual model of a hydrologic system. Knowledge of recharge rates is important

for determining sustainable yields in many groundwater systems (Sophocleous *et al.*, 2004; Devlin and Sophocleous, 2005), but recharge rates by themselves are not sufficient for determining sustainability (Bredehoeft *et al.*, 1982; Bredehoeft, 2002). The effects of changes in groundwater levels on groundwater discharge rates and aquifer storage must also be considered. From a hydrologic perspective, sustainable yield is best studied within the context of the entire hydrologic system of which the aquifer is a part, but decisions as to what constitutes a sustainable yield often involve more than just hydrologic considerations. Ecological, cultural, economic, and other considerations should help to determine the acceptability of any effects related to groundwater development (Alley and Leake, 2004).

1.4.1 Spatial and temporal variability in recharge

Recharge rates vary in space in both systematic and random fashions. This is true for both focused and diffuse recharge. Systematic trends often are associated with climatic trends, but land use and geology are also important. Statewide maps of estimated annual recharge for Texas (Figure 1.3; Keese *et al.*, 2005) and Minnesota (Lorenz and Delin, 2007) both

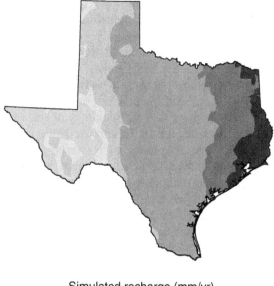

Simulated recharge (mm/yr)

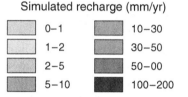

0–1		10–30
1–2		30–50
2–5		50–00
5–10		100–200

Figure 1.3 Map of average annual recharge rate for the state of Texas (Keese *et al.*, 2005).

display trends similar to those in statewide maps of annual precipitation. The concept of recharge rates increasing with increasing precipitation rates is certainly intuitive – recharge cannot occur if water is not available. The random factor in recharge variability can be viewed as local-scale variability that can be attributed, for example, to natural heterogeneity in permeability in surface soils or variability in vegetation. Any of the factors addressed below can contribute to apparent random variability. Delin *et al.* (2000) found that annual recharge varied by more than 50% within what appeared to be a uniform 2.7-hectare agricultural field simply because of slight differences in surface topography; the total relief in the field was less than 1.5 m. It could be argued that this difference in topography was not random; indeed, distinguishing between systematic and random patterns of recharge is sometimes a matter of scale. In the context of the entire upper Mississippi River valley, the topographic

differences in this field are minute, apparently random; to someone standing in the field during a rain storm, the systematic pattern in recharge is obvious.

Recharge also varies temporally. Seasonal, multiyear, or even long-term trends in climate affect recharge patterns. Because of its close link to climate, temporal variability of recharge is addressed more thoroughly in Section 1.4.2. Changes in land use or in vegetation type and density can also result in large changes in recharge rates over time.

The importance of spatial and temporal variability of recharge must be considered within the context of study objectives. Spatial variability may not be critical for groundwater resource evaluation if an average rate of recharge can be determined for an entire aquifer. Spatial variability is important, though, for assessing aquifer vulnerability to contamination; therefore, methods that provide point estimates of recharge may be appropriate. Historically, many groundwater-flow models were developed under the assumption that recharge was constant in time. Current model applications typically allow recharge to vary over time but hold it constant for periods of months or years. Recent advances in incorporating landscape features into combined surface-water and groundwater flow (Section 3.6) will allow impacts of climate, land-use, and vegetation change on water resources to be examined at unprecedented levels of temporal and spatial variability.

1.4.2 Climate

Climate variability is often the most important factor affecting variability in recharge rates. Precipitation, the source of natural recharge, is the dominant component in the water budget for most watersheds. The relation between spatial trends of precipitation and recharge has been noted in Section 1.4.1. Temporal variability in precipitation also is important. Seasonal, year-to-year, and longer-term trends in precipitation, as well as frequency, duration, and intensity of individual precipitation events also affect recharge processes. Conditions are most favorable for water drainage through

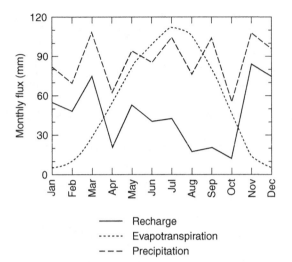

the unsaturated zone to the water table when precipitation rates exceed evapotranspiration rates. In regions outside of the tropics, evapotranspiration rates follow a seasonal trend, with highest rates occurring during summer months and lowest rates in winter months. If precipitation rates are fairly uniform throughout the year, the most likely time of the year for drainage to occur is winter through spring, when precipitation rates exceed evapotranspiration rates. At a site in the eastern United States, Rasmussen and Andreasen (1959) estimated that 62% of recharge over a 2-year period occurred in the months of November through March (Figure 1.4); precipitation was relatively uniform throughout the year, but evapotranspiration rates were lowest during these months.

Duration and intensity of individual precipitation events can have a large influence on recharge in some settings. On the humid, windward side of the Hawaiian Islands, precipitation and evapotranspiration rates are relatively uniform throughout the year. Recharge occurs at any time of the year in response to intense rain storms, when the total precipitation for a day exceeds the daily evapotranspiration rate (Ahuja and El-Swaify, 1979).

In arid regions, focused recharge from ephemeral streams and playas is often the dominant form of recharge. The frequency and duration of streamflow play important roles in the recharge process. The frequency of streamflow in Rillito Creek in Tucson, Arizona, coincides with the frequency of recharge events. Pool (2005) showed that interannual variability in recharge from the creek is linked to the El Niño/Southern Oscillation climate trend. Years dominated by El Niño conditions (high winter precipitation rates) produced significantly higher streamflow and recharge rates than years dominated by La Niña conditions.

1.4.3 Soils and geology

Permeabilities of surface and subsurface materials can greatly affect recharge processes. Recharge is more likely to occur in areas that have coarse-grained, high-permeability soils as opposed to areas of fine-grained, low-permeability soils. Coarse-grained soils have a relatively high permeability and are capable of transmitting water rapidly. The presence of these soils promotes recharge because water can quickly infiltrate and drain through the root zone before being extracted by plant roots. Finer-grained sediments are less permeable, but are capable of storing greater quantities of water. Thus, in areas of finer-grained sediments, one would expect decreased infiltration, enhanced surface runoff, increased plant extraction of water from the unsaturated zone, and decreased recharge relative to an area of coarser-grained sediments. Permeability also is important in terms of focused recharge. High-permeability streambeds facilitate the exchange of surface water and groundwater. In the Black Hills of South Dakota, most recharge to the Madison Limestone aquifer occurs at high elevations as focused flow from streams that cross rock outcrops (Swenson, 1968; Downey, 1984). In karst regions, dissolution cavities or sinkholes that have developed in the geologic material can rapidly channel streamflow directly to an aquifer; these cavities also facilitate ground-water discharge in the form of springs.

Subsurface geology influences discharge processes as well as recharge processes. If the rate of discharge from an aquifer is less than the recharge rate, water storage within the aquifer increases. Aquifer storage can reach a maximum at which point additional recharge cannot be accepted, regardless of the amount of precipitation. This condition typically leads to enhanced runoff.

Geophysical techniques have a wide range of uses in geologic and hydrologic studies, providing information on the electrical, physical, and chemical properties of surface and sub-surface sediments. In regards to quantifying groundwater recharge, geophysical methods are most useful for determining soil-water content (Section 5.2.1) and changes in subsurface water storage (Section 2.3.3). However, information obtained from the application of geophysical techniques is also useful in a qualitative sense. Geophysical techniques can be used to infer aquifer geometry and hydraulic properties, important information for shaping conceptual models of hydrologic systems and for constructing computer models of groundwater flow (Robinson *et al.*, 2008a).

1.4.4 Surface topography

Land-surface topography plays an important role for both diffuse and focused recharge. Steep slopes tend to have low infiltration rates and high runoff rates. Flat land surfaces that have poor surface drainage are more conducive to diffuse recharge; these conditions also favor flooding. Small, often subtle depressions can have a profound influence on infiltration rates. Delin *et al.* (2000) showed that, even with highly permeable soils, slight depressions in an apparently uniform agricultural field caused runoff to be focused in certain areas, with the result that infiltration (and recharge) in those areas was substantially greater than that in the rest of the field. Even with uniform surface characteristics, apparent infiltration rates increase in the downslope direction along a long hill slope (Dunne *et al.*, 1991) because downslope portions of the hill are exposed to runoff from

upslope portions as well as precipitation. Local relief, orientation, and altitude of mountain ranges are additional topographic factors that can affect recharge processes (Stonestrom and Harrill, 2007).

1.4.5 Hydrology

A conceptual model of recharge processes needs to consider the surface-water and groundwater flow systems and how they are linked. Are streams in the area perennial or ephemeral? Are streams gaining (receiving groundwater discharge) or losing (providing recharge)? A single stream could conceivably be losing water to an aquifer in one reach, but gaining water in another reach; the difference between ground-water and surface-water elevations, according to Darcy's law, determines whether water is moving to or from the subsurface. These are key questions, the answers to which will help shape the conceptual model.

The depth to the water table also is important. If the unsaturated zone is thin, infiltrating water can quickly travel to the water table; recharge may be largely episodic, occurring in response to any large precipitation event. However, shallow water tables are also susceptible to groundwater discharge by plant transpiration. Therefore, water that recharges shallow subsurface systems may only reside in the saturated zone for a short time before it is extracted by plant roots and returned to the atmosphere. Thick unsaturated zones are less likely to have episodic recharge events; recharge would be expected to be seasonal or quasi-steady because wetting fronts moving through the unsaturated zone tend to slow with depth and multiple fronts may coalesce and become indistinguishable from each other.

1.4.6 Vegetation and land use

Vegetation and land use can have profound effects on recharge processes. Types and densities of vegetation influence patterns of evapotranspiration. A vegetated land surface typically has a higher rate of evapotranspiration (and, hence, less water available for recharge) than an unvegetated land surface under similar conditions. The depth to which plant roots extend influences the efficiency with which plants can extract water from the subsurface. Trees, for example, are capable of drawing moisture from depths of several meters or more. In contrast, shallow-rooted crops cannot access soil water that penetrates to those depths. Thus, enhanced recharge rates in areas with shallow-rooted vegetation are seen in some semiarid regions when native perennial vegetation is replaced by shallow-rooted crops (Allison et al, 1990; Scanlon et al., 2005; Leblanc et al, 2008). Nonirrigated agricultural crops can have higher or lower evapotranspiration rates than native plants; therefore, it is difficult to generalize as to whether the potential for recharge will increase or decrease due to changes in vegetation alone. In most settings, the influence of vegetation is seasonal; in periods of senescence, the presence of plants can actually promote recharge. Decay or shrinkage of roots can expose cavities that can act as preferential flow channels and enhance infiltration. Plowing and tilling in agricultural fields can have opposing effects – breaking up surface crusts, thus increasing the potential for infiltration – and destroying preferential flow channels, thus decreasing infiltration potential. Satellite remote sensing (Section 2.5) can provide information on surface characteristics, such as vegetation type and percent coverage, leaf area index, and land use that can be useful in formulating a conceptual model (Brunner et al., 2007).

In the Murray Basin of Australia, native eucalyptus trees were gradually replaced with nonirrigated agricultural crops through the 1900s. Allison and Hughes (1983) estimated natural recharge rates under native vegetation to be less than 0.1 mm/yr. After clearing and subsequent cropping, estimated recharge rates increased by up to two orders of magnitude (Allison et al., 1990). Unfortunately, the increased recharge has led to increased leaching of salts to groundwater and subsequently to the Murray River and its tributaries.

Conversion from natural savannah to nonirrigated millet crops over large areas of southwest Niger since the 1950s has produced soil crusting on slopes, resulting in increased runoff and focused recharge beneath ephemeral ponds that collect runoff (Leblanc et al., 2008; Favreau et al., 2009). Increased recharge rates arising from the land-use change can explain the paradoxical relationship between rising groundwater levels (about 4 m between 1963 and 2007) and decadal droughts (23% average annual decline in precipitation from 1970 to 1998 relative to the previous two decades). Areally averaged recharge rates are estimated to have increased from 2 to 25 mm/yr (Favreau et al., 2009).

Irrigation can play an important role in groundwater recharge. *Irrigation return flow* is any excess irrigation water that drains down beneath the root zone or is captured in drainage ditches. It constitutes a significant amount of recharge in many areas, especially in arid or semiarid regions where natural recharge rates are low. Fisher and Healy (2008) studied recharge processes at two irrigated agricultural fields in semiarid settings; virtually all of the annual recharge occurred during the irrigation season and was attributed to irrigation return flow. Faunt (2009) used a complex groundwater-flow model to show that, in addition to the natural recharge that occurs during winter to aquifers in California's Central Valley, recharge also occurs during the growing season as a result of irrigation return flow. Within the United States, flood irrigation has gradually been replaced with more efficient sprinkle or drip irrigation systems; conversion to these new methods has reduced return flows substantially in some areas (McMahon et al., 2003).

Urbanization brings about many landsurface changes that can have significant ramifications for recharge processes. Roads, parking lots, and buildings all provide impervious areas that can inhibit recharge. Runoff diversions are common features in urban landscapes. Diversions may lead to surface-water bodies or to infiltration galleries. In the former case, overall recharge for the area is reduced. In the latter case, recharge may not necessarily be reduced, but at the very least it is redirected and may change from a diffuse source to a focused source. Runoff diversions may be important in terms of aquifer vulnerability to contamination because they can quickly funnel contaminants to the subsurface.

Delivery systems for water supply and treatment are additional artifacts of urbanization that can affect recharge processes, both in terms of water supply and potential for contamination. These systems consist of open channels or water pipes and sewers. Invariably, there is leakage associated with any delivery system. This leakage, a form of potential recharge, may become actual recharge. Norin *et al.* (1999) estimated that 26% of water transmitted through water mains in Göteborg, Sweden, was lost to leakage.

1.4.7   Integration of multiple factors

A conceptual model of recharge processes is formed by integrating the above factors, and perhaps other factors as well, into hypotheses on where, when, and why recharge occurs. For example, the timing and location of recharge in high mountainous valleys is often controlled by geology, climate, and hydrology. Snowfall in the mountains from late fall through early spring is the source of recharge water. Water is stored in the snowpack until late spring, when it is released to streams as rising air temperatures melt snowpacks. As swollen streams flow to valleys, water seeps downward, recharging underlying aquifers. Variations of this predictable pattern of seasonal recharge occur in many mountainous regions.

Numerical or analytical models of climatic conditions, watershed processes, surface-water flow, groundwater flow, or unsaturated-zone flow are useful tools for integrating the factors affecting the conceptual model of recharge. The suggestion of using a numerical model in the initial stages of a recharge study may seem unusual because oftentimes the specific goal of a recharge study is to develop estimates for use in groundwater-flow models or for comparison with model results. Nonetheless, a simple numerical model can be a useful tool for identifying important mechanisms, evaluating hypotheses included in a conceptual model, and determining optimum locations

and timing for data collection. Application of numerical or analytical models provides benefits at all stages of a recharge study. The conceptual and numerical models are both part of an iterative process, whereby the models are continually refined and revised as new data, interpretations, and simulation results become available.

1.4.8   Use of existing data

Construction of a conceptual model should make use of all available data for the study area and surrounding areas. Many of the recharge estimation methods described in the following chapters, including watershed and groundwater-flow models, can be applied without collecting any new data. Careful analysis of all existing data precedes any decisions on the collection of new data. Existing databases contain climatological data, surface-water flow data, land-use data, groundwater levels, chemistry of surface water and groundwater, physical and hydraulic properties of soils, and land-use characteristics. Pertinent data sources are described more thoroughly in Chapters 2 and 3.

1.4.9   Intersite comparison

As a first estimate of recharge for a particular study area, one might use an estimate derived from a site with similar climate, land use, and other features. A review of literature for similar sites is always a worthwhile endeavor. Such a review would benefit from a common classification scheme for climatic/hydrologic/geologic provinces. Such a scheme would facilitate intersite comparisons and would also be useful in the construction of conceptual models and selection of appropriate techniques. Currently (2010), there is no such scheme in widespread use, although classification schemes suggested by Salama *et al.* (1994b) for hydrogeomorphic units and by Winter (2001) for hydrologic landscapes hold promise. For discussion of generic recharge processes, we resort to the groundwater regions of the United States defined by Thomas (1952); this discussion is provided in Chapter 9 so that concepts and specific methods can be discussed in complementary fashion.

1.5 | Challenges in estimating recharge

1.5.1 Uncertainty in recharge estimates

Accurate estimates of recharge are always desired; yet it is beyond our current capabilities to determine, with any degree of confidence, the uncertainty associated with any recharge estimate, let alone claim that an estimate is accurate. Actual recharge rates are unknown; therefore, there are no standards that can be used to evaluate the accuracy of recharge estimates. The most serious errors are those associated with an incorrect conceptual model. An incorrect conceptual model can lead to the application of inappropriate estimation techniques and meaningless recharge estimates. Any estimates based on an incorrect conceptual model are inherently unreliable. Additional sources of error arise from improper application of methods and measurement errors.

Improper application of a method can result from lack of understanding of the method or from failure to adequately account for spatial and temporal variability. Errors related to the latter arise from an inability to measure at enough locations or failure to make measurements for a sufficient length of time. Spatial and temporal variabilities of recharge cannot be determined exactly, but they can be examined in some detail with numerical models. A map of the annual recharge for Texas (Figure 1.3) was developed by Keese *et al.* (2005) by combining a one-dimensional variably saturated water-flow model with spatially variable soil, vegetation, and climate properties. Techniques for the upscaling of point estimates of recharge to large areas are discussed more fully in Chapter 3.

Measurement errors relate to inaccuracies in data collection. Complications arise because the magnitude of a recharge flux is generally quite small and often cannot be measured directly. Most methods presented in this text are indirect methods that rely on more readily measured parameters, such as changes in water storage or tracer concentrations, to make inferences on recharge rates. Measurement errors are the only type of errors that are conducive to classical error analysis (Lerner *et al.*, 1990). Approaches for such an analysis for water-budget methods are described in Chapter 2.

Concerns about inaccuracy in recharge estimates should not deter application of any reasonable method for estimating recharge. Simple techniques, applied with careful consideration of conceptual models, can be not only useful, but enduring. Theis (1937) used a simple application of the Darcy equation to estimate natural recharge rates of between 3 and 7 mm/yr for the southern High Plains aquifer, values that fall midway in the range of estimates generated in subsequent years with more sophisticated techniques (Gurdak and Roe, 2009). White (1930) used a salt tracer to estimate subsurface flow into a portion of the Mimbres River watershed in southwestern New Mexico; the groundwater-flow model of Hanson *et al.* (1994) corroborated that estimate.

Because much of the error associated with a recharge estimate is not quantifiable, it is wise to apply multiple methods for estimating recharge in any study (Lerner *et al.*, 1990; Simmers, 1997; Scanlon *et al.*, 2002b). Estimates from multiple methods may not quantitatively reduce uncertainty; consistency in results, while desirable, may not be a reliable indicator of accuracy. Application of different approaches may have qualitative benefits, however; inconsistencies in estimates may provide insight into measurement errors or the validity of assumptions underlying a method and, thus, may provide direction for revising the conceptual model.

1.5.2 Spatial and temporal scales of recharge estimates

The concept of spatial scale is important in terms of selecting appropriate methods. Different methods provide estimates that are integrated over various spatial scales. Some methods provide essentially point estimates; these methods are useful for evaluating aquifer vulnerability to contamination in support of land-use decisions, for example; however, application at many points may be required to determine an areally averaged recharge rate. Other techniques provide an average estimate for an entire aquifer or watershed or for a stream reach, but they may provide little insight into recharge rates at specific locations. The user must reconcile project objectives and the spatial scale of a field site of interest with the spatial scale inherent in the techniques to be applied. The spatial scale of each method is discussed in some detail in this text.

Methods for estimating recharge are associated with temporal scales as well as spatial scales. Some methods, such as the water-table fluctuation method, can provide an estimate of recharge for each individual precipitation event. Most tracer methods, on the other hand, provide a single estimate of recharge that is averaged over the time period between tracer application and tracer sampling. That time period can extend from a few days to years for applied tracers to decades or centuries or even millennia for naturally occurring tracers. In regions with thick unsaturated zones, recharge is sometimes assumed to be constant in time (some methods for estimating recharge are based on this assumption). On the basis of the chloride mass-balance method, Scanlon (1991) determined that there has essentially been no recharge since the Pleistocene in interdrainage areas of the Chihuahuan Desert in the southwestern United States. As with spatial variability, the importance of identifying temporal variability can be determined only by the user. The user must take care that the methods that are selected will provide results over the time frame of interest.

Water-budget methods, Darcy methods, and other methods can be applied over a variety of time intervals, for example, with daily, monthly, or annual data; however, results from application of the same method over different time intervals can vary. Consider a simple water-budget approach for a watershed in a subhumid climate, where recharge can occur only when precipitation exceeds evapotranspiration. On a monthly basis, evapotranspiration totals exceed precipitation totals for the months of May through August; therefore, no recharge would be predicted. Within one of these months, though, there could be days when precipitation exceeds evapotranspiration. Water-budget calculations, when performed with daily values, could conceivably calculate recharge on those days.

1.5.3 Expense

Expense is a common limitation for applying some recharge-estimation methods. Some methods can be applied by means of a single field trip to collect and analyze soil and water samples (this is sometimes the case with use of tracers); other methods require continuous monitoring over the course of a year or more. Analytical costs for tracers such as carbon-14 may appear to be beyond the means of many recharge studies. But the costs of collecting and analyzing a single set of samples may be less than that required for continuous monitoring. It should be kept in mind that "more expensive" does not always mean "better or more accurate." Methods that require large expenses or hard-to-collect data are seldom applied outside of research studies. In discussions in the following chapters, the relative cost of each method is addressed. The user must balance cost against expected improvements in recharge estimates and attempt to answer the question: how much will my knowledge of the system be improved and at what cost?

1.6 | Discussion

The selection of methods for estimating recharge is largely driven by the goals of a study and the financial and time constraints placed on that study. Examples of study goals might be to:

- obtain a long-term average rate of recharge for an aquifer,
- determine point estimates of recharge in space for assessing aquifer vulnerability to contamination,
- estimate recharge at specific points in time and space to serve as observations for calibration of a model for simulating combined surface-water/groundwater flow,

• assess the effects of land-use or climate change on past and future patterns of recharge.

Satisfying these objectives requires recharge estimates that span a wide range of space and time scales; therefore, it is unlikely that a single approach would work for all studies. Spatial and temporal scales associated with each method are discussed in the following chapters to facilitate matching methods with study goals. Also discussed are expenses, in terms of data requirements and manpower, for each method, so the reader can determine whether study constraints will permit application of a particular method.

The selection of methods also must be tied to the conceptual model of the hydrologic system under study. Throughout this first chapter, the importance of building a sound conceptual model has been expounded. The final chapter contains a more detailed discussion of conceptual models, along with examples for typical systems in generic groundwater regions of the United States. Chapters 2 through 8 provide analysis of individual methods for estimating recharge. Each of these methods is based on a set of assumptions. Assumptions can relate to the mechanism of recharge (diffuse or focused), the timing and the location of recharge, the importance of other components in the water budget of an aquifer or watershed, the uniformity of properties (such as hydraulic conductivity), and various other aspects or features. Assumptions inherent to each method are discussed. The importance of these assumptions needs to be assessed in the context of each application, and the reader must decide whether inherent assumptions are consistent with the conceptual model of the hydrologic system under consideration.

Estimation of recharge is an iterative process with continual refinement. The conceptual model can help guide selection of suitable methods and indicate where and when the methods might best be applied. Recharge estimates obtained early in the course of a study may lead to refinement of the conceptual model, which, in turn, could lead to the application of alternative estimation techniques. Many methods described in the following chapters can be applied with a minimal amount of effort by using existing data.

Water-budget methods

2.1 | Introduction

A water budget is an accounting of water movement into and out of, and storage change within, some control volume. *Universal* and *adaptable* are adjectives that reflect key features of water-budget methods for estimating recharge. The universal concept of mass conservation of water implies that water-budget methods are applicable over any space and time scales (Healy *et al.*, 2007). The water budget of a soil column in a laboratory can be studied at scales of millimeters and seconds. A water-budget equation is also an integral component of atmospheric general circulation models used to predict global climates over periods of decades or more. Water-budget equations can be easily customized by adding or removing terms to accurately portray the peculiarities of any hydrologic system. The equations are generally not bound by assumptions on mechanisms by which water moves into, through, and out of the control volume of interest. So water-budget methods can be used to estimate both diffuse and focused recharge, and recharge estimates are unaffected by phenomena such as preferential flow paths within the unsaturated zone.

Water-budget methods represent the largest class of techniques for estimating recharge. Most hydrologic models are derived from a water-budget equation and can therefore be classified as water-budget models. It is not feasible to address all water-budget methods in a single chapter. This chapter is limited to discussion of the "residual" water-budget approach, whereby all variables in a water-budget equation, except for recharge, are independently measured or estimated and recharge is set equal to the residual. This chapter is closely linked with Chapter 3, on modeling methods, because the equations presented here form the basis of many models and because models are often used to estimate individual components in water-budget studies. Water budgets for streams and other surface-water bodies are addressed in Chapter 4. The use of soil-water budgets and lysimeters for determining potential recharge and evapotranspiration from changes in water storage is discussed in Chapter 5. Aquifer water-budget methods based on the measurement of groundwater levels are described in Chapter 6.

Water budgets are fundamental to the conceptualization of hydrologic systems at all scales. Building a preliminary water budget from existing data is an easy and logical first step in any study, regardless of whether water-budget methods will actually be used to estimate recharge. Initial analysis of a water budget can provide insight as to the suitability of any recharge estimation technique. Refinement of the water budget and the conceptual model of the system throughout the duration of a study can further guide study efforts. Evaluation of the water budget at the conclusion of a study is a particularly useful exercise, serving as a postaudit or a check of study results.

Inconsistencies between the conceptual model of a hydrologic system and recharge estimates generated during the study may indicate the need for further evaluation of both the model and the estimates.

The following section presents a general analysis of water-budget equations. Methods for measuring and estimating components in typical water-budget equations, including precipitation, evapotranspiration, runoff, and water storage, are presented in subsequent sections. Temporal and spatial scales of interest play an important role in selecting appropriate methods. For convenience, methods are organized by spatial scales: local, corresponding to a small field plot where point measurements of individual components of the water-budget equation are sufficient – measurement techniques for the individual components are addressed in Section 2.3; mesoscale, corresponding to a watershed where either upscaling of multiple point measurements is required or estimating equations must be used for each component – estimation techniques are stressed in Section 2.4; and macroscale, corresponding to regional, continental, or global scales where remote-sensing tools are usually applied – these tools are reviewed in Section 2.5.

2.2 | Water budgets

The first step in a water-budget analysis is to select one or more control volumes for study. A control volume can be a volume of earth or atmosphere or a hydrologic structure – a laboratory soil column, a lake or stream, an aquifer, a watershed, a country or state, and the Earth itself are all examples of control volumes for which water budgets can and have been constructed. Any control volume whose water-budget equation contains recharge as a component can be used to estimate recharge. Insight on the hydrology of the system is useful in selecting a control volume. Where and when recharge occurs, whether recharge is focused or diffuse, what types of data are available – these are important considerations in selecting a control volume. Locations where fluxes are

known or can be easily measured or estimated make good boundaries for a control volume. For purposes of estimating recharge, watersheds, aquifers, and one-dimensional soil columns, or some part of these, are the most widely used control volumes; and for that reason, they are emphasized in this chapter.

A simple water-budget analysis used in many hydrological studies is based on a soil column that extends downward from land surface to some depth, L (Figure 2.1) as the control volume:

$$P = ET + \Delta S + R_{off} + D \qquad (2.1)$$

where P is precipitation; ET is evapotranspiration, which includes evaporation and plant transpiration; ΔS is change in water storage in the column; R_{off} is direct surface runoff (precipitation that does not infiltrate); and D is drainage out of the bottom of the column. All components are given as rates per unit surface

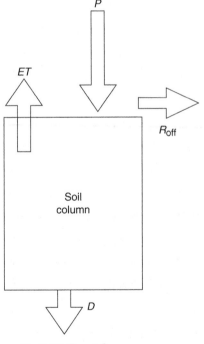

$$D = P - ET - R_{off} - \Delta S$$

Figure 2.1 Schematic diagram showing water budget of a one-dimensional soil column. D is drainage out the bottom of the column, P is precipitation, ET is evapotranspiration, R_{off} is runoff, and ΔS is change in storage.

area, such as millimeters per day. Equation (2.1) states that the change in storage within the column is balanced by the difference between flow into the column, from precipitation, and flow out of the column, by either evapotranspiration or drainage through the bottom of the column.

Drainage, D, is equivalent to recharge, R, only if the bottom of the column extends to the water table. According to the definitions given in Chapter 1, draining water is not referred to as recharge until it actually arrives at the water table. Some methods described in this and following chapters produce estimates of drainage; others produce estimates of actual recharge. The distinction between the two phenomena is a matter of timing; if the soil column extends beyond the bottom of the root zone, water draining out the bottom will eventually become recharge, but the time it takes for that water to move to the water table is not known. Approaches for describing this travel time are discussed in Chapter 3. Distinguishing between drainage and recharge may be important if temporal trends of recharge are of interest. Land-use and climate changes can result in current drainage rates that are substantially different from historical rates of recharge. But if recharge is being estimated for input to a steady-state groundwater flow model, the two terms can be considered synonymous. It is not always clear whether a particular method is providing an estimate of drainage or recharge; many authors do not even make this distinction. In this chapter, and in the remainder of this book, if it is clear that a method estimates drainage and not actual recharge, then the estimate is referred to as drainage (D). If it is clear that a method estimates actual recharge or if there is any ambiguity as to what type of estimate is provided, the estimate is referred to as recharge (R).

The water budget equation for a watershed and the underlying unsaturated and saturated zones can be written in a form similar to that of Equation (2.1):

$$P + Q_{on} = ET + \Delta S + Q_{off} \tag{2.2}$$

where Q_{on} is surface and subsurface water flow into the watershed; and Q_{off} is surface and

subsurface water flow out of the watershed. Equation (2.2) can be easily expanded and refined, as the following discussion illustrates. Precipitation can occur in the form of rain, snow, dew, or fog drip; irrigation could also be explicitly included. Water flow into the watershed can be written as the sum of surface-water flow, Q^{sw}_{on} (which could include diversion from another watershed), and groundwater flow, Q^{gw}_{on} (which could include subsurface injection):

$$Q_{on} = Q^{sw}_{on} + Q^{gw}_{on} \tag{2.3}$$

Evapotranspiration can be divided on the basis of the source of evaporated water:

$$ET = ET^{sw} + ET^{gw} + ET^{uz} \tag{2.4}$$

where ET^{sw} is evaporation or sublimation of water stored on land surface, ET^{uz} is bare soil evaporation and plant transpiration of water stored in the unsaturated zone, and ET^{gw} is evapotranspiration of water stored in the saturated zone. (Although distinguishing among ET^{sw}, ET^{gw}, and ET^{uz} can be accomplished in theory, in practice, it may be problematic because standard measurement techniques determine the flux of water vapor from land surface to the atmosphere and are essentially blind to the source of the evaporating water. Available measurement techniques, their accuracy, and their cost are important considerations when fashioning a water-budget equation.)

Storage of water in a watershed occurs in surface-water reservoirs, in ice and snowpacks, in the unsaturated zone, and in the saturated zone. Change in water storage can be written as:

$$\Delta S = \Delta S^{sw} + \Delta S^{snow} + \Delta S^{uz} + \Delta S^{gw} \tag{2.5}$$

where the superscripts refer to the aforementioned compartments. ΔS^{sw} can include storage in surface depressions, in plants, and in the plant canopy (due to interception of precipitation). ΔS^{uz} refers to changes in water storage in the unsaturated zone at depths equal to or less than the zero-flux plane, ZFP, which is a plane at some depth in the subsurface where the magnitude of the vertical hydraulic gradient is 0 (Figure 2.2; Section 5.3). Water above that plane moves upward in response to the

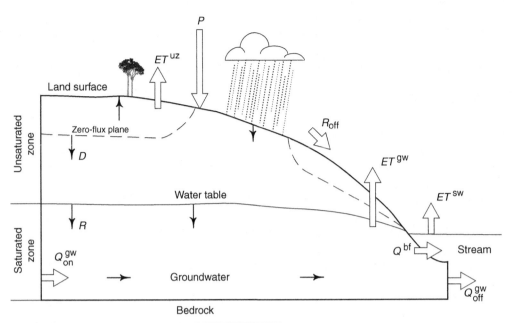

Figure 2.2 Schematic diagram of a vertical cross section through a watershed showing water-budget components (large arrows) and directions of water movement (small arrows). Recharge, R, occurs at the water table. Drainage, D, is water movement between the zero-flux plane and the water table.

evapotranspirative demand of the atmosphere. Water below that plane moves downward to eventually recharge the groundwater system. The zero-flux plane is sometimes equated with the bottom of the root zone. ΔS^{gw} refers to changes in storage in the saturated zone; as written here, it actually includes all storage changes that occur at depths greater than that of the zero-flux plane.

Flow out of the watershed, Q_{off}, can be divided into surface-water flow, Q^{sw}_{off}, and groundwater flow, Q^{gw}_{off} (which could include extraction of groundwater by pumping wells or groundwater flow to adjacent watersheds). In addition, Q^{sw}_{off} can be split into direct runoff, R_{off}, which now includes water that may have infiltrated and traveled through a part of the unsaturated zone but was diverted to land surface before reaching the saturated zone, and base flow, Q^{bf}, which is water that has been discharged from the saturated zone to springs and streams and is sometimes referred to as groundwater runoff:

$$Q_{off} = Q^{gw}_{off} + R_{off} + Q^{bf} \qquad (2.6)$$

Incorporating Equations (2.3 through 2.6) into Equation (2.2), the water-budget equation for the watershed can be expressed as (Figure 2.2):

$$P + Q^{sw}_{on} + Q^{gw}_{on} = ET^{sw} + ET^{gw} + ET^{uz} + \Delta S^{sw} \\ + \Delta S^{snow} + \Delta S^{uz} + \Delta S^{gw} + Q^{gw}_{off} + R_{off} + Q^{bf}$$

$$(2.7)$$

Equation (2.7) can be refined further, if need be. On the other hand, for some watersheds, many of the terms in the equation will be negligible in magnitude and can be ignored.

Several of the parameters described in this section are also components of the water budget of an aquifer. A water-budget equation for an aquifer equates change in groundwater storage with the difference between flow into and out of the aquifer. Inflow is in the form of recharge and groundwater flow; outflow is by groundwater flow, base flow, and evapotranspiration. The equation can be written as:

$$R + Q^{gw}_{on} = \Delta S^{gw} + Q^{bf} + ET^{gw} + Q^{gw}_{off} \qquad (2.8)$$

Equation (2.8) is a form of the groundwater-flow equation that is solved by numerical groundwater-flow models.

Combining Equations (2.7) and (2.8) results in a new water-budget equation for the control volume consisting of the watershed and the underlying unsaturated zone:

$$P + Q^{sw}_{on} = R + ET^{sw} + ET^{uz} + \Delta S^{sw}$$
$$+ \Delta S^{snow} + \Delta S^{uz} + R_{off} \qquad (2.9)$$

Watershed models (Section 3.4) are usually based on some form of Equation (2.7) or (2.9).

For residual water-budget methods, all terms in a water-budget equation, except recharge, are independently measured or estimated and recharge is set equal to the residual value. Equations (2.1), (2.8), or (2.9), or some variation of them, serve as the basis for most water-budget methods for estimating recharge. However, other equations can also be used. The residual water-budget approach is also used to estimate other water-budget components, such as evapotranspiration (e.g. Allen *et al.*, 1998; Wilson *et al.*, 2001).

Water-budget components can display markedly different trends. Precipitation can be highly episodic, and rates can range from 0 upward toward 100 mm/h; evapotranspiration, on the other hand, usually follows a distinct seasonal trend with rates ranging between 0 and 1 mm/h. Rates of water movement through the unsaturated zone vary considerably, being affected by soil properties, climate, land use, and depth to water table. Depending on conditions, rainfall that infiltrates might move to the water table (and thus become recharge) over the course of 1 day. In another system, infiltrating water may take several months to reach the water table. Because of these different temporal trends, recharge calculations are sensitive to the time step over which water budgets are tabulated. Recharge generally is predicted to occur only over a time interval in which precipitation exceeds evapotranspiration. This condition can occur on any day in virtually any environment. On a monthly basis, however, evapotranspiration exceeds precipitation for summer months in most environments. Hence, monthly time steps are likely to produce recharge estimates that are less than those produced by using daily time steps; longer time steps tend to dampen

out differences between daily precipitation and evapotranspiration. If data permit, daily time steps are recommended for water-budget tabulations.

2.2.1 Uncertainty in water budgets

Uncertainty in recharge estimates generated by water-budget methods can arise from inaccuracy in the underlying conceptual model, inadequate accounting of spatial and temporal variability of water-budget components, and uncertainty in measuring or estimating values for those components. Incorrect conceptual models of the hydrologic system introduce the most serious errors in using a water-budget method, or any other approach, to estimate recharge. An incorrect conceptual model can result in an incorrect water-budget equation, one with missing or superfluous terms. Conceptual models are addressed in Chapters 1 and 9. The effects of other sources of error on recharge estimates are discussed herein.

A simple example can be used to illustrate the effects of water-budget uncertainties on recharge estimates. Consider a soil column with no surface storage and no surface-water flow onto it, as described in Equation (2.1). Precipitation falling on the soil column runs off, is evapotranspired, remains in storage in the column, or drains out of the bottom of the column. Considering only measurement error, classical statistics can be applied to define the variance of the drainage estimate (σ^2_D) as:

$$\sigma^2_D = \sigma^2_P + \sigma^2_{ET} + \sigma^2_{\Delta S} + \sigma^2_{Roff} \qquad (2.10)$$

where σ^2_j is the variance of component j (Lerner *et al.*, 1990). This analysis requires assumptions that the water-budget components are independent of each other and that the measurement errors for all methods are unbiased and normally distributed with a mean of 0. If these assumptions are valid and the variances were known, confidence intervals could be calculated for drainage estimates. However, this approach cannot account for any spatial or temporal trends in water-budget components.

Dages *et al.* (2009) analyzed the water budget for a small Mediterranean watershed, considering only the spatial variability of water-budget variables. Variances were calculated for total precipitation depth, for example, by using the standard statistical equation:

$$\sigma^2_P = \sum_{i=1}^{n} \frac{(Pi - P^*)^2}{(n-1)} \qquad (2.11)$$

where P_i is precipitation at rain gauge i, P^* is average precipitation for all gauges, and n is the number of rain gauges in the watershed. The variances were used to calculate confidence intervals. This analysis required that water-budget variables be independent, normally distributed, and unbiased. These requirements are not easily verified, and Dages et al. (2009) made no attempt to do so. Geostatistical techniques such as kriging can account for both measurement error and spatial variability of water-budget components (Journel and Huijbregts, 1978). These techniques are discussed in Chapter 3.

Informal, empirical approaches to water budget uncertainty analyses are relatively easy to apply and can provide useful information. A common approach is to assume that the measured value of a component is equal to the true value plus some value of uncertainty. For example:

$$P = \underline{P} + \varepsilon_P \qquad (2.12)$$

where P is the measured or estimated value of precipitation, $\underline{P}$ is the true value, and ε_P is uncertainty associated with measurement error, spatial variability, or other unspecified factors. Modifying the other terms in Equation (2.1) accordingly, the water-budget equation for the soil column can be rewritten as:

$$D = (\underline{D} + \varepsilon_D) = (\underline{P} + \varepsilon_P) - (\underline{ET}^{uz} + \varepsilon_{ET}) \\ - (\underline{\Delta S}^{uz} + \varepsilon_{\Delta S}) - (\underline{R}_{off} + \varepsilon_{Roff}) \qquad (2.13)$$

The magnitude of the uncertainty in the drainage estimate, ε_D, cannot be determined exactly, but it must be less than or equal to the sum of the magnitudes of the other error terms:

$$|\varepsilon_D| \leq |\varepsilon_P| + |\varepsilon_{ET}| + |\varepsilon_{\Delta S}| + |\varepsilon_{Roff}| \qquad (2.14)$$

Insight on the magnitude of ε_D can be obtained if uncertainty terms on the right-hand side of Equation (2.14) can be determined or assumed (e.g. it might be assumed that ε_P was ±10% of the measured P). Drainage rates are then recalculated a number of times with Equation (2.13), each time varying uncertainty terms for one component at a time or for all components simultaneously. The effect of the variation on the calculated drainage can be used to predict a range of drainage values. Variations can be systematic, such as varying each component by a fixed amount or a fixed percentage of a measured value (Engott and Vana, 2007; Izuka et al., 2005). Alternatively, a Monte Carlo type analysis can be performed with component values randomly selected from an assumed distribution of values (Leake, 1984; Dages et al., 2009). This informal error analysis has two benefits: providing a range of possible drainage rates and identifying variables that most affect calculated drainage rates. Reducing uncertainty in the most influential variables should improve the reliability of drainage estimates. Sensitivity analyses such as this are also used in conjunction with numerical simulation models (Chapter 3).

When recharge is small relative to other water-budget components, the uncertainty in water-budget estimates of recharge can easily exceed 100% of the true value (Gee and Hillel, 1988). Consider the water budget of a soil column in a semiarid setting (Equation (2.1)), where drainage is assumed equal to the difference between precipitation and evapotranspiration (i.e. no surface runoff and negligible annual change in storage). Suppose true $\underline{P} = 200$ mm/yr and true $\underline{ET}^{uz} = 190$ mm/yr, but that measurement uncertainties for each were 10% of actual totals (not an unreasonable supposition). The true value of $\underline{D}$ is 10 mm/yr, but calculated values of D would fall in the range of –29 to +49 mm/yr, a large range of uncertainty by any measure. This simple analysis does not imply that all water-budget methods are inappropriate in arid regions. Intermittent focused recharge from ephemeral streams is often the dominant recharge mechanism in arid and semiarid regions, and methods based on the water budget of a stream are often used for quantifying stream loss (Chapter 4).

The magnitude of water-budget components can be readily assessed, at least in a preliminary sense, by using available data sources, such as those shown in Table 2.1 or in the Hydrometeorlogical Networks internet site of the National Center for Atmospheric Research (http://www.eol.ucar.edu/projects/ hydrometnet/; accessed August 17, 2009). These preliminary estimates may indicate that one component is very small relative to the others so that a large relative uncertainty in that component will ultimately contribute little to the overall uncertainty in recharge. For example, groundwater recharge from a lake might be addressed by examining the lake's water budget. If that water budget is dominated by surface-water inflow and outflow and lake level is fairly constant, then initial analysis might indicate that ΔS and ET are minor components. The uncertainty in the estimates of ΔS and ET would have little impact on the overall water budget and on the estimate for recharge. This implies that there would be little value in investing resources to improve upon the initial estimates of ΔS and ET. Resources are generally best directed at accurately quantifying the largest components in the water-budget equation, in this case surface-water flow into and out of the lake, because uncertainties in these components are generally the largest source of error in recharge estimates.

For ease of reading, uncertainty terms are not explicitly included in all water-budget equations in this text; however, the reader should understand that uncertainty terms are inherent to and implicitly included in every water-budget equation. Examination of uncertainties should be an integral part of any water-budget analysis, even though many studies reported in the literature neglect to do this.

2.3 | Local-scale application

Working at the local scale permits use of the most accurate methods for measuring individual components of the water-budget equation. Discussion in this section centers on point measurement techniques for each of the individual components. The control volume corresponds to a field plot or perhaps a small watershed with a length scale typically in the range of 1 m to 1 km.

2.3.1 Precipitation

Hourly and daily precipitation data are available from tens of thousands of weather stations within the United States maintained by the National Oceanic and Atmospheric Administration (NOAA) (Table 2.1). Analogous data are available in many other countries. However, it may be necessary or beneficial to directly measure precipitation in specific study areas, especially in regions with few gauges or where precipitation is dominated by convective thunderstorms or orographic effects. Most readers should be familiar with techniques for measuring precipitation; therefore, these techniques are only described briefly. More details can be obtained from the World Meteorological Organization (1983).

A number of gauges can be used to measure precipitation. The most common are the standard NOAA gauge and the tipping-bucket gauge. Considerations when selecting a gauge include accuracy, desired frequency of data collection (e.g. hourly or daily), operating temperatures (if snow and freezing conditions are a concern), and cost of purchase and maintenance. Most gauges have some type of analog or electronic recording capability.

The standard NOAA precipitation gauge consists of a 200 mm diameter cylinder with the top open to allow rain and snow to accumulate within the cylinder. The height of water in the cylinder is measured with a graduated rod. Advantages include easy calibration, ability to measure total precipitation under any intensity, and ability to measure snowfall. Antifreeze must be added to the collector during periods of freezing temperatures. During periods of warmer temperatures a small amount of oil is sometimes added to reduce water loss due to evaporation. Disadvantages of this type of gauge include limited accuracy and poor resolution of precipitation intensity (Strangeways, 2004). Automatic recording is possible with an electronic balance, a float,

Table 2.1. Sources of data for water-budget components within the United States. (H is hourly; D is daily; W is weekly; M is monthly; NOAA is National Oceanic and Atmospheric Administration; NASA is the National Aeronautics and Space Administration; USBR is the US Bureau of Reclamation; CIMIS is the California Irrigation Management Information System; USGS is the US Geological Survey; NRCS is the Natural Resources Conservation Service.) All websites accessed August 25, 2009.

Component	Freq.	Locations	Source	Description and reference
Precipitation	HDM	Gauge sites, US	NOAA	Current/historical: http://www.ncdc.noaa.gov/oa/ncdc.html
Precipitation	D	4-km grid, US	NOAA	Current/historical (2005–present). MPE interpolates using surface gauge and radar data: http://water.weather.gov
Precipitation	D	1-km grid, US	Daymet	Historical (1980–1997). Interpolated from NOAA surface data. Temperature, humidity, and radiation also available: http://www.daymet.org
Precipitation	M	4-km grid, US	Prism	Historical: Interpolated from NOAA surface data accounting for elevation. Temperature also available: http://www.prism.oregonstate.edu
Precipitation	DM	25-km grid, worldwide	NASA	Current/historical (2003–present). AMSR-E radiometer aboard AQUA satellite: http://nsidc.org/data/amsre/
Evapotranspiration	D	Gauge sites	NOAA	Current/historical. Pan evaporation: http://www.ncdc.noaa.gov/oa/ncdc.html
Evapotranspiration	D	Gauge sites, northwestern US	USBR	Current/historical. PET and other climate data: http://www.usbr.gov/pn/agrimet/
Evapotranspiration	D	Gauge sites, CA	CIMIS	Current/historical. PET and other climate data: http://www.cimis.water.ca.gov/
Evapotranspiration	D	2-km grid, FL	USGS	Current/historical (1995-present). PET and ET_o: http://fl.water.usgs.gov
Streamflow/stream stage	HD	Stream gauges, US	USGS	Current/historical: http://water.usgs.gov
Snow depth/water content	D	Gauge sites western US	NRCS SNOTEL	Current/historical: http://www.wcc.nrcs.usda.gov/snow/
Snow depth/water content	D	25-km grid, worldwide	NASA	Current/historical. AMSR-E radiometer aboard AQUA satellite: http://nsidc.org/data/amsre/
Soil-water content	W	Gauge sites, worldwide	Global Soil Moisture Bank	Historical. Compendium of measurements at worldwide locations: http://climate.envsci.rutgers.edu/soil_moisture/
Soil-water content	HD	Gauge sites, US	NRCS SCAN	Current/historical: http://www.wcc.nrcs.usda.gov/scan/
Groundwater levels	D	Gauge sites, US	USGS	Current/historical: http://water.usgs.gov

or a pressure transducer connected to an electronic recorder.

Tipping-bucket rain gauges are often used in hydrologic studies. This gauge contains two small buckets that fill with water from a collector. When one bucket fills, it drops, triggering a switch, emptying that bucket, and exposing the other bucket. Advantages of tipping-bucket gauges are their ability to measure precipitation intensity and small quantities (0.25 mm or less). Disadvantages include a limitation on the maximum rate of precipitation that can be measured (this depends on the size of the gauge, but is usually about 250 mm/hr), the need for calibration, and the necessity of heating the gauge to collect data in subfreezing temperatures (Molini *et al.*, 2005).

Precipitation gauges should be placed at locations that are representative of the field site and mounted so that the collectors are horizontal and not shielded by vegetation, fences, or other obstructions that could intercept precipitation. Ideally, the collector would be at a height equal to that of the vegetative canopy in the immediate surroundings. Wind shields can help mitigate the distortion of measurements caused by high winds if the gauge must be located above the canopy.

Snow pillows, or lysimeters, that can record small changes in weight (Johnson and Schaefer, 2002) are gradually replacing older snow boards and snow stakes for monitoring snow depth. Proper location and maintenance of snowfall gauges are important because wind can redistribute snow after it has fallen. In some parts of the world, dew or condensation is an important form of precipitation; it can be measured with a screen fog collector (Schemenauer and Cereceda, 1994).

2.3.2 Evapotranspiration

Evapotranspiration can be measured or estimated at the local scale by one or more of the *micrometeorological* methods described by Rosenberg *et al.* (1983). These methods can provide accurate estimates of evapotranspiration over small areas. Requirements for proper application of the methods can be onerous, however. The measurement site should be relatively flat and contain uniform vegetation (in terms of density, plant type, and height) for a distance at least 100 times instrument height in the predominant wind direction. Instruments mounted at a height of 2 m above the vegetative canopy would provide an estimate of evapotranspiration that was integrated over the horizontal surface extending approximately 200 m upwind. Instrumentation is expensive and can require frequent maintenance. The most popular of the micrometeorological methods are the energy-balance Bowen ratio (EBBR) and the eddy correlation methods.

The EBBR method is based on an energy-balance equation written for a unit area of land surface (Figure 2.3):

$$R_n = H + G + \lambda ET \qquad (2.15)$$

where R_n is net radiation, H is sensible-heat flux, G is soil heat flux, λ is latent heat of vaporization of water or the amount of energy required to evaporate a unit mass of water, and the product λET is referred to as the latent-heat flux and represents the energy required to evaporate water at the specified ET rate; units are usually

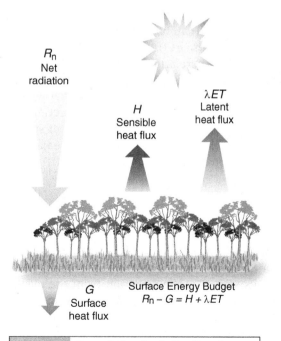

R_n
Net radiation

λET
Latent heat flux

H
Sensible heat flux

G
Surface heat flux

Surface Energy Budget
$R_n - G = H + \lambda ET$

Figure 2.3 Energy budget for the Earth's surface.

in terms of energy flux density, such as W/m^2. Equation (2.15) simply states that the net radiative input at land surface is used to heat the air, warm the soil, and evaporate water. Energy and water budgets are directly linked because ET is a component in each. Net radiation can be measured with a net radiometer. G is measured with soil-heat-flux plates and/or soil temperature probes (Sauer, 2002). Because of the limited accuracy with which sensible-heat flux can be directly measured, attempts to determine evapotranspiration from Equation (2.15) by the residual method are susceptible to large errors. Bowen (1926) proposed using the ratio $\beta = H/\lambda ET$ to solve Equation (2.15). β can be calculated as a constant times the vertical gradient in air temperature divided by the vertical gradient of water-vapor pressure. Then Equation (2.15) can be rewritten as:

$$ET = (R_n - G)/(\lambda(1 + \beta)) \qquad (2.16)$$

Measurements of air temperature and vapor pressure are made at two heights, usually between 0.5 and 2.0 m above the top of the vegetation canopy. Measurements are made at intervals of a minute or less, and Equation (2.16) is applied with values that are averaged over periods of 15 to 60 minutes. Additional details on this method can be found in Rosenberg et al. (1983) and Brutsaert (1982).

The eddy correlation method directly measures vertical fluxes of heat (H) and water (ET). The method is based on the concept that water vapor is transported in the vertical direction by the upward and downward movement of small parcels of air, or eddies (Lee et al., 2004). The relevant equations are:

$$H = \rho c_p \; w'T' \qquad (2.17)$$

$$ET = w'q' \qquad (2.18)$$

where ρ is the density of air, c_p is specific heat of air, w' is the deviation from the mean vertical wind speed, T' is the deviation from the mean air temperature, and q' is the deviation from the mean water-vapor density. Precise and rapid (10 Hz) measurement of w, T, and q requires highly specialized instruments (Figure 2.4).

A sonic anemometer, fine-wire thermocouple, and a krypton hygrometer are commonly used. These are expensive instruments; recent innovations have substantially improved their durability and reliability relative to early designs (Lee et al., 2004). The eddy correlation method is the most popular land-based point or ground-truth method for large-scale regional and global evapotranspiration and trace-gas flux experiments, such as FLUXNET (Baldocchi et al., 2001) and EUROFLUX (Aubinet et al., 2000). However, it is well known that direct application of the method systematically underestimates both H and ET and that corrections are required to ensure the closure of the energy budget (Wolf et al., 2008). Twine et al. (2000) showed that these corrections can be as high as 10 to 30% of available energy, $R_n - G$.

Additional approaches for measuring evapotranspiration include the zero-flux plane method and lysimeters (both described in Chapter 5). Sap-flow meters (Wilson et al., 2001) provide estimates of transpiration by sensing water velocity in plant stems. Small static chambers (Stannard and Weltz, 2006) can be used to estimate evapotranspiration rates for areas of about 1 m^2.

2.3.3 Change in storage

Change in storage and its evaluation should be considered with respect to the time scale and spatial scale of interest. Changes occur in response to daily and seasonal weather patterns. Annual cycles are of interest in many, if not most, hydrologic studies. Natural hydrologic systems typically display little change in storage from year to year, so the simple assumption that storage change is 0 when averaged over 1 year may be appropriate for some systems (Healy et al., 1989). However, decadal and longer term changes can occur in response to changing climates or land use.

Equation (2.5) includes change in storage for four distinct compartments: surface water, snow, unsaturated zone, and saturated zone. Change in storage between times t_1 and t_2 in any of these compartments is usually calculated as the difference in total water storage in the compartment at the two observation

Figure 2.4 Tower to measure evapotranspiration by the eddy correlation method within a forest and close up of a sonic anemometer, fine-wire thermocouple, and krypton hygrometer.

times; $\Delta S > 0$ indicates an increase in storage from time t_1 to time t_2. Water storage in surface-water bodies is usually determined by measurement of stage and use of a stage-volume relationship. This relationship can be determined for a lake, for example, by preparing a bathymetric map of the lake bottom and then calculating the volume of water present at a given stage level. Lake stage can be measured manually, with staff or wire-weight gauges, or recorded automatically with float gauges or pressure transducers.

Water storage in snowpacks has historically been measured with snow surveys, whereby manual measurements of snow depth and density were made at fixed locations on transects at specific time intervals through winter and spring. In recent years, the Natural Resources Conservation Service (NRCS) has set up the SNOTEL network of snow monitoring sites across the western United States (Table 2.1). Snow depth, snow-water equivalent, and other

climatic parameters are electronically recorded. The data are available in real time (via satellite telemetry) to the general public. The network is particularly useful for watershed managers who use the data in streamflow and reservoir forecasting models to optimize water use and storage.

Water storage within the subsurface, S^{subs}, is described by the following general equation:

$$S^{subs} = \int_{z_{base}}^{0} \theta dz \qquad (2.19)$$

where θ is volumetric soil-water content, the upper limit in the integral (0) represents land surface, z_{base} is depth to the base of the aquifer, and storage due to compression of water and solids is ignored. Equation (2.19) is written for a soil column of unit surface area, so S^{subs} has units of length (e.g. mm). Storage within individual depth intervals of the subsurface can be represented with Equation (2.19) by adjusting the limits on the integral (Figure 2.5). For calculating S^{uz}, the amount of water stored between land surface and the zero-flux plane, the equation becomes:

$$S^{uz} = \int_{z_{ZFP}}^{0} \theta dz \qquad (2.20)$$

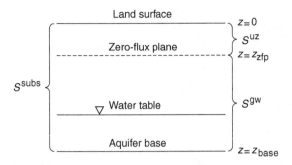

Figure 2.5 Schematic of soil column extending from land surface to the base of the aquifer showing different storage compartments. S^{subs} represents all water stored in column, S^{uz} is water stored between land surface and the zero-flux plane, and S^{gw} is water stored between the zero-flux plane and the base of the aquifer. z_{ZFP} is depth to zero-flux plane; z_{base} is depth to base of aquifer.

where z_{ZFP} is the depth to the zero-flux plane. The relevant equation for calculating ΔS^{uz}, as defined in Equation (2.5), is:

$$\Delta S^{uz}_i = (S^{uz}_i - S^{uz}_{i-1})/(t_i - t_{i-1}) \qquad (2.21)$$

where i is an index of time, t. Water-budget methods for estimating evapotranspiration are often based on Equations (2.20) and (2.21). For determining the amount of storage underlying the zero-flux plane, S^{gw}, the equation becomes:

$$S^{gw} = \int_{z_{base}}^{z_{ZFP}} \theta dz \qquad (2.22)$$

And an equation analogous to Equation (2.21) is used to define ΔS^{gw}.

There are many different bookkeeping schemes for subsurface water-storage accounting, so some clarification is in order. Our approach divides the soil column into two compartments separated by the zero-flux plane. A problem with this approach (and, indeed, a problem with many other water-budget approaches) is that time lags may not be properly taken into account. Water moves at different rates through different parts of the hydrologic cycle; it can take weeks or months for precipitation from a day-long storm to move from land surface to the water table. If a water-budget equation such as Equation (2.9) is tabulated on a weekly

basis, water draining beneath the zero-flux plane that has not yet reached the water table would be counted as recharge, instead of what it actually should be called, which is drainage. By our definition (Equations (2.5) and (2.8)), both ΔS^{gw} and R include actual recharge (indicated by change in water table height) and drainage (indicated by a change in water storage within the interval between the zero-flux plane and the water table – that water does not become actual recharge until it arrives at the water table). Different boundaries for describing subsurface storage are used in different studies, so it is important that those boundaries be clearly defined.

Estimates of drainage produced by the zero-flux plane method (Section 5.3) are based on changes in storage within some interval between the zero-flux plane and the water table; changes in water-table height have no effect on those estimates if the water table does not intersect the measurement window. On the other hand, the water-table fluctuation method (Section 6.2) ignores any change in storage between the zero-flux plane and the water table; estimates of recharge are obtained by looking only at the changes in groundwater storage that occur with rises and falls of the water table:

$$\Delta S^{gw} = S_y \Delta H / \Delta t \qquad (2.23)$$

where S_y is specific yield and ΔH is change in water-table height over time interval Δt.

Measurement of soil-water content by gravimetric and geophysical techniques and calculation of ΔS^{uz} are described in some detail in Chapter 5. Soil-water content data are not as widely available as precipitation and other climatological data, but there are some sources of data. The Global Soil Moisture Data Bank (Robock et al., 2000; Table 2.1) contains soil-water content data obtained at regular time intervals, to depths of about 1 m, at over 600 fixed locations worldwide. These data have been used for assessing regional trends over decades (e.g. Robock and Li, 2006) and for evaluating simulation results of general circulation and land surface models (e.g. Fan et al., 2006). The NRCS

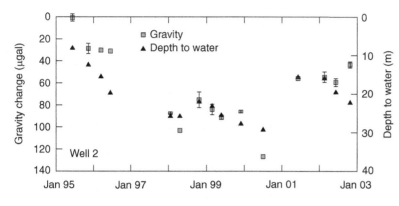

Figure 2.6 Gravity and groundwater-level relations for an unconfined aquifer, 1995 to 2002 (after Pool, 2008). Error bars represent standard deviation of gravity survey. Gravity values without error bars had insufficient survey data to calculate standard deviation.

maintains the Soil Climate Analysis Network (SCAN) of sites across the United States where soil-water content and temperature are measured hourly at five depths to a maximum of 1 m (Table 2.1). These databases are still sparse in terms of areal coverage, but as more sites are added, they should become more useful for water-budget studies. Groundwater levels for application of Equation (2.23) can often be found in databases maintained by state and federal agencies. Within the United States, the US Geological Survey maintains the largest database on real-time and historic groundwater levels (Table 2.1).

Temporal gravity method for determining change in subsurface storage

Gravity measurements made from land surface are reliable indicators of total subsurface mass. Measurements have historically been used to map spatial variability in mass for purposes of defining geologic structure. Change over time in the amount of water stored in the subsurface is usually the dominant factor in temporal changes in total subsurface mass. So changes in subsurface mass at a particular location, as determined with gravity measurements made at different points in time, can be used to infer changes in subsurface water storage. Measurements of gravity are made at specific points on land surface; these points may lie on a transect or be distributed across a watershed. Measurements are made at time intervals that can range from days to months. After readings have been corrected for tidal and instrument effects, changes in gravity between successive readings are then used

to calculate changes in total storage, ΔS (Pool and Eychaner, 1995):

$$\Delta S = 1.99 \Delta g \qquad (2.24)$$

where ΔS is in millimeters of water and g is gravitational acceleration in μgals. Equation (2.24) is based on an assumption of areally uniform change in storage. Pool and Eychaner (1995) found standard deviations of gravity changes to be between 1.6 and 5.9 μgals, equivalent to 3.3 to 11.7 mm of water. Chapman *et al.* (2008) estimated the accuracy of gravity measurements at ± 5 μgals in their study.

The temporal gravity method determines total water storage, as opposed to soil-water content. The method cannot distinguish between changes in storage within the unsaturated zone and saturated zone; this can limit the method's applicability in some recharge studies. If a well is near each measurement location, changes in gravity can be correlated with changes in groundwater levels. Pool and Schmidt (1997) and Pool (2008) used such a correlation to determine aquifer specific yield. Pool (2008) found reasonable agreement between the measured depth to the water table and gravity change in an unconfined aquifer in southeastern Arizona (Figure 2.6). Gravity measurements can be obtained quickly without the need for a borehole, but gravimeters are relatively expensive. The method works best when fluctuations in groundwater levels are in the order of a few meters or more.

2.3.4 Surface flow

Hourly and daily streamflow data for thousands of stream gauges within the United

States are available through the US Geological Survey (Table 2.1). However, direct measurement of surface flow may be necessary in some studies. Judicious selection of boundaries of a control volume or study area can facilitate measurement of surface flow. Small watersheds with no surface inflow make good control volumes; in this sense, only streamflow out of the watershed would need to be measured. In semiarid and arid regions with permeable soils, runoff is often assumed to be negligible. Surface flow can occur in stream channels or as overland flow (sheet flow). Measurement of surface flow onto and off a site is straightforward if flow occurs only in stream channels. Overland flow is much more difficult to directly measure (the estimation methods discussed in Section 2.4.4 are more commonly applied). So it is difficult to construct an accurate water budget for a control volume (such as a small field plot) where overland flow across boundaries is substantial.

Streamflow can be measured directly with an acoustic Doppler current profiler (Olson and Norris, 2007); this device can provide discharge data in real time if it is maintained at a single location. Because of the high cost of the Doppler profiler, few studies can afford to install one permanently. Streamflow at gauging stations is more commonly determined by measuring stream stage and converting that stage to a discharge by means of a rating table. Rating tables can be developed theoretically or empirically. Hydraulic structures, such as flumes, culverts, and weirs, have theoretical rating curves (Kennedy, 1984; Chow et al., 1988; Maidment, 1993). Installation of a hydraulic structure in a small stream channel removes the need for direct discharge measurements. In the absence of a Doppler profiler, streamflow can be determined by careful measurement of the cross-sectional area of the stream (normal to the direction of flow) and measurement of flow velocity at a number of locations in the cross section (Olson and Norris, 2007). Velocity is measured with a mechanical current meter. Stream stage can be monitored with a float or pressure transducer; real time data on stream stage are available for many stream gauge sites

in the United States (Table 2.1). Stream discharge determined from a Doppler profiler or from a rating table is generally thought to be, at best, within 5% of actual discharge (Rantz, 1982; Oberg et al., 2005). Additional discussion on streamflow measurement is contained in Section 4.2, in which the stream water-budget method is described. Steep channels and ice-affected streams are not amenable to standard methods of discharge measurement; tracer injection methods (discussed in Section 4.6.2) can be used in these streams (Kilpatrick and Cobb, 1985; Clow and Fleming, 2008).

Overland flow can be measured on small plots, but it is a tedious procedure, and the measurement technique may alter natural flow rates. Impermeable edging can be pressed into the ground surface so as to funnel overland flow to a collector. A tipping-bucket mechanism, similar to that used in precipitation gauges, can be used at the collection point to measure flow (Wilkison et al., 1994).

2.3.5   Subsurface flow

Quantifying groundwater inflow to and outflow from a small study area is problematic because of the three-dimensional nature of groundwater flow and the inherent variability in properties of subsurface materials (Alley et al., 2002). Knowledge of hydraulic conductivity and hydraulic head at multiple locations and depths is required for determining groundwater flow rates. As with surface flow, selection of boundaries for control volumes will influence the accuracy with which boundary fluxes can be determined. Groundwater divides provide natural boundaries because there is no flow across them. For some small basins, all groundwater discharges to a stream near the mouth of the watershed, thus simplifying the water-budget study.

One-dimensional groundwater fluxes can be estimated with the Darcy equation:

$$q = -K_s \, \partial H / \partial x \qquad (2.25)$$

where q is specific flux (flux per unit cross-sectional area), K_s is saturated hydraulic conductivity in direction x, and H is total head. Data on hydraulic conductivity and water levels can be expensive to obtain. Hydraulic conductivity of geologic materials varies over a wide range. Hydraulic conductivity of gravels, fractured basalts, and karst limestones can be as high as 1000 m/d, whereas values for shales, marine clays, and glacial tills can be as low as 10^{-7} m/d (Heath, 1983). Variability in hydraulic conductivity is typically the largest source of uncertainty in estimating groundwater flow. Hydraulic conductivity can be measured in the laboratory on undisturbed soil-core samples or repacked samples (Reynolds *et al.*, 2002); these samples usually have lengths of tens to hundreds of millimeters. Hydraulic conductivity can be determined over larger spatial scales by applying field techniques such as single well slug tests (Butler, 1997) and multiwell aquifer pumping tests (Walton, 2007).

Groundwater flow models (Section 3.5) are often used as an alternative to Equation (2.25) for estimating groundwater flow rates because of the complex nature of many groundwater systems. Prior to the development of numerical flow models, groundwater flow rates were estimated with flow nets (Section 6.3.3). Additional information on groundwater flow is available in texts such as Freeze and Cherry (1979) and Todd and Mays (2005).

Groundwater withdrawal or injection may be an important part of the water budget of a watershed or aquifer. Reporting rules for groundwater users vary by governing entity. In some areas, data on water use are easily obtained. In other areas, with many domestic or irrigation wells, data on water use may be difficult to obtain. Most of these wells are not equipped for monitoring flow rates. For large capacity irrigation wells, pumping rates can be estimated on the basis of power consumption (Hurr and Litke, 1989).

Example: Northwestern Illinois
The water budget for an 8-hectare waste-disposal site in northwestern Illinois was studied from 1982 to 1984 (Healy *et al.*, 1989). A modified form of Equation (2.9) was used to determine drainage below the zero-flux plane (D):

$$D = P - ET^{uz} - \Delta S^{uz} - R_{off} \qquad (2.26)$$

Precipitation was measured with two tipping-bucket gauges and one weighing gauge. Evapotranspiration was estimated by the energy-balance Bowen ratio method. Change in unsaturated-zone storage was determined by borehole logging with a gamma-attenuation tool at three locations on a weekly to biweekly schedule. Runoff to three intermittent streams was measured with a Parshall flume or a V-notch weir. Drainage was calculated by the residual method, using Equation (2.26). Drainage was also determined directly by applying the unsaturated-zone Darcy method (Section 5.4) at two of the three logging locations. Therefore, this study provides an opportunity to compare the residual water-budget method with another method.

Results for the first year of the study (Figure 2.7) illustrate the variability in time of individual water-budget components as well as the interdependency among all components. Total precipitation of 927 mm is similar to the long-term average of 890 mm for the site; it was distributed fairly evenly across the year with slightly lower amounts in winter months. Evapotranspiration rates followed the seasonal pattern in net radiation, highest in late spring and summer and close to 0 during winter months. Runoff occurred only after large precipitation events. Drainage, as calculated from Equation (2.26), was episodic, occurring in response to large rainfalls, mostly in spring and early summer when evapotranspiration was low and soil-water contents were high. Negative values for drainage were calculated for many days. These values do not indicate groundwater discharge (the water table was at a depth of 10 m or more). Instead, they are attributed to the following factors: the time lag between measurements of ΔS^{uz} (the rate of change was assumed uniform between measurements), instrument inaccuracy, and an insufficient number of sampling points particularly for ΔS^{uz} and runoff. Drainage calculated for the year by Equation (2.26) was 126 mm, whereas that calculated by the Darcy method was 216 mm (Healy *et al.*, 1989).

Several factors may have contributed to the discrepancy between the two drainage

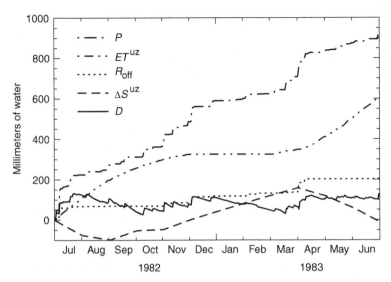

Figure 2.7 Cumulative water-budget components for July 1982 through June 1983 for a site in northwestern Illinois (after Healy et al., 1989). P is precipitation, ET^{uz} is evapotranspiration from the unsaturated zone, R_{off} is runoff, ΔS^{uz} is unsaturated-zone storage change, and D is drainage.

estimates. The discrepancy could be explained by a 10% inaccuracy in precipitation measurements or a 15% inaccuracy in evapotranspiration estimates. There is also the question of spatial variability in processes and properties and of measurement scale for determining water-budget components. Runoff estimates are integrated over the entire area that drained to the stream gauges (more than 80% of the site). Likewise, evapotranspiration rates are values integrated over a fairly large portion of the site. Precipitation did not vary considerably across the site. So the water-budget estimate of drainage can be viewed as a value integrated over the entire site. The Darcy method produced point estimates at two discrete locations. Perhaps more accurate sitewide estimates would have been obtained with additional measurement locations. Uncertainty in hydraulic conductivity can also contribute to inaccuracy in Darcy method estimates. In any regard, the accuracy of either drainage estimate cannot be quantified. This example illustrates the importance of applying multiple methods for estimating recharge.

Example: Eastern Long Island

Steenhuis et al. (1985) estimated groundwater recharge at a site on eastern Long Island by analyzing the water budget of a column of soil extending from land surface to the water table.

Owing to the permeable soils at their study site, surface-water flow, direct runoff, change in unsaturated-zone storage, and travel time for water draining through the unsaturated zone were assumed to be negligible. Recharge was calculated as the difference between precipitation and evapotranspiration:

$$R = P - ET \qquad (2.27)$$

By our definition (Section 2.2), Equation (2.27) actually calculates drainage, but Steenhuis et al. (1985) referred to this as recharge. Again, the distinction between drainage and recharge is one of lag time. An energy-budget approach was used to estimate evapotranspiration. Micrometeorological measurements made at the site included precipitation (three gauges), wind speeds and air temperatures at 1 and 3 m heights, dew point temperatures at 1.68 and 2.68 m heights, and net radiation. Data were recorded at hourly intervals.

Monthly estimates of recharge and precipitation for 8 months in 1980 are shown in Table 2.2. Negative recharge rates reflect decreases in soil-water storage that were not accounted for in this approach. Also shown in Table 2.2 are recharge estimates obtained with the unsaturated-zone Darcy method. For most months, these values were less than the estimates obtained by the water-budget method. The discrepancies are again attributed to soil-water

Table 2.2. Monthly precipitation and estimated recharge in millimeters by the water-budget method, R1, and the Darcy method, R2, for a site on eastern Long Island in 1980 (after Steenhuis et al., 1985).

	Jan	Feb	Mar	Apr	May	Jun	Nov	Dec
R1	56	29	153	93	−17	−16	81	35
R2	31	4	137	114	24	9	16	24
P	42	27	176	139	49	68	92	13

storage changes that were unaccounted for. The two methods produced similar estimates of recharge totals for the 8-month period, 414 mm for the water-budget method and 359 mm for the Darcy method. These estimates correspond to 68% and 59% of the 606 mm of precipitation for this period.

Example: Southwest India

Langsholt (1992) conducted a soil water-budget study in the state of Kerala, southwest India, during the monsoon months of June through August of 1988 and 1989. Permeable lateritic soils, 10 m thick, overlie low-permeability bedrock. Heavy monsoon rains typically produce water-level rises of several meters in a perched water table within these soils. The water budget for the 600 m² field site is given as:

$$R = P - ET^{uz} - \Delta S^{uz} - R_{off} \qquad (2.28)$$

Instrumentation included two precipitation gauges, a V-notch weir located in the drainage channel that bisected the site, water-level recorders in observation wells, and a neutron probe for monitoring changes in soil-water content. The Morton (1978) equation was used to estimate evapotranspiration. Air temperature, humidity, and sunlight duration data were measured at a station 1 km from the field site. The hydrology of this system is rather unique; the soils have a high permeability and a very large storage capacity. In spite of the high precipitation rates (Figure 2.8), runoff from the site was essentially negligible. The magnitude of recharge (70% of precipitation) greatly exceeds that of both runoff and evapotranspiration; therefore, the water-budget method is well suited for application at this site.

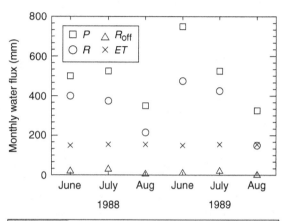

Figure 2.8 Water-budget components for June, July, and August 1988 and 1989 for site in southwest India (after Langsholt, 1992). Data for June and August 1988 are for partial months. P is precipitation, R_{off} is runoff, R is recharge, and ET is evapotranspiration.

2.4 | Mesoscale application

The term mesoscale refers, herein, to a medium to large watershed with a length scale of 1 to 1000 km. At this scale, measurement of water-budget components at a single location may not produce values that are representative of the entire basin. Accurate characterization of the water budget requires measurements or estimates at multiple locations in conjunction with a scheme for integrating or upscaling the information from these locations to the basin as a whole. Fiscal and time constraints and limited availability of data often dictate the application of simplified estimation equations in place of the direct measurement techniques discussed in Section 2.3. This section provides a review of methods for estimating individual water-budget components. The methods presented are straightforward and inexpensive to apply and, in general, require only data that are available from standard sources, such as National Weather Service weather stations and US Geological Survey stream-gauging stations. Although discussed here in terms of mesoscale studies, these methods may also be applied in smaller scale studies. In particular, estimating equations are often preferred to the expensive and time consuming micrometeorological methods for measuring evapotranspiration. Watershed models are commonly used for studying water budgets at this scale (Section 3.4). Discussion of upscaling procedures, a key part of mesoscale watershed studies, is deferred to Section 3.7.

2.4.1 Precipitation

In the past, maps of mean annual precipitation and precipitation frequency distribution for the United States were published by the US Department of Commerce (e.g. Karl and Knight, 1985; Barnston, 1993). These widely used maps were generated from data collected at surface-based precipitation gauges. Newer techniques that use radar as well as gauge data provide more accurate data at finer resolutions. Since 2005, NOAA's Multisensor Precipitation Estimation (MPE) program (Seo, 1998; Seo et al., 1999; 2000) has provided estimates of hourly and daily precipitation on a 4-km grid for the entire conterminous United States by using surface-based Nexrad Doppler radar in conjunction with real-time standard rain gauges and satellite imagery (Table 2.1). These estimates are used for predicting streamflow and providing warnings of severe weather and floods, especially for large watersheds; the estimates are also useful in water-budget studies for estimating recharge. Refined temporal and spatial resolutions of the MPE estimates should be available in the future. The Daymet database (Table 2.1; Thornton et al., 1997) is another source of precipitation data. It contains daily precipitation, air temperature, and radiation data for the United States from 1980 to 1997 generated on a 1-km grid using data from NOAA's surface-based weather stations. The Prism modeling system (Table 2.1, Daly et al., 1994; 2008) provides monthly precipitation and air temperature on a 4-km grid;

values are determined on the basis of surface-based weather stations and elevation.

2.4.2 Evapotranspiration

Because of the limited availability of evapotranspiration information at the mesoscale, values of evapotranspiration are often estimated. Many estimating equations have been developed, and although they may lack the accuracy of the micrometeorological methods, they are easier and less expensive to apply. Rosenberg et al. (1983) referred to these techniques as *climatological* methods. Jensen et al. (1990) reviewed several of these methods. Brief descriptions of some of the more widely used equations follow.

Most of the techniques described in this section were developed to predict potential evapotranspiration, *PET*. According to Rosenberg et al. (1983), "Potential evapotranspiration is the evaporation from an extended surface of a short green crop which fully shades the ground, exerts little or negligible resistance to the flow of water, and is always well supplied with water. Potential evapotranspiration cannot exceed free water evaporation under the same weather conditions." Conditions within a watershed will not always be consistent with these requirements, so actual evapotranspiration is often less than *PET*. Actual values can be estimated from values of *PET* if the ratio of *ET/PET*, called the crop coefficient (K_c), is known. Doorenbos and Pruitt (1975) and Jensen et al. (1990) provide tables of K_c values for various crops under different climate conditions and stages of growth. Both actual and potential evapotranspiration are important in hydrologic studies. Potential evapotranspiration calculations are used in irrigation management systems. Many hydrologic models require potential, as opposed to actual, evapotranspiration data (or the data from which *PET* can be calculated).

Pan evaporation is measured at many weather stations (Table 2.1). *PET* can be estimated from recorded pan evaporation using the formula:

$$PET = k_p E_{pan} \qquad (2.29)$$

where k_p is the pan coefficient, which is a function of pan type, climate, fetch, and the pan environment relative to nearby surfaces, and E_{pan} is mean daily pan evaporation in mm/d. Values of k_p for the standard Class A pan are given in Doorenbos and Pruitt (1975); they range from 0.35 to 1.10 and decrease with increasing wind speed and decreasing humidity. The method is simple to apply and data are available from many weather stations. Pan measurements themselves, however, can be problematic; birds and other animals are often attracted to the pans (Jensen et al., 1990).

The Thornthwaite (1948) method for estimating PET requires only air temperature data. To estimate monthly PET in mm/month, the following formula was proposed:

$$PET = 16(l_1 / 12)(N / 30)(10\ T_a / I)^{a1} \qquad (2.30)$$

where l_1 is length of day in hours, N is number of days in month, T_a is mean monthly air temperature (°C), $I = \sum_{i=1}^{12} (T_a)_i^{1.514}$ is the heat index (i is an index to months), and $a1 = 6.75 \times 10^{-7}I^3 - 7.71 \times 10^{-5}I^2 + 0.0179I + 0.49$. According to Rosenberg et al. (1983), this method can be successful on a long-term basis, although PET tends to be underestimated during summer periods of peak radiation.

The Jensen and Haise (1963) method produces daily estimates of PET from daily air temperature and solar radiation:

$$PET = R_s(0.025T + 0.08) \qquad (2.31)$$

where PET is in mm/d, R_s is daily total solar radiation in units of millimeters of water/day, and T is mean daily air temperature (°C). According to Rosenberg et al. (1983), the Jensen-Haise method seems to give good results under nonadvective conditions, but may underestimate PET under highly advective conditions (i.e. when high wind carries warm dry air, and hence energy, into the study area).

Penman's (1948) formula, which requires net radiation, soil-heat flux, and humidity, was the first widely used method for estimating PET:

$$PET = (s(R_n - G) + \gamma E_a) / (s + \gamma) \qquad (2.32)$$

where s is the slope of the water vapor saturation curve, γ is the psychrometric constant, and E_a is a function of wind speed and water vapor pressure deficit that has taken several forms. Thom and Oliver (1977) proposed the formula:

$$E_a = 0.037(1 + U / 160)(e_s - e_a) \qquad (2.33)$$

where U is the windrun in km/d at a 2 m height and e_s and e_a are saturated and actual vapor pressure, respectively, expressed in kPa.

Monteith (1963) modified the Penman equation for vegetated areas to account for the fact that specific humidity at leaf surfaces is usually less than saturation. The Penman–Monteith equation predicts actual ET by including resistance terms:

$$ET = \{r_a s(R_n - G) + \rho c_p(e_s - e_a)\} / \lambda\{r_a s + \gamma(r_a + r_c)\} \qquad (2.34)$$

where r_a and r_c are the aerial boundary layer and the total canopy resistance, respectively, in units of time over length. The canopy resistance depends on leaf area and environmental factors such as solar radiation and temperature in ways that vary from species to species. So while Equation (2.34) is general, application to specific sites is hampered by a lack of knowledge of these resistance terms for a wide variety of species.

The Priestley and Taylor (1972) equation is another widely used variation of the Penman equation:

$$PET = \alpha s(R_n - G) / (s + \gamma) \qquad (2.35)$$

where α is an empirically derived constant that was originally set at 1.26.

Allen et al. (1998) generalized the Penman equation for a well-watered reference crop of short grass to obtain what the authors refer to as reference evapotranspiration, ET_0, in mm/d:

$$ET_0 = \{0.408s(R_n - G) + 900\gamma u_2(e_s - e_a) / (T + 273)\} / \{s + \gamma(1 + 0.34u_2)\} \qquad (2.36)$$

where R_n and G are in MJ/m²/day, u_2 is wind speed at the 2 m height in m/s, and s and γ are in kPa/°C. The concept of reference evapotranspiration is similar to that of potential evapotranspiration; actual crop evapotranspiration

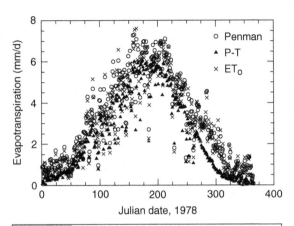

Figure 2.9 Daily estimates of potential evapotranspiration by the Penman (Equation (2.32)) and Priestley–Taylor (P-T) (Equation (2.35)) methods and reference evapotranspiration (ET_o) by Equation (2.36) for Kimberly, Idaho, for 1978 as calculated by the computer program Ref-ET (University of Idaho, 2003). Data used in the calculations included daily precipitation, minimum and maximum air and soil temperatures, relative humidity, and solar radiation. Similar trends are apparent over the course of the year for all three estimates, although considerable variability exists on a daily basis.

can be calculated by multiplying ET_0 by the crop coefficients listed in Allen et al. (1998).

An attractive feature of these climatological methods is that they can usually be applied with data that are available at standard NOAA weather stations (e.g. air temperature, precipitation, relative humidity, wind speed, and solar radiation); no additional data collection is required. Some parameters in the above equations, such as net radiation and soil-heat flux, are seldom measured at NOAA stations, but equations for estimating these variables can be found in Allen et al. (1998), Jensen et al. (1990), and Rosenberg et al. (1983). The similarities in the above equations facilitate calculation of multiple estimates of PET or ET_0 in a spreadsheet or in computer programs such as Ref-ET (University of Idaho, 2003). Ref-ET will calculate PET or ET_0 by as many as 15 different equations (Figure 2.9) using available data and estimating values for parameters that were not measured.

No national network of evapotranspiration sites exists within the United States, but stations are maintained in certain regions to assist water managers and irrigation schedulers.

Daily estimates of reference evapotranspiration are available at approximately 120 locations in the California Irrigation Management Information System (CIMIS) network (Table 2.1) and at 70 locations in the Pacific Northwest of the United States as part of the Agrimet network operated by the Bureau of Reclamation, in conjunction with the Agricultural Research Service (Table 2.1).

2.4.3 Change in storage

Much of the discussion contained in Section 2.3.3 for measuring changes in storage at the local scale also applies at the mesoscale. Snow courses are used to determine water storage in transects of snowpacks. Aircraft-mounted gamma-ray spectrometry (Glynn et al., 1988) and microwave measurements (Chang et al., 1997) offer more rapid means to conduct snow surveys; 500 km transects can typically be covered in a single day. In addition, Fassnacht et al. (2003) demonstrated how to interpolate snowpack data from SNOTEL sites onto a 1-km grid. The temporal gravity technique described in Section 2.3.3 has been applied over transects as long as 20 km; longer transects are feasible. Thornthwaite and Mather (1955; 1957) proposed a bookkeeping technique for determining total change in storage in a watershed. Their approach is illustrated in a following example.

2.4.4 Surface flow

Estimates of surface-water flow may be needed in recharge studies for predicting direct runoff in the absence of any streamflow gauges or data. A common technique for predicting direct runoff is the US Soil Conservation Service curve number method (Natural Resources Conservation Service, 2004). The dimensionless curve number, CN, is determined on the basis of soil type, land use, and antecedent soil-water content. Direct runoff, R_{off}, in inches, is estimated as:

$$R_{off} = (P - .2S)^2 / (P + .8S) \qquad (2.37)$$

where P is precipitation (in inches) and $S = 1000 / CN - 10$. Numerous variations of Equation (2.37) appear in the literature (Garen and Moore, 2005a, b).

Estimates of streamflow at ungauged sites or for future times are often required in studies examining the effects of changes in land use, climate, or aquifer management. Watershed models are often used for this purpose (Micovic and Quick, 1999; Merz and Bloschl, 2004). These models (Section 3.4) are water-budget models of a watershed, incorporating precipitation and other climatic data along with characteristics such as surface elevation, soil properties, and vegetation into their calculations. Less complex, nonparametric models, such as regression equations, are also used for predicting streamflow at ungauged sites (Schilling and Wolter, 2005).

2.4.5 Subsurface flow

Estimations of inflow and outflow of groundwater at the watershed scale can be obtained by applying the Darcy equation (Equation (2.25)), as discussed in Section 2.3.5, but there will always be large uncertainties associated with those estimates. Historically, flow nets have been used to estimate groundwater flow rates (Cedergren, 1988; Section 6.3.3). Groundwater-flow models are probably the most useful tools for determining groundwater flow rates at the mesoscale. These models and their use for estimating recharge are discussed in Section 3.5. Zheng et al. (2006) reviewed internet sources of data for groundwater flow models.

Example: Panther Creek watershed
Schicht and Walton (1961) developed water budgets for Panther Creek and two other watersheds in Illinois for the years 1951 and 1952. Precipitation, groundwater levels, and stream stage and discharge were monitored. The Panther Creek watershed occupies approximately 246 km² in north central Illinois. The groundwater budget was described by a modified form of Equation (2.8); there was no groundwater flow into or out of the watershed:

$$R = Q^{bf} + ET^{gw} + \Delta S^{gw} \qquad (2.38)$$

Base flow, Q^{bf}, and evapotranspiration from groundwater, ET^{gw}, were determined by developing rating curves for streamflow vs.

groundwater level for different times of the year. Net change in saturated zone storage, ΔS^{gw}, was determined by using the water-table fluctuation method described in Chapter 6.

The water budget for the Panther Creek watershed is given by:

$$P = ET^{uz} + R_{off} + R \qquad (2.39)$$

where direct runoff, R_{off}, is determined by subtracting base flow from measured stream discharge. ET^{uz} was taken as the residual of Equation (2.39). Figure 2.10 shows cumulative monthly values for all components of Equation (2.39) for 1951 and 1952. Precipitation was near normal for 1952; recharge was about 25% of precipitation. The year 1951 was very wet; recharge was 19% of precipitation. For both years, most recharge occurred during spring months.

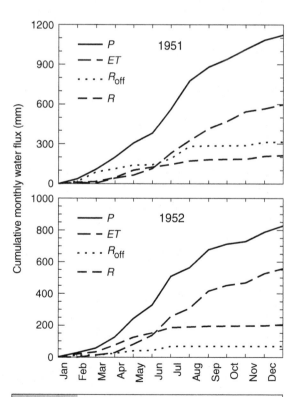

Figure 2.10 Cumulative monthly values of the water-budget components for Panther Creek, north central Illinois, 1951 and 1952 (after Schicht and Walton, 1961). P is precipitation, ET is evapotranspiration, R_{off} is runoff, and R is recharge.

Example: Thornthwaite and Mather method

In the late 1940s and through the 1950s C. W. Thornthwaite and colleagues at the Laboratory for Climatology of Drexel University developed a systematic approach to the study of watershed water budgets. The objective of this work was to identify relations among precipitation, potential evapotranspiration, and actual evapotranspiration worldwide. The approach laid the foundation for the development of watershed models in the following decades. Groundwater recharge was not an explicit component in these water budgets, but with some simple assumptions, recharge estimates can be obtained.

The procedure of Thornthwaite and Mather (1955) requires only measurements of air temperature and precipitation, although data on soil texture and the thickness of the soil zone could also be used. The following equation was solved for a watershed on a monthly basis:

$$P = ET + \Delta S^{uz} + Q^{sw}_{off} \qquad (2.40)$$

where ΔS^{uz} is the change in soil-water storage and Q^{sw}_{off} is the sum of runoff, R_{off}, and base flow, Q^{bf}. If evapotranspiration from groundwater and change in groundwater storage are negligible, recharge is often equated with base flow (a more detailed discussion of base flow is given in Section 4.1.2). PET was estimated by Equation (2.30), and a bookkeeping procedure was used for determining actual ET, ΔS^{uz}, and Q^{sw}_{off}. Soil-water storage, S^{uz}, was assumed to have a maximum value of $S^{uz}max$ (100 mm was originally suggested, but this can be estimated from soil texture and thickness). If P was greater than PET, then ET was set to PET and the excess amount of precipitation was added to S^{uz}. If S^{uz} exceeded $S^{uz}max$, then S^{uz} was set equal to $S^{uz}max$ and the remaining excess was added to the moisture surplus, MS. The user decides what percentage of the moisture surplus constitutes direct runoff. Thornthwaite and Mather (1955) suggested 50% for large basins as a general rule, so as a simple first approximation, the remaining 50% of the MS for each month can be designated

as base flow (storage in other reservoirs could also be considered). Alley (1984) and Steenhuis and van der Molen (1986) proposed variations to this approach that more accurately portray recharge processes. McCabe and Markstrom (2007) describe an easy-to-use computer program that determines the water budget in a fashion similar to that of Thornthwaite and Mather (1955).

Thornthwaite and Mather (1955) applied the model with average monthly data from Seabrook, NJ (Table 2.3). It was assumed that moisture surplus was split evenly between direct runoff and base flow. $S^{uz}max$ was set to 300 mm. Precipitation varied little from month to month, whereas PET and ET followed a distinct seasonal pattern being greatest in summer months and essentially 0 during winter months. Moisture surplus and base flow were greater than 0 only during winter and early spring months. Total base flow for the year was 189 mm or 17% of precipitation.

Example: West Maui, Hawaii

A water-budget approach was used to estimate historic rates of diffuse drainage to the unconfined aquifers underlying western Maui, the second largest of the Hawaiian Islands (Engott and Vana, 2007). Of concern were falling groundwater levels and increasing chloride concentrations in groundwater. Rising to a height of 1760 m, West Maui Mountain is the dominant topographic feature in the approximately 1000-km² study area. The climate is tropical, and precipitation varies with elevation and orientation. Average rainfall is about 8900 mm/yr at the highest elevations but less than 250 mm/yr along the southwestern coast of the island. Fresh groundwater occurs in lens systems and in dike-impoundment systems. Irrigation water is primarily obtained by surface-water diversion.

A variation of the Thornthwaite–Mather approach was applied in conjunction with a Geographical Information System (GIS). The water-budget equation for a column of soil extending from land surface to the bottom of the root zone was written as:

$$S^{uz}temp = P_i + I_{rri} + FD_i - R_{offi} - ET_i + S^{uz}_{i-1} \qquad (2.41)$$

Table 2.3. Average monthly water budget in millimeters for Seabrook, NJ (Thornthwaite and Mather, 1955).

	Jan	Feb	Mar	April	May	June	July	Aug	Sept	Oct	Nov	Dec	Year
P	87	93	102	88	92	91	112	113	82	85	70	93	1108
PET	1	2	16	46	92	131	154	136	97	53	19	3	750
ET	1	2	16	46	92	129	147	130	92	53	19	3	730
ΔS^{uz}	0	0	0	0	0	−38	−35	−17	−10	32	51	17	0
MS	86	91	86	42	0	0	0	0	0	0	0	73	378
Q^{bf}	43	46	43	21	0	0	0	0	0	0	0	37	189

If S^{uz}temp $< S^{uz}$ max, then $S^{uz}_i = S^{uz}$temp, and $D_i = 0$

If S^{uz}temp $> S^{uz}$ max, then $S^{uz}_i = S^{uz}$ max, and

$$D_i = S^{uz}\text{temp} - S^{uz} \text{ max}$$

where S^{uz}temp is the preliminary calculated soil-water storage, the index i indicates values for the current day, i-1 represents the previous day's value, I_{rr} is irrigation, FD is fog drip, D is drainage beneath the root zone, and S^{uz} and S^{uz}max are again soil-water storage and maximum soil-water storage, respectively.

Water budgets were calculated for six historical periods, which extended from 1926 through 2004. Within each period, it was assumed that there was no change in land use. The first period ran from 1926 to 1979; subsequent periods were 5 years in duration. The amount of land in cropped agriculture decreased with each period. Historical precipitation data from 33 locations were used. Irrigation rates were calculated as potential evapotranspiration minus rainfall infiltration (precipitation minus direct runoff); frequency of irrigation application was based on crop type. Fog drip was assumed to occur only on the windward (east) side of West Maui Mountain, where it was set equal to 20% of rainfall based on the findings of Scholl et al. (2005). Direct runoff, R_{off}, was determined by using streamflow hydrograph separation techniques (described in Chapter 4) and calculating monthly ratios of runoff to precipitation for four runoff regions. PET rates were determined from pan evaporation data (Equation (2.29)). Crop coefficients for sugar cane were varied throughout the growing cycle, but coefficients for all other vegetation were held constant. Actual evapotranspiration,

ET_i, was set equal to PET for each day, if precipitation and soil-water storage were sufficiently high; otherwise, ET_i was calculated on the basis of S^{uz}_i and a threshold parameter. S^{uz}max was set equal to the depth of rooting multiplied by the available water capacity (AWC). AWC is the difference in water content between field capacity and wilting point and varies with soil texture. Values of AWC were obtained from the Natural Resources Conservation Service SSURGO database (http://soils.usda.gov/survey/geography/ssurgo/; accessed July 27, 2009) and other published reports for the 65 mapped soil types in the study area. Rooting depth varied by land cover class (there were 26 such classes), and was held constant throughout the year.

Mean annual drainage rates for the 1926 to 1979 period varied widely (Figure 2.11), ranging from less than 250 mm/yr to more than 7500 mm/yr. The pattern of drainage matches that of precipitation, with increasing rates with increasing elevation. Drainage rates were highest in the 1926 to 1979 period and showed a general decreasing trend in subsequent periods (Figure 2.12). The estimated drainage for the 2000 to 2004 period was 44% less than that for the first period. The decreased drainage was attributed mainly to a reduction in the amount of land in agricultural use (the amount of land planted in sugar cane decreased by 22% between 1979 and 2004), improved irrigation efficiency (furrow irrigation of sugar cane was replaced with drip irrigation during the 1980s), and decreased rainfall rates from 1990 to 2004. Engott and Vana (2007) conducted a sensitivity analysis to identify parameters that most

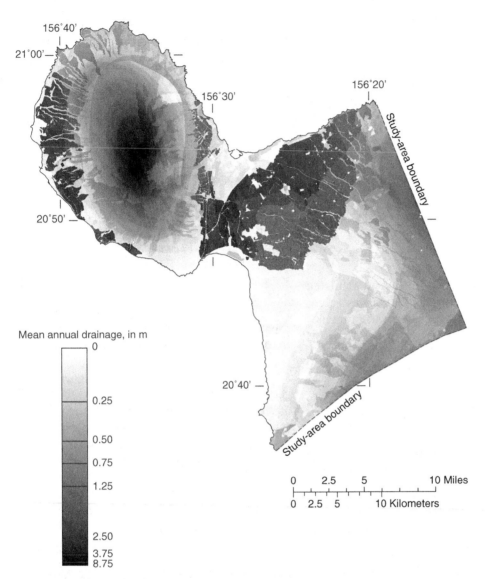

Figure 2.11 Distribution of mean annual drainage, central and west Maui, Hawaii for the period 1926 to 1979 (Engott and Vana, 2007).

affect drainage rates. The effect of future land-use and climate scenarios on predicted drainage rates was also examined.

2.5 | Macroscale application

Macroscale refers to length scales that exceed 1000 km. Studies conducted at this scale are of a regional, continental, or global nature and typically make use of remotely sensed data. The feasibility of accurately estimating recharge at this scale is questionable because of the large number of groundwater basins that are contained within such an area and the inherent variability among the different basins. There are no direct methods for estimating recharge with remotely sensed data, but the data can be useful for constructing water budgets. Sheffield *et al.* (2009) noted substantial closure errors for a water budget of the Mississippi River watershed, but water-budget methods based on remotely sensed data may eventually prove

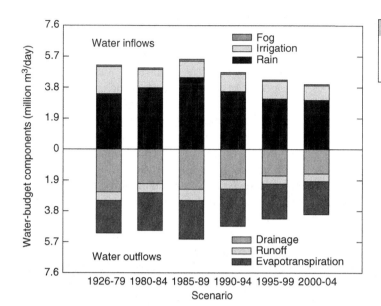

Figure 2.12 Estimated water-budget components for six historical scenarios, 1926 to 2004, central and west Maui, Hawaii (Engott and Vana, 2007).

useful for developing recharge estimates. The abundance of satellite-based data, the continual improvement of satellite-mounted sensors, and the large research effort invested in analysis of remotely sensed data provide promise for future utility (Schmugge *et al.*, 2002; Entekhabi and Moghaddam, 2007). In addition, in underdeveloped parts of the world, remote sensing may provide the only approach for estimating water-budget components (Brunner *et al.*, 2007).

Data on energy emissions and reflections from Earth have been collected for decades by Earth-orbiting satellites. Translation of these data, which are collected in a variety of wavelength intervals, into reliable estimates of water and energy fluxes has evolved slowly over the years. Estimation of precipitation, evapotranspiration, and water-storage changes over large areas can benefit from remote-sensing techniques. Remote sensing is not commonly used to estimate streamflow (although satellite-based data have been used to delineate flooded areas). Satellite remote sensing is attractive because of the broad spatial coverages that can be obtained. The frequency with which images are obtained may be a concern. Some sensors repeat measurements hourly or daily; for others, the interval between images can be as long as 26 days. Each image provides an instantaneous set of data, whereas what is needed for

recharge studies, and water-budget studies in general, are data that are integrated over the time between images.

Models are used to relate sensor readings to actual water fluxes. These models include land surface models (LSMs), such as NOAH (Ek *et al.*, 2003) and Interaction Soil Biosphere Atmosphere (ISBA; Noilhan and Planton, 1989), and Soil-Vegetation-Atmosphere Transfer (SVAT) models, such as the Surface Energy Balance Algorithm for Land (SEBAL), the Two Source Energy Budget model (TSEB) (Timmermans *et al.*, 2007), and the Community Land Model (CLM) (Bonan and Levis, 2006). These models vary in complexity, but they are all water-budget and energy-budget models of one type or another, and they are largely based on equations presented in this chapter. They relate surface energy state and temperature to water and energy exchanges among the soil surface, subsurface, vegetation canopy, and atmosphere. In addition to their use in analyzing remotely sensed data, these models are important components in atmospheric general circulation models. More details on remote sensing and water budgets can be found in Schmugge *et al.* (2002) and Krajewski *et al.* (2006).

In addition to estimating water-budget components, remote sensing provides useful information on surface characteristics, such as temperature, vegetation type and cover, leaf area index, and land use. This information can have important indirect uses, such as defining geologic structure, identifying areas that have similar hydrologic characteristics (Brunner *et al.*, 2007), and providing a basis for the upscaling of point measurements of evapotranspiration (Smith *et al.*, 2007).

2.5.1 Precipitation

Satellite-based estimates of daily precipitation (such as those from NASA's AQUA or Tropical Rainfall Measurement Missions) are generated worldwide on an approximately 25-km grid (Table 2.1). The spatial resolution is not as fine as the 4-km grid provided for the United States by MPE (Section 2.3.1, Table 2.1), but unlike ground-based radar, satellite measurements are not adversely affected by complex terrain, such as mountainous areas, or areas with no precipitation gauges. Continued refinement and integration of ground-based and satellite-based methods for estimating precipitation rates may result in the widespread use of these methods in future recharge studies at the watershed scale.

2.5.2 Evapotranspiration

Although currently (2010) there is no database of evapotranspiration estimates generated from satellite data, this is an area of active research. Most methods for estimating latent-heat flux, λET, are based on the surface energy-balance equation (Equation (2.15)). Remotely sensed data can be used to estimate net radiation, R_n, sensible-heat flux, H, and soil-heat flux, G; and λET can be determined as the residual in the equation (Timmermans *et al.*, 2007). Estimation of net radiation as the sum of net shortwave and net longwave radiation is fairly straightforward. Net shortwave radiation requires estimates of incoming radiation and surface albedo or reflectance; these can be obtained on a 250-m grid with microwave readings from an instrument such as NASA's moderate resolution imaging spectroradiometer, MODIS (http://aqua.nasa.gov/science/; accessed August 25, 2009). Net longwave radiation can

be estimated with knowledge of surface and atmospheric temperatures, which can be determined from radiation readings in the infrared wavelength range. Soil-heat flux is usually assumed to be a percentage of calculated net radiation. Estimation of sensible-heat flux is more problematic; H is often estimated with a temperature gradient/resistance formula, where the gradient is a function of the surface and atmosphere temperatures and the aerodynamic resistance term is a function of vegetation density and height (Norman *et al.*, 1995). Unfortunately, cloud cover interferes with some imagery and thus may limit the applicability of some methods.

Liu *et al.* (2003) estimated daily evapotranspiration on a 1-km grid for all of Canada for 1996 using the Penman–Monteith method (Equation (2.34)). The required model input was land cover, leaf area index, soil-water holding capacity (AWC), and daily meteorological data (net radiation, minimum and maximum air temperature, humidity, precipitation, and snowpack). Ten classes of land cover were defined (e.g. coniferous forest, cropland, and urban area) using data from the Advanced Very High Resolution Radiometer (AVHRR) on the NOAA 14 satellite. Leaf area index was determined with Landsat Thematic Mapper (TM) images and surface measurements at eight sites. AWC was obtained from the Soils Landscapes of Canada database (http://sis.agr. gc.ca/cansis/nsdb/slc/intro.html; accessed July 21, 2009). Daily meteorological data for 1996 were obtained from the medium-range forecast Global Flux Archive of the National Center for Environmental Prediction (Kalnay *et al.*, 1990). Bilinear interpolation was used to assign data values to each point on the 1-km grid.

Daily Priestley–Taylor potential (Equation (2.35)) and reference (Equation (2.36)) evapotranspiration are available on a 1-km grid for the State of Florida from 1995 to present (Table 2.1). The calculations are made with a combination of satellite-derived and land-based data. Daily solar radiation is calculated from hourly readings from NOAA's Geostationary Operational Environmental Satellite (GOES) system. Surface albedo was also obtained from the GOES system. Incoming and outgoing longwave radiation was estimated from air and soil-surface temperatures. Temperature, relative humidity, and wind speed at each grid point were determined by interpolation (inverse-distance weighting) of data obtained from land-based NOAA and University of Florida weather stations.

2.5.3 Change in storage

Passive and active microwave sensors are the most widely used satellite instruments for estimating surface-water storage and soil-water content (Krajewski et al., 2006). The emission (passive sensing) or backscattering (active sensing) of energy in the microwave region is a function of the dielectric constant of the soil and water. The dielectric constant varies with soil-water content. As part of NASA's AQUA satellite mission, readings from the Advanced Microwave Sensing Radiometer (AMSR-E) provide daily estimates of snow depth, snow-covered area, snow-water equivalent, and soil-water content (top 50 mm) worldwide on a 25-km grid (http://nsidc.org/data/amsre/; accessed August 25, 2009). The RADARSAT-1 satellite's active microwave sensor is another source for soil-water content data (Shi and Dozier, 1997; http://www.asc-csa.gc.ca/eng/satellites/radarsat1/; accessed August 25, 2009). Microwave readings are not affected by weather, but they are affected by land cover, soil properties, surface slope, and aspect. If compensations can be made for these interferences, reliable estimates of soil-water content in the top 50 mm of soil can be obtained (Makkeasorn et al., 2006). Land-surface models, such as NOAH, are used to relate changes in water storage at the land surface to changes throughout the entire soil profile.

The Gravity Recovery and Climate Experiment (GRACE) satellites provide monthly measurements of the Earth's gravity field. As discussed in Section 2.3.3, changes in the gravity field over land surface are mainly related to changes in the amount of water stored on the surface and in the subsurface. Therefore, data from GRACE can be used to estimate water storage changes, although the data cannot be used to distinguish between water on land surface and water in the subsurface. In areas where surface-water storage varies slightly over time, Rodell et al. (2004; 2007) and Strassberg et al. (2007) demonstrated that GRACE satellite data could be used to monitor changes in groundwater storage. Rodell et al. (2004) used a water-budget method, with change in storage determined from GRACE data, to generate estimates of evapotranspiration rates for the Mississippi River watershed.

2.5.4 Indirect use of remotely sensed data

In addition to supporting water-budget analyses, remotely sensed data are useful in a number of indirect ways, such as locating areas of ground- and surface-water exchange (Becker, 2006), mapping areas of groundwater recharge and discharge (Tweed et al., 2007), and identifying areas of apparently uniform evapotranspiration rates (Smith et al., 2007). As part of a water-budget study of the Basin and Range carbonate-rock aquifer system, estimates of evapotranspiration were developed for an area of approximately 37 500 km² in Nevada and Utah with the aid of satellite imagery (Moreo et al., 2007; Laczniak et al., 2008). Areas of similar vegetation type and density and soil characteristics, called evapotranspiration units, were identified by using the Landsat Thematic Mapper. The Landsat images have a resolution of approximately 30 m and cover six spectral bands. The identification procedure included, among other criteria, a soil-vegetation index (modified from Qi et al., 1994). The 10 evapotranspiration units ranged from areas of no vegetation (including open water, dry playa, and moist bare soil) to areas of increasing density of grasses, rushes, and phreatophytic shrubs (Smith et al., 2007). Field measurements of evapotranspiration rates were obtained over 1 year by the eddy correlation method in representative locations in each unit. Evapotranspiration rates for the entire study area were determined by multiplying measured evapotranspiration rates for each unit by the mapped area of each unit.

2.6 | Discussion

Water budgets are the fundamental building blocks upon which conceptual models of hydrologic systems are constructed. Water-budget equations for an aquifer, a watershed, a stream, even a column of soil offer more than just an approach for estimating recharge. Water-budget equations provide insight to the entire hydrology of a system, supplying information not only on recharge, but also on the interrelationship between recharge and other water-budget components. Water budgets serve another important role; as will be shown in following chapters, water budgets provide a framework within which the assumptions and limitations of other techniques for estimating recharge can be analyzed.

Many, if not most, methods for estimating recharge are based on some form of a water-budget equation. The universal nature of water budgets allows water-budget methods to be applied over the wide range of space and time scales encountered in hydrologic studies. The lack of assumptions on the mechanisms that drive the individual components in a water-budget equation provides these methods with additional flexibility. Models are linked with water-budget methods at multiple levels. Most hydrological models are water-budget models, being based on some variation of Equations (2.1), (2.8), or (2.9). In many water-budget studies, models are used to evaluate one or more components in a water-budget equation. As such, there is a certain amount of overlap between this chapter on water budgets and the following chapter on modeling methods.

The major limitation of the residual water-budget approach for estimating recharge is that the accuracy of the recharge estimate is dependent on the accuracy with which the other components in the water-budget can be determined. This limitation is important when the magnitude of recharge is small relative to that of the other variables. In this case, small inaccuracies in values for those variables may result in large uncertainties in recharge.

Future applications of water-budget methods should see improved accuracy as new tools for measuring and estimating water-budget components evolve. New developments in remote sensing, geophysical techniques, and modeling should be particularly helpful.

Modeling methods

3.1 | Introduction

Simulation models are widely used in all types of hydrologic studies, and many of these models can be used to estimate recharge. Models can provide important insight into the functioning of hydrologic systems by identifying factors that influence recharge. The predictive capability of models can be used to evaluate how changes in climate, water use, land use, and other factors may affect recharge rates. Most hydrological simulation models, including watershed models and groundwater-flow models, are based on some form of water-budget equation, so the material in this chapter is closely linked to that in Chapter 2. Empirical models that are not based on a water-budget equation have also been used for estimating recharge; these models generally take the form of simple estimation equations that define annual recharge as a function of precipitation and possibly other climatic data or watershed characteristics.

Model complexity varies greatly. Some models are simple accounting models; others attempt to accurately represent the physics of water movement through each compartment of the hydrologic system. Some models provide estimates of recharge explicitly; for example, a model based on the Richards equation can simulate water movement from the soil surface through the unsaturated zone to the water table. Recharge estimates can be obtained indirectly from other models. For example, recharge

is a parameter in groundwater-flow models that solve for hydraulic head (i.e. groundwater level). Recharge estimates can be obtained through a model calibration process in which recharge and other model parameter values are adjusted so that simulated water levels agree with measured water levels. The simulation that provides the closest agreement is called the best fit, and the recharge value used in that simulation is the model-generated estimate of recharge.

Both diffuse and focused recharge can be estimated with models. When selecting a model it is important to ensure that assumptions inherent in the model are consistent with the user's conceptual model of the recharge process. As with other estimation methods, space and time scales of recharge estimates vary for different models and model applications. Applications of watershed or groundwater-flow models may assume that recharge is uniform over the simulated domain, or recharge may be allowed to vary in space within the domain. In terms of time scales, recharge can be estimated on a daily, monthly, or annual basis; or recharge may be assumed constant over time.

This chapter provides descriptions of generic categories of models: unsaturated zone water-budget models, watershed models, and groundwater-flow models. Models within each category share a common approach, but complexity and features of models can vary substantially. An overview of models within each category is presented, representative models are described, and major assumptions underlying the models are analyzed; examples of model applications are also presented. Model calibration and use of inverse modeling techniques, topics that apply to all models, are described in Section 3.2. Section 3.7 describes modeling tools for upscaling point estimates of recharge to obtain integrated estimates at the watershed or regional scale. Aquifer vulnerability to contamination is addressed in Section 3.8 because estimates of recharge used in vulnerability assessments are often generated by one of the modeling approaches described in the preceding sections; the link between model recharge estimates and aquifer vulnerability studies is discussed. Proper model application and interpretation of results require a sound understanding of assumptions inherent to that particular model. It is not possible to include in this chapter all the details of every model discussed. Readers considering application of a model are directed to cited references for exact model details.

3.1.1 Data sources

Table 2.1 provides a list of sources for real-time and historical hydrologic data. Data obtained from those sources can be used as input for many of the models discussed herein. Additional data on soil properties, land-surface elevation, land cover, and other features may be required for some models, such as watershed and soil-water-budget models; these data are also useful in upscaling procedures, for example, in development of regression equations to relate recharge estimates to climate, land use, and soils. Sources for these additional data are briefly described in this section.

The US General Soil Map (STATSGO2) is a digital dataset of soil information for the United States. The 2006 release of STATSGO2 supersedes the State Soil Geographic (STATSGO) dataset published in 1994. The dataset was developed by the National Cooperative Soil Survey and is distributed by the Natural Resources Conservation Service (http://soils.usda.gov/survey/geography/statsgo/description.html; accessed September 22, 2009). The dataset consists of georeferenced digital data on physical and chemical properties of soils. The data were collected in

1° by 2° topographic quadrangle units and are distributed by state or territory.

The Natural Resources Conservation Service has digitized soil maps for most of the United States and compiled that information into the Soil Survey Geographic (SSURGO) dataset (http://soils.usda.gov/survey/geography/ssurgo/description.html; accessed September 22, 2009). SSURGO data are available at much finer spatial resolution (scales of 1:12 000 to 1:63 360) than STATSGO2 data. In addition to soil descriptions, information such as available water capacity and land-use category are included in the SSURGO dataset.

Elevation data for the United States are available from the US Geological Survey (USGS). The National Elevation Dataset (NED) is a seamless digital raster product derived primarily from USGS 30 m digital elevation models (DEMs) along with higher resolutions where available (http://eros.usgs.gov/; accessed December 15, 2009). Global elevation data are also available from the USGS at a horizontal grid spacing of approximately 1 km (http://eros.usgs.gov/; accessed December 15, 2009).

The National Land Cover Data 1992 (NLCD 92) is a 21-category land-cover classification scheme that has been applied over the United States. It is primarily based on Landsat Thematic Mapper images from 1992 (http://eros.usgs.gov/; accessed December 15, 2009). In addition to the land-cover classification, data on topography, census, agricultural statistics, and soil characteristics are also available. The classification is provided as raster data at a spatial resolution of 30 m. The classification is currently (2010) being revised on the basis of 2001 imagery by the Multi-Resolution Land Characteristics (MRLC) consortium (http://eros.usgs.gov/; accessed December 15, 2009). The USGS also provides global land-cover characterization at a 1-km scale (http://eros.usgs.gov/; accessed December 15, 2009). The characterization is based on imagery from 1992 and 1993 from the Advanced Very High Resolution Radiometer (AVHRR); ancillary data include digital elevation data, ecoregions interpretation, and vegetation and land-cover maps.

3.2 | Model calibration and inverse modeling

Hydrologic models form the basis of this chapter. Although there are a variety of modeling approaches, common model-calibration practices can be applied to most models. Model calibration is the process of adjusting model input data (parameters) so that model output matches independent observations or measurements (Figure 3.1). For illustrative purposes, consider groundwater-flow models. Groundwater-flow models generate estimates of hydraulic head and possibly base flow or spring discharge. Field measurements of any of these phenomena constitute the observations to which model results are compared. Model parameters consist of properties of aquifer materials (such as hydraulic conductivity and storage coefficient), initial water levels, boundary conditions, and sources or sinks. Recharge can be represented as a boundary condition or an areal or point source. During the calibration process, multiple model runs are made; values for one or more parameters are adjusted between model runs, and simulated results are compared with measured data to see whether the parameter adjustment resulted in an improved model fit. Calibration can be done in a manual trial-and-error manner or in a formal manner by using statistical techniques such as nonlinear regression (Cooley and Naff, 1990; Hill, 1998; Hill and Tiedeman, 2007). The latter approach is referred to as *inverse modeling* or *parameter estimation* and is gradually becoming a standard practice in hydrologic modeling (Poeter and Hill, 1997; Carrera *et al.*, 2005; Doherty, 2005; Poeter *et al.*, 2005).

Inverse modeling techniques seek to determine the model parameters that produce the best agreement between measured and simulated data. An objective function is a measure of this agreement or model fit. Typically, the objective function, $S(b)$, is defined as the sum of squared weighted differences between observed and simulated heads and fluxes:

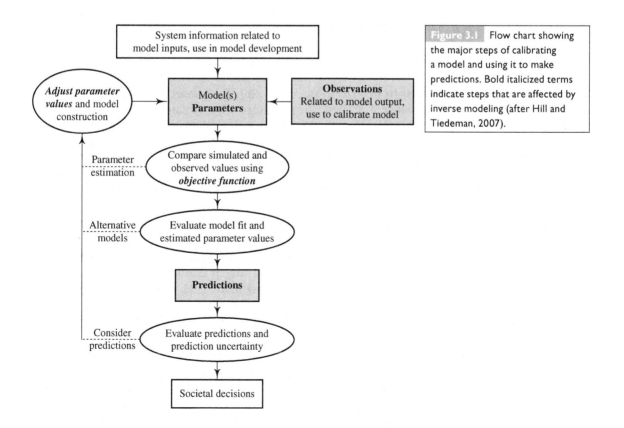

Figure 3.1 Flow chart showing the major steps of calibrating a model and using it to make predictions. Bold italicized terms indicate steps that are affected by inverse modeling (after Hill and Tiedeman, 2007).

$$S(b) = \sum_{i=1}^{N_{obs}} \omega_i [y_i - y_i'(b)]^2 \qquad (3.1)$$

where b is a vector of the parameter values that are being estimated, N_{obs} is the number of measured heads and fluxes, ω_i is the weight for the ith observation, y_i is the observed head or flux, and y_i' is the simulated head or flux. Inverse models automatically adjust parameter values so as to minimize the objective function. The weights, ω_i, serve a dual purpose. They convert all observations to a uniform set of units, and they can be used to more heavily weight the observations that are the most accurate or the most important (e.g. a measured groundwater level is essentially a point estimate in space and time; base flow, on the other hand, reflects a response that is integrated over some larger space and time scales, so in a groundwater-flow model, measurements of base flow may warrant larger weights than measurements of groundwater levels). Use of nonuniform weights in inverse modeling adds a degree of subjectivity to the calibration process, so this needs to be done with careful consideration.

Inverse modeling is a powerful tool for any hydrologic modeling. Automatic calibration can save substantial amounts of time relative to manual model calibration, especially for large, complex simulations. In addition to determining parameter values that produce the best model fit, inverse modeling programs can generate diagnostic statistics that quantify the quality of the calibration, point out data shortcomings and needs, determine the sensitivity of model results to individual parameters, and quantify reliability of parameter estimates and predictions (Hill, 1998).

Potential problems with using inverse modeling include insensitivity, nonuniqueness, and instability. Insensitivity occurs when the observations do not contain enough information to support estimation of all of the parameters. Nonuniqueness implies that different combinations of parameter values produce equally good matches to observed data. This occurs when one model parameter is highly correlated with another; a steady-state groundwater-flow model that is calibrated only with water level data will have a nonunique best fit because hydraulic conductivity and recharge will be highly correlated. In this case, the inverse model will actually be determining the ratio of recharge to hydraulic conductivity that provides the best model fit; any simulation that uses that same ratio will produce an identical model fit. Including base-flow or other flux measurements along with water levels in the calibration process alleviates this nonuniqueness problem (Sanford, 2002). Instability refers to the situation where slight changes in parameter values or observed data cause large changes in simulation results.

PEST (Doherty, 2005) and UCODE_2005 (Poeter et al., 2005) are two widely used universal model-calibration programs. These programs can be applied with virtually any hydrologic model; both of these programs use nonlinear regression techniques to minimize the objective function. Alternative inverse-modeling approaches are available. Adjoint-state methods (Carrera and Neuman, 1986; Xiang et al., 1993; Tarantola, 2005) rely on the derivative of the objective function with respect to parameter value. Global optimization methods, such as simulated annealing, genetic algorithms (Wagner, 1995), and tabu search (Zheng and Wang, 1996), are useful for problems with highly irregular objective function surfaces. These problems may not be amenable to solution by nonlinear regression.

The ability of inverse models to quantify uncertainty in estimated recharge rates is a benefit not available from other recharge-estimation techniques. The uncertainties are valid only if the model accurately represents the actual hydrologic system. Two methods for quantifying parameter uncertainty are inferential statistics and Monte Carlo methods (Hill and Tiedeman, 2007). Inferential statistics can be used to calculate linear and nonlinear confidence intervals for parameter values. An individual linear confidence interval for a parameter estimate, such as recharge, has a given probability of containing the true value of that parameter regardless of whether confidence intervals on other model parameters include their true values. Linear confidence intervals of parameter estimates can be calculated and printed by PEST and UCODE_2005; the accuracy of any of these calculations depends on the validity of the assumptions invoked for a particular application (Hill and Tiedeman, 2007). Nonlinear confidence intervals (Vecchia and Cooley, 1987) offer improved accuracy and can be calculated with the above mentioned programs, but only at considerable computational cost.

Monte Carlo methods use a known or assumed distribution of one model parameter value to infer information on the distribution of model results or other model parameters. This is accomplished through repeated application of the model with values of the known parameter randomly selected from the given distribution. In groundwater-flow modeling, for example, a log normal distribution, mean, and standard deviation of hydraulic conductivity may be assumed for the simulated domain. Values of hydraulic conductivity for the entire domain (referred to as a realization) are randomly drawn from that distribution. The groundwater-flow model is solved for each realization to produce estimates of recharge. If model runs are made for 1000 realizations of hydraulic conductivity fields, then the mean and standard deviation of the 1000 calculated recharge fields can be determined at specific points in space and for the entire simulated domain.

Monte Carlo methods can produce results similar to those from inferential statistical methods, and they have some advantage over inferential methods when dealing with highly nonlinear models (Hill and Tiedeman, 2007). Monte Carlo methods have been used with all types of hydrologic models, including soil water-budget models (Bekesi and McConchie, 1999; Baalousha, 2009), watershed models (Bogena et al., 2005), and groundwater-flow models (Hunt et al., 2001; Bakr and Butler, 2004). Monte Carlo methods are not limited to application to models; they are also useful for assessing uncertainty in recharge estimates derived by other techniques, including water-budget methods (Leake, 1984; Dages et al., 2009).

3.3 | Unsaturated zone water-budget models

Diffuse recharge occurs as water moving through the unsaturated zone to underlying aquifers. It is the dominant form of recharge in many areas. Unsaturated zone water-budget models can be useful tools for generating estimates of diffuse recharge. There is great variability among these models, but a common assumption is one-dimensional vertical water flow through the unsaturated zone, so that a study site can be represented by a single soil column or by multiple noninteracting soil columns. Two types of models are considered: soil water-budget models and models based on the Richards equation. The former type of model is based on a water budget of the soil or root zone, whereas the latter considers the water budget over the entire thickness of the unsaturated zone.

3.3.1 Soil water-budget models

Soil water-budget models for estimating recharge usually consist of two components: a water budget for the soil or root zone that estimates the amount of drainage through the base of that zone and a submodel that transports that drainage to the water table. Drainage from the soil zone, D, is typically determined from a water-budget equation such as:

$$D = P - ET^{uz} - \Delta S^{uz} - R_{off} \qquad (3.2)$$

where, following the notation of Chapter 2, P is precipitation plus irrigation, ET^{uz} is evapotranspiration from the soil zone, ΔS^{uz} is change in storage in the soil zone, and R_{off} is runoff. Implicit in this approach is the assumption that the maximum amount of storage within the root zone is fixed, or that the depth of the base of the root zone (more properly referred to as the zero-flux plane, Figure 2.2) does not vary with time. Equation (3.2) is usually solved for daily values; the equation is valid for any time-step size, including weekly, monthly, and annual, but estimates of drainage may be sensitive to time-step size. As discussed in Section 2.2, annual drainage rates calculated with weekly or monthly time steps are often less than those calculated by using daily time steps.

Typical model input consists of daily precipitation and actual or potential evapotranspiration. Because measured values of evapotranspiration are often not available, some models allow input of air temperatures in place of evapotranspiration; one of the equations described in Section 2.4.2 is then used to estimate potential evapotranspiration. Runoff is often estimated with the US Soil Conservation Service curve number method (see Section 2.4.4), and change in storage is determined by a bookkeeping procedure based on field capacity, wilting point capacity, and initial amount of water in storage. Additional data may be required in models that use more detailed water-budget equations. Soil–vegetation–atmosphere transfer (SVAT) models, for example, may explicitly account for plant growth as well as water uptake from soil by plant roots and release of that water to the atmosphere through leaf stomata (e.g. Fischer et al., 2008). This more sophisticated approach requires data such as plant type, stage of growth, slope and aspect of land surface, relative humidity, solar radiation, and wind speed. Storage can be simulated more accurately if information on water retention properties of the soil is available. Similarly, precipitation intensity and surface roughness information can improve runoff predictions. Land-surface models (Section 2.5) are forms of soil water-budget models that act as subcomponents in large-scale atmospheric general circulation models.

Most models consider the unsaturated zone to be a transition zone within which water moves downward to the water table. Some modeling approaches do not consider water movement through the unsaturated zone; they simply refer to drainage from the soil zone as potential recharge or net infiltration (Flint and Flint, 2007). Others assume that drainage from the soil zone arrives instantaneously at the water table (Batelaan and De Smedt, 2007; Dripps and Bradbury, 2007), an assumption appropriate for sites with shallow water tables or for studies that span long periods, such as decades or centuries, where travel time to the water table is negligible relative to the length of the study period.

Water movement through the unsaturated zone has been simulated with approaches that range from bucket models to numerical models that solve the Richards equation. The more detailed models offer insights into physical processes that affect water movement, but they require information on model parameters and can be computationally intensive. In contrast, bucket models require little information on the unsaturated zone, and their application is relatively easy and straightforward. These simpler models are amenable to use with geographical information system (GIS) applications, in which many one-dimensional simulations are used to represent the response of a watershed to changes in factors such as land use, climate, or water use.

Simple bucket models represent the surface and subsurface by a series of storage reservoirs that have been referred to as buckets, tanks, compartments, or layers (Figure 3.2). The shallowest subsurface bucket is filled with infiltration from precipitation or irrigation. As one bucket is filled, excess water flows to the next deepest bucket, and so on. Water can be extracted from shallow buckets by evaporation, plant transpiration, interflow, and other phenomena. Flow of water to or from the deepest bucket is assumed to be recharge. Ragab et al. (1997) and Finch (1998) proposed models that had four layers of buckets within the root zone.

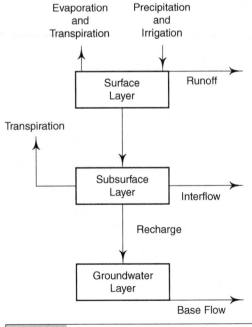

Figure 3.2 Schematic diagram of simple tank model.

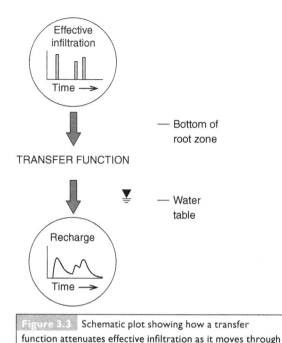

Figure 3.3 Schematic plot showing how a transfer function attenuates effective infiltration as it moves through the unsaturated zone (after O'Reilly, 2004).

The HELP3 model (Schroeder *et al.*, 1994), which allows the user to select the number of layers, has been used to estimate recharge in a number of studies (Jyrkama *et al.*, 2002; Risser, 2008; and Risser *et al.*, 2009). Bucket models are easy to apply, but they provide little information on the physics of water movement.

Besbes and de Marsily (1984) proposed use of a transfer function for transporting drainage from the bottom of the root zone to the water table. Recharge is determined according to the following equation:

$$R(t) = \int_0^t I_{net}(t - \tau)\Phi'(\tau)d\tau \qquad (3.3)$$

where $R(t)$ is cumulative recharge arriving at the water table between times 0 and t; I_{net} is net infiltration or drainage from the base of the root zone; Φ' is a linear transfer function; and τ is the variable of integration that represents the time lag of the transfer function measured backwards in time from t. Transfer-function models can be viewed as black boxes relative to the physics of water movement; drainage is delayed and smoothed before arriving at the water table

(Figure 3.3). Transfer functions have proved useful in several areas in hydrology including surface-water flow (Dooge, 1959), stream/aquifer interaction (Hall and Moench, 1972), and solute transport through the unsaturated zone (Jury, 1982). Morel-Seytoux (1984), Bierkens (1998), O'Reilly (2004), and Berendrecht *et al.* (2006) used transfer-function models to simulate the movement of drainage from the root zone to the water table. Transfer models can be calibrated with measured groundwater levels and independent recharge estimates (such as from the water-table fluctuation method, Section 6.2). Model coefficients are adjusted, either manually or automatically, so that model output is in agreement with the independent estimates (Berendrecht *et al.*, 2006).

Example: Orange County, Florida

A water-budget transfer-function (WBTF) model was developed and applied at a site in west Orange County, Florida (O'Reilly, 2004). The site contained herbaceous vegetation with rooting depth generally less than 0.3 m. Runoff from the site was negligible because of the high permeability of the sandy soils. Water-table depths (2–3.5 m) were deemed too great for plants to draw water from the aquifer. Evapotranspiration at the site was determined for a 1-year period in 1993 and 1994 by the eddy-correlation method (Sumner, 1996). Daily precipitation and water-table depth were also measured. The WBTF model calculates the water budget (as described in Equation (3.2)) for the soil zone with input data consisting of daily precipitation and evapotranspiration, initial and maximum root-zone and vegetation storage, and three parameters of the transfer-function model. (In areas where runoff is nonnegligible, the difference between daily precipitation and runoff is entered in place of precipitation.) The model delays and attenuates the drainage signal, but the transfer-function model is conservative in that all drainage calculated in the water budget is ultimately delivered to the water table.

Optimum values for the three model parameters were determined with the parameter estimation program, UCODE (Poeter and Hill, 1998), by minimizing the differences between simulated

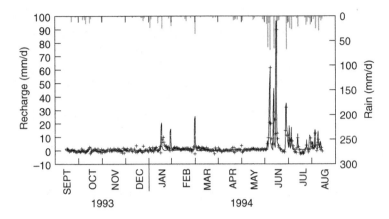

Figure 3.4 Measured daily rainfall and estimated recharge by the water-table fluctuation method (+) and the water-budget transfer-function model (dark line) for a site in west Orange County, Florida (after O'Reilly, 2004).

daily recharge rates estimated with WBTF and those calculated by using the water-table fluctuation method (Section 6.2) on the basis of measured groundwater levels. Simulated recharge rates were in good agreement with estimates from the water-table fluctuation method (Figure 3.4). If daily precipitation and actual evapotranspiration data are available, the WBTF model is easy to use and can be readily applied to multiple sites within a watershed by using a GIS. The need for calibration of transfer-function models at every application site remains an unresolved question. O'Reilly (2004) provides estimates of the WBTF model parameters for variable soil textures and water-table depths.

3.3.2 Models based on the Richards equation

Flow models based on the Richards equation, such as VS2DI (Hsieh et al., 1999), HYDRUS (Simunek et al., 1999), and UNSAT-H (Fayer, 2000), provide a physically based approach for simulating water movement through the unsaturated zone. Estimates of travel times for wetting fronts moving from the soil zone to the water table can be generated, and physical features that affect water movement can be identified. The Richards equation is derived from a water conservation equation and the Darcy equation; it can be written for one-dimensional vertical flow as:

$$\partial\theta / \partial t = \partial K_s K_r(h)\partial H / \partial z^2 + Q'$$ (3.4)

where θ is volumetric water content, K_s is saturated hydraulic conductivity in the vertical direction, K_r is relative hydraulic conductivity, h is pressure head, H is total head, Q' accounts for sources and sinks, t is time, and z is the vertical coordinate. Models based on the Richards equation are sometimes used in the transport phase of soil water-budget models, but they are actually self-contained water-budget and transport models. The Richards equation is, after all, a statement of a water budget. The land surface, unsaturated zone, and saturated zone are treated as a continuum through which water is allowed to move upward as well as downward. These models have capabilities of allowing water for evapotranspiration to be withdrawn at any depth and allowing the locations of the zero-flux plane and the water table to fluctuate.

Application of a Richards equation-based model requires data on water content as a function of pressure head (the water-retention curve) and relative hydraulic conductivity as a function of pressure head (the unsaturated hydraulic conductivity curve). These curves can be determined in the field or in the laboratory (Section 5.2), but measurements are expensive and time consuming (Dane and Topp, 2002). Empirical equations, such as those of van Genuchten (1980) and Brooks and Corey (1964), have been derived to represent these curves. Use of these equations with parameters that have been developed for generic soil types provides an alternative to direct measurements of the curves. The van Genuchten equations are given by:

$$\theta(h) = (\theta_s - \theta_r)[1 + (\alpha h)^n]^{-m} + \theta_r$$ (3.5)

Table 3.1. Average hydraulic conductivity (K_s), saturated water content (θ_s), residual water content (θ_r), and van Genuchten α and n for generic soil texture classes as determined by Carsel and Parrish (1988).

Soil texture	K_s (cm/hr)	θ_s	θ_r	α (cm^{-1})	n
Clay	0.20	0.38	0.068	0.008	1.09
Clay loam	0.26	0.41	0.095	0.019	1.31
Loam	1.04	0.43	0.078	0.036	1.56
Loamy sand	14.59	0.41	0.057	0.124	2.28
Silt	0.25	0.46	0.034	0.016	1.37
Silt loam	0.45	0.45	0.067	0.020	1.41
Silty clay	0.02	0.36	0.070	0.005	1.09
Silty clay loam	0.07	0.43	0.089	0.010	1.23
Sand	29.70	0.43	0.045	0.145	2.68
Sandy clay	0.12	0.38	0.100	0.027	1.23
Sandy clay loam	1.31	0.39	0.100	0.059	1.48
Sandy loam	4.42	0.41	0.065	0.075	1.89

$$K_r(h) = \frac{\{1-(\alpha h)^{n-1}[1+(\alpha h)^n]^{-m}\}^2}{[1+(\alpha h)^n]^{m/2}} \qquad (3.6)$$

where θ is volumetric water content, h is pressure head, θ_s is saturated water content, θ_r is residual water content, α and n are referred to as the van Genuchten parameters, and m is taken equal to $1 - 1/n$. On the basis of thousands of published measurements, Carsel and Parrish (1988) determined average values of saturated water content, residual water content, α, and n for 12 soil textural classifications (Table 3.1). Models such as VS2DI and HYDRUS offer users the option of using these values. Pedotransfer functions, such as ROSETTA (Schaap *et al.*, 2001), have also been developed for predicting van Genuchten parameters on the basis of soil texture.

A typical model scenario consists of a column of soil, with the top representing land surface and the bottom representing a fixed depth that could be above, equal to, or below the water-table depth. The column is divided into a number of computational cells or layers. Input data commonly consist of measured daily precipitation and potential evapotranspiration; the top boundary is treated as an infiltration or an evaporation boundary depending on whether precipitation occurs during the current day. A fixed pressure-head or free-drainage boundary condition is assigned to the bottom of the column. Flow out of the bottom of the column represents recharge unless the depth of the bottom column is above the water table, in which case the flow represents drainage. Multiple soil types can be represented in the column to simulate water movement through layered media.

The physically realistic approach for representing flow through the unsaturated zone offered by the Richards equation provides a useful tool for identifying controls on groundwater recharge and examining the effects that climate and land-use change may have on recharge. Richards-equation models can also account for flow upward from the water table to the soil zone. Numerical solution of Equation (3.4) can be computationally intensive, however, because of the nonlinear nature of the water-retention and hydraulic conductivity terms. A number of studies have used a Richards equation-based model for estimating recharge or drainage, including Fayer *et al.* (1996), Keese *et al.* (2005), Smerdon *et al.* (2008), and Webb *et al.* (2008).

The Richards equation can be simplified by application of the kinematic wave approximation, in which drainage within the unsaturated zone is assumed to be driven solely by gravity (Colbeck, 1972):

$$\partial\theta/\partial t = \partial K_s K_r(\theta)/\partial z + Q' \qquad (3.7)$$

where relative hydraulic conductivity is written as a function of volumetric water content instead of pressure head. This form of the equation can be solved by the method of characteristics (Smith, 1983; Charbeneau, 1984; Niswonger *et al.*, 2006) with much less computational effort than that required for Equation (3.4). This approach cannot account for layered soils, however; the simulated soil column must have uniform properties.

Example: Maryland agricultural field

Webb *et al.* (2008) described a hypothetical study of recharge and pesticide leaching at a field in Maryland that was planted with a corn and soybean rotation. Recharge was calculated for 1995 through 2004 with LEACHM, a one-dimensional unsaturated-zone flow model based on the Richards equation (Hutson and Wagenet, 1992). Model input included daily precipitation (measured at the site in 2003–2004 and from a nearby National Weather Service site for 1995–2002) and potential evapotranspiration (calculated by the Hamon (1963) equation on the basis of daily air temperature).

Simulations were run to determine the effect of unsaturated-zone thickness on recharge rates and on travel times for water movement from land surface to the water table. Three sets of simulations were run, with unsaturated-zone thicknesses of 1, 5, and 10 m that span the range of thicknesses actually measured in the field. The top of the simulated soil column represented land surface and the bottom represented the water table; computational cells within the column were 100 mm in height. Optimum van Genuchten parameters for the water-retention and hydraulic-conductivity curves (Equations (3.5) and (3.6)) were determined by simulating a tracer experiment that was conducted at the site and matching observed and simulated rates of tracer movement. As expected, average recharge rates decreased with increasing depth to the water table. Annual rates of recharge were 159, 124, and 111 mm for unsaturated-zone thicknesses of 1, 5, and 10 m, respectively. These estimates range from 9 to 13% of the annual precipitation (about 1200 mm). Thicker unsaturated zones also led to longer lags

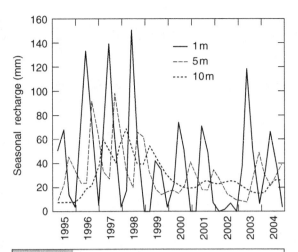

Figure 3.5 Simulated recharge for unsaturated zone thicknesses of 1, 5, and 10 m at 3-month intervals for an agricultural field in Maryland from 1995 to 2004 (Webb *et al.*, 2008).

between infiltration and arrival of recharge at the water table (Figure 3.5); peak rates of recharge occurred in winter for the 1-m thick unsaturated zone, but not until the following fall for the 10-m thick unsaturated zone.

3.4 | Watershed models

Watershed models were developed to evaluate the effects of climate and land use on watershed hydrology. Affected processes include evapotranspiration, surface and subsurface water storage, and groundwater recharge and discharge; but for many if not most applications, the process of primary interest is streamflow. The first watershed model, the Stanford Watershed Model (Crawford and Linsley, 1962), was developed to aid design of flood-control projects. Newer watershed models have improved upon early models in terms of sophistication with which processes are represented and GIS input and output processing capabilities. These improved models, along with increased availability of online data, have led to an expansion in the types of studies in which watershed models are used. Watershed models are now used in studies of groundwater recharge, surface-water chemistry, groundwater chemistry, biology, and ecology. Results of watershed models are used in resource management decisions, risk and uncertainty analyses, and environmental impact studies.

Among the more widely used watershed models are SWAT (Arnold *et al.*, 1998), PRMS (Leavesley *et al.*, 1983), and MIKE-SHE (Graham *et al.*, 2006). Singh (1995) and Singh and Frevert (2006) provided reviews of these and many other watershed models. Watershed models are all based on a water-budget equation, but there are important differences among models in terms of spatial and temporal scales of applicability, processes that are accounted for (e.g. interception, runoff, infiltration, evapotranspiration, water movement through the unsaturated zone, and groundwater flow), techniques for representing those processes (e.g. physically based vs. empirical), data input requirements, and output options. Groundwater recharge and discharge are components of a watershed water budget; hence, watershed models can provide estimates for these components. Applications of watershed models to estimate recharge include Weeks *et al.* (1974), Arnold and Allen (1996), Sami and Hughes (1996), Bauer and Mastin (1997), Desconnets *et al.* (1997), Laenen and Risley (1997), Zhang *et al.* (1999), Arnold *et al.* (2000), Steuer and Hunt (2001), Eckhardt and Ulbrich (2003), and Cherkauer (2004).

A watershed model assumes a series of reservoirs (Figure 3.6), representing water storage on land surface, in the plant canopy, within the unsaturated zone (soil zone and subsurface), and in groundwater. These reservoirs fill and drain in response to atmospheric phenomena (e.g. precipitation, snowmelt, evaporation, and transpiration) and to the current state of other reservoirs. The complexity with which water movement within a reservoir or among different reservoirs is represented varies among models. In the simplest case, these reservoirs can be thought of as tanks, buckets, or layers, as depicted in Figure 3.2, that fill by precipitation or drainage from other tanks. Drainage from tanks can be at a constant rate, a rate dependent on the height of water currently in the tank, or a rate dependent on external parameters, such as air temperature and humidity.

More sophisticated representations of processes include routing of surface-water flow over land or through channels and surface reservoirs and simulating water movement through the unsaturated zone by an approach based on the Richards equation.

Climatic and hydraulic data requirements for watershed models are similar to those of soil water-budget models (Section 3.3). Minimum data requirements typically include average daily precipitation and temperature (for estimating potential evapotranspiration) and soil water-storage capacity. Additional information (such as a digital elevation map, storage and conveyance properties of surface-water bodies, depth to water table, hydraulic properties of unsaturated and saturated zone materials, and types of vegetation) can be used by many models. Recent advances allow watershed models to work with GIS datasets, thus facilitating a highly distributed representation of parameters within a watershed. Watershed models have historically been calibrated by comparing simulated streamflow with measured streamflow. It is not uncommon in current studies to include measurements of evapotranspiration, water storage in the unsaturated zone, and groundwater levels in the calibration process.

An attractive aspect of using a watershed model to estimate recharge is the power to predict how future changes in climate and land use may affect recharge patterns (Zhang *et al.*, 1999; Steuer and Hunt, 2001). Physically based models can provide insight into mechanisms that influence the hydrologic response of a basin. Disadvantages to using watershed models include the need for training in model use and the large number of parameters that may need to be determined or estimated. The common assumption made in many applications of watershed models that groundwater-flow boundaries coincide with surface-water flow boundaries (it is often assumed that there is no groundwater inflow to or outflow from the watershed) should be fully evaluated. Because watershed models are water-budget methods, the caveats expressed in Chapter 2 deserve consideration. Most importantly, if the magnitude of recharge is only a small fraction of

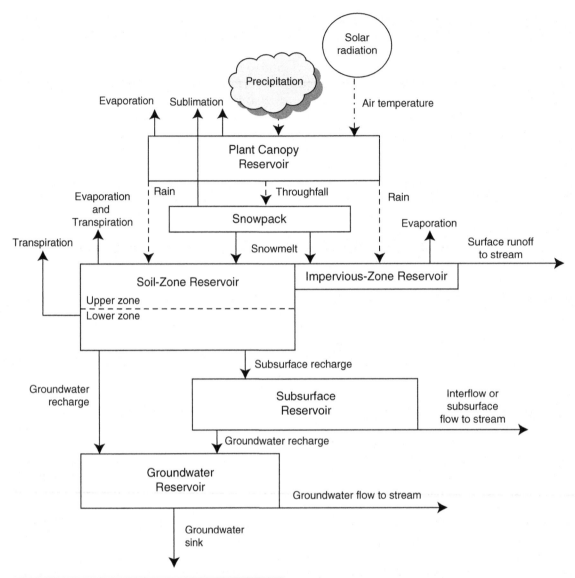

Figure 3.6 Flow chart for PRMS model (Markstrom et al., 2008).

that of other components in the water-budget equation, small inaccuracies in the other components may produce large uncertainties in recharge estimates.

3.4.1 Precipitation runoff modeling system (PRMS)

Details of a single watershed model are provided here to illustrate the manner in which watershed models typically represent hydrological processes. The Precipitation Runoff Modeling System, PRMS (Leavesley et al., 1983), allows the watershed to be partitioned into a set of spatial units of uniform characteristics such as soil and vegetation type, elevation, and slope. For each of these units, called Hydrologic Response Units (HRUs), daily water and energy budgets are calculated; fluxes to and from the atmosphere and fluxes between all reservoirs depicted in Figure 3.6 are determined, as are storage changes within the reservoirs. The daily watershed response is equal to the sum of the responses of all HRUs. Required climate data include daily precipitation, minimum and maximum air temperature, and solar radiation (if available). Incoming precipitation can be intercepted by vegetation, contribute to the snowpack, fall on impervious material, or infiltrate the soil. Evapotranspiration may occur from water in each of these four reservoirs. Evapotranspiration is calculated on the basis of water availability and potential evapotranspiration as determined by the Jensen-Haise method (Equation (2.31)) or by two alternative methods.

Precipitation falling on impervious areas becomes direct runoff. For other areas, direct runoff is calculated by an empirical method, similar to the SCS curve number method (Equation (2.37)). Accretion or depletion of the snowpack is governed by air temperature as well as precipitation. Infiltration is determined as the difference between water (either precipitation, snowmelt, or surface run on from adjacent areas) arriving at the soil surface and runoff. (An alternative approach taken in other watershed models is to first calculate infiltration with the Green-Ampt equation (Singh, 1995), for example, and to then set runoff equal to the difference between water arriving at the soil surface and infiltration.) Drainage from the soil-zone reservoir occurs when its storage capacity is exceeded. A portion (determined by an empirical factor) of that excess water goes to the subsurface reservoir; the remainder goes directly to the groundwater reservoir. Water can also drain from the subsurface reservoir to the groundwater reservoir. There is no lag time in the exchange of water among these reservoirs; drainage is assumed to be instantaneous. Recharge for each HRU consists of all drainage to the groundwater reservoir. Total recharge for the watershed is the sum of recharge for all HRUs. Water in the groundwater reservoir can discharge to a stream (base flow) or flow out of the watershed as groundwater flow; groundwater discharge rates vary linearly with groundwater reservoir storage.

Since its initial release (Leavesley *et al.*, 1983) PRMS has undergone multiple revisions that have added alternative options for calculating fluxes and have made the model more user friendly. Improvements include incorporation of a modular modeling system to link available databases (Leavesley *et al.*, 1996) and a GIS tool for delineating HRUs on the basis of elevation, topography, climate and vegetation zones, and depth to water table (Viger and Leavesley, 2007).

Example: Puget Sound Lowland
Bauer and Mastin (1997) used the Deep Percolation Model, DPM (Bauer and Vaccaro, 1987; Vaccaro, 2007), to construct detailed water budgets for three small watersheds in glacial-till mantled terrains in the southern Puget Sound Lowland of Washington. The goal was to estimate the amount of recharge that penetrated the till. The DPM is a distributed parameter, physically based watershed model developed specifically for estimation of groundwater recharge. Input requirements and water exchange calculations are similar to those of PRMS.

The surface of most of the Puget Sound Lowland is characterized by rolling, hilly glacial-till mantled areas and level glacial-outwash bench land. At the three study sites, the aquifer in the glacial outwash deposits is overlain by 3 to 20 m of till. From October 1991 through September 1993, discharge was measured at the mouth of each catchment using a flume or culvert and a stage recorder. Piezometers were installed within each catchment to monitor groundwater levels. Soil-water content was measured at three depths with time-domain reflectometry probes at the piezometer locations. Precipitation was measured in each basin; daily maximum and minimum air temperatures were obtained from nearby National Weather Service stations. Solar radiation was measured at one location and assumed to be uniform for all watersheds.

Most precipitation and recharge occurred in winter at the three sites; little rain fell during summer (Figure 3.7). Average annual recharge calculated by the DPM ranged from 37 to 172 mm (Table 3.2). Thicker till is the apparent reason for the relatively small amount of recharge at the Clover site. A unique aspect of this study was the use of groundwater-level and water-content data in model calibration; simulated and measured soil-water storage were in good agreement (Figure 3.8). Independent estimates of recharge were obtained by analyzing soil cores for tritium concentrations and

Table 3.2. Annual average recharge estimated for three small watersheds in Puget Sound Lowlands, Washington, by the Deep Percolation Model (DPM) and by the tritium mass balance method (Bauer and Mastin, 1997).

Watershed	DPM estimated recharge (mm)	Ratio of DPM recharge to precipitation	Recharge from tritium mass balance (mm)
Clover	37	0.04	42–53
Beaver	138	0.14	104–134
Vaughn	172	0.17	169–200

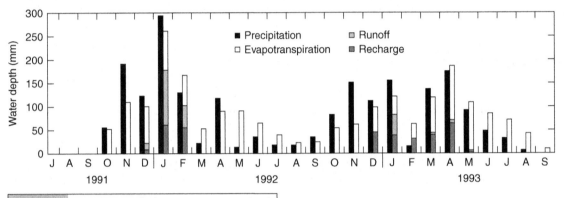

Figure 3.7 Monthly water-budget components simulated with the Deep Percolation Model during the study period for the Vaughn catchment (after Bauer and Mastin, 1997).

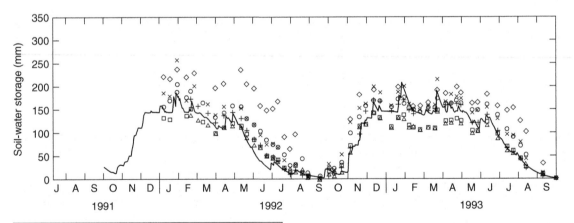

Figure 3.8 Observed (symbols) soil-water storage at six locations and Deep Percolation Model simulated (line) soil-water storage in millimeters above wilting point during the study period for the Vaughn catchment (after Bauer and Mastin, 1997).

applying the tritium mass-balance method (Section 7.2.3); these estimates were similar to the DPM determined values (Table 3.2).

Example: Monavale catchment, Australia
Zhang *et al.* (1999) described the application of the watershed model, TOPOG_IRM, for estimating recharge and changes in soil-water content within the 161-ha Monavale catchment near Wagga Wagga, New South Wales, Australia. TOPOG_IRM is a physically based, distributed parameter model that predicts water movement within a soil–vegetation–atmosphere

system. It explicitly accounts for topography and soil and vegetation characteristics and is driven by conventional meteorological data (Dawes *et al.*, 1997). Bare soil evaporation and plant transpiration are calculated separately by using Penman–Monteith type equations. Runoff is generated when precipitation exceeds the infiltration capacity of the surface soils. Water movement through soils is assumed to be vertical and is simulated by using the one-dimensional Richards equation. Drainage from the base of the simulated soil zone was assumed to be recharge.

The field study was conducted from May 1992 through December 1994. A meteorological station was established at the study site to measure air temperature, relative humidity, wind speed, radiation, and rainfall. Leaf-area index was measured on three 1-m^2 plots. Soil-water content was measured biweekly at four depths with time-domain reflectometry probes at nine sites. Water levels in three wells were measured at the same frequency. Hydraulic properties of the sediments were determined from analysis of soil cores. Streamflow from the watershed was measured with a V-notch weir with an automatic stage recorder.

Model input data consisted of daily values of precipitation, air temperature, vapor pressure deficit, and solar radiation. Land-surface contours and catchment boundaries were derived with a digital elevation model. The watershed was divided into 1977 computational elements with an average area of 835 m^2. Drainage was calculated for each individual element and ranged from 10 to 200 mm for the period of simulation. Sensitivity analysis indicated that soil type was the most important parameter influencing recharge because vegetation was fairly uniform. Average drainage for the watershed was 98 mm, about 5% of total rainfall. Drainage typically occurred in late winter and spring (June through November); within that period, large rainfall events triggered brief sharp increases in drainage rates (Figure 3.9). Little recharge occurred in late 1994 because of a severe drought. Simulated soil-water contents compared favorably with measured data (Figure 3.10); simulated streamflow and evapotranspiration were also similar to measured values. Zhang *et al.* (1999) used the model to predict how planting trees on all or parts of the watershed would affect recharge rates.

3.5 | Groundwater-flow models

Groundwater-flow models are used to predict aquifer response, in terms of head (groundwater level) and fluxes into and out of an aquifer, to natural and human-induced stresses; they are important tools for resource and environmental management, and they provide groundwater velocities needed for simulation of subsurface contaminant transport. A common form of the groundwater-flow equation (Harbaugh, 2005) can be written as:

$$\partial(K_{xx}\partial H / \partial x) / \partial x + \partial(K_{yy}\partial H / \partial y) / \partial y \\ + \partial(K_{zz}\partial H / \partial z) / \partial z + Q' + R = S_s \partial H / \partial t \tag{3.8}$$

where K_{xx}, K_{yy}, and K_{zz} are values of saturated hydraulic conductivity along the x, y, and z coordinate axes (and it is assumed that the major axes of the hydraulic conductivity tensor are aligned with the coordinate axes), H is total head, Q' represents all sources and sinks except recharge, including water withdrawn from wells and groundwater evapotranspiration, and S_s is specific storage. Equation (3.8) is derived from the continuity equation and the Darcy equation and is, therefore, a water-budget equation.

Groundwater-flow modelers may be reading this book for insight on independent estimates of recharge for use as input for a groundwater-flow model, but estimates of recharge can be obtained indirectly from a groundwater-flow model through the calibration process if actual measurements of water levels and groundwater discharges are available (Section 3.2). Values of recharge used in the model are adjusted to bring simulated heads and fluxes (such as base flow) into agreement with measured values. The best estimates of recharge rates are those that produce the best model results, in terms of minimized objective function. This approach

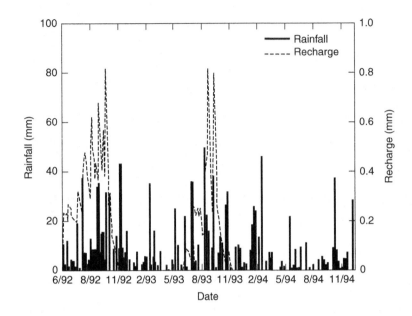

Figure 3.9 Rainfall and calculated recharge for one computational element, site L (Zhang *et al.*, 1999).

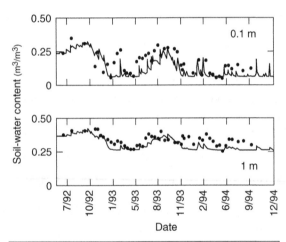

Figure 3.10 Observed (dots) and simulated (lines) soil-water contents at site P, depths of 0.1 and 1 m (Zhang *et al.*, 1999).

has been used in many studies, including Cooley (1979; 1983), Boonstra and Bhutta (1996), Tiedeman *et al.* (1997), Sanford *et al.* (2004), and Dripps *et al.* (2006).

Most groundwater-flow models solve Equation (3.8) numerically using either the finite-difference or finite-element method. The simulated domain is discretized into a grid of computational cells and nodes (Figure 3.11). The groundwater system may be represented by a single layer of cells, in which case two-dimensional horizontal flow is simulated, or by multiple layers of cells for simulation of three-dimensional flow. An approximation of Equation (3.8) is developed for each cell, and the set of equations for all cells is solved simultaneously to determine head values at each node. Depending on the size of the domain, there may be thousands, even millions, of computational cells. Required input data include hydraulic conductivity, recharge and other sources or sinks, aquifer geometry, initial head values, and boundary conditions. No-flow boundaries are the usual default boundary type; constant head boundaries and boundaries where flux is constant or varies over time or as a function of head can also be used, as explained in documentation for MODFLOW-2005 (Harbaugh, 2005).

Groundwater-flow models use different approaches for simulating diffuse and focused recharge. Diffuse recharge is usually assigned as a constant flux (i.e. a volumetric flow per unit horizontal surface area, L/T) to a cell representing the water table; within MODFLOW-2005, diffuse recharge is simulated by using the Recharge Package. Focused recharge from a stream or lake or other surface-water body occurs to model cells that underlie those surface features and is typically simulated with fixed head or head-dependent boundaries;

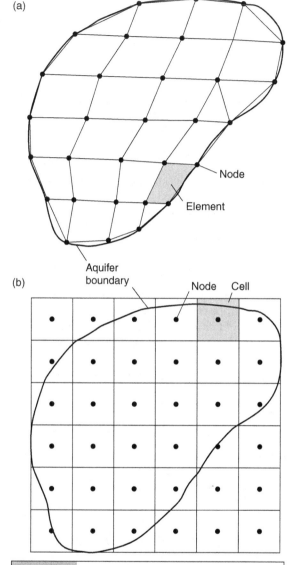

(a)

Node

Element

Aquifer boundary

(b)

Node   Cell

Figure 3.11 Example of areal finite element grid (a) and finite difference grid (b) overlain on an aquifer. Approximations to Equation (3.8) are solved for values of head at each node. Multiple grid layers are required for simulating three-dimensional groundwater flow.

A uniform value of diffuse recharge can be assigned to the entire model domain, or values can be assigned on a cell-by-cell basis. The simulated domain is sometimes divided into a number of regions in which recharge is assumed uniform (Cooley, 1979); these regions are somewhat analogous to the HRUs used in watershed models. Historically, many groundwater-flow models invoked a steady-state assumption, wherein recharge was assumed constant in time, because there seldom were sufficient data to calibrate a transient model. Recent advances in linking watershed and groundwater-flow models (Section 3.6) allow recharge to vary over time, even to the point of having daily values of recharge.

Focused recharge can be simulated by using the River Package in MODFLOW-2005. For each finite-difference cell, the user must specify stream stage and a streambed conductance term that accounts for streambed thickness and hydraulic conductivity and stream area (length times width) within the cell (Harbaugh, 2005). Exchange of water between the stream and aquifer is calculated according to the Darcy equation. If stream stage for a model cell is greater than calculated head for that cell, water flows from the stream to the aquifer at a rate dictated by model input parameters. If calculated head exceeds stream stage for a model cell, the stream acts as a sink for groundwater discharge.

Recharge to confined aquifers occurs largely in areas where the aquifers are unconfined, but many aquifers receive substantial amounts of replenishment in the form of vertical flow through confining beds. Flow through overlying confining beds was the largest source of water for the Ozark aquifer under predevelopment conditions (Czarnecki et al., 2009). As confined aquifers are developed by humans, the importance of flow through confining beds as a source of aquifer replenishment is enhanced because declining water levels in the aquifers induce additional flow into the system through these beds.

Overlying confining layers may or may not be included as active model layers when simulating flow in a confined aquifer. If the confining

recharge fluxes from these boundaries are determined by the model on the basis of calculated heads and therefore are not known a priori. Focused recharge can also be represented as a point source with a specified volumetric flow (e.g. by using the MODFLOW Well Package) for sources such as intermittent streams.

beds are represented by active model layers, flow through the beds to the aquifer is calculated by the model according to Equation (3.8). When confining beds and other overlying layers are not actively represented in a model, vertical flux to the aquifer may be represented as a constant or head-dependent flux, as described in this section. Groundwater modelers commonly refer to this flux as recharge. In the terminology introduced in Section 1.2, flow through confining layers to an aquifer is referred to as inter-aquifer flow instead of recharge. Regardless of terminology, this flux is often an important source of water for the aquifer. For modeling purposes, flux through a confining layer can be simulated as diffuse recharge by using the Recharge Package of MODFLOW-2005. Inverse groundwater-flow modeling is a useful tool for estimating interaquifer flow.

Stoertz and Bradbury (1989) described a procedure that uses a groundwater-flow model to calculate recharge at every water-table cell. A water-table map was drawn by hand for the model domain and superimposed on a two-dimensional horizontal grid. The mapped water-table elevation at each nodal location was assigned to that node. All nodes were set as constant head boundaries. MODFLOW was then used to simulate groundwater flow. Because all cells were constant head, no new head values were calculated, but fluxes between cells were calculated by the model. Net flux through each cell is equal to the difference between outflow to adjacent cells and inflow from adjacent cells. The net flux must be balanced by flow into the cell from the constant head boundary, which is by definition recharge. In addition to water levels, the method requires estimates of hydraulic conductivity and aquifer thickness. The method is perhaps of most use in a qualitative sense, identifying areas of recharge and discharge, and in that sense the method may be useful for aquifer vulnerability studies. Quantitative estimates of recharge by this method are problematic because calculated recharge rates are linearly related to hydraulic conductivity; if hydraulic conductivity is doubled, the calculated recharge is also doubled. This method has been used in a number of studies (Hunt *et al.*,

1996; Gerla, 1999; Lin and Anderson, 2003). Lin *et al.* (2009) developed a GIS application that uses a similar approach for estimating recharge and discharge rates; their approach has more flexibility because it is not constrained by a MODFLOW grid.

The analytic-element method (AEM) is an alternative to the finite-difference and finite-element methods for solving the groundwater-flow equation (Haitjema, 1995). The AEM uses superposition of an analytical solution to Equation (3.8) to determine head values at any point in the simulated domain; hence, a spatial grid is not required. The AEM is generally not capable of capturing the fine-scale complexities that can be incorporated into a model such as MODFLOW, but AEM models are relatively easy to set up, and they can serve as useful tools for calibrating more complex models (Hunt *et al.*, 1998). Dripps *et al.* (2006) demonstrated how an AEM model can be used in conjunction with a parameter estimation code to derive estimates of recharge in a small watershed in northern Wisconsin. Model calibration was based on matching simulated base flow with annual average base flow as determined by streamflow hydrograph separation (Section 4.5) for four streamflow gauge sites for 5 years. Steady-state conditions were assumed for each year. Using best estimates of recharge, simulated base flows were within 5% of measured values.

Example: Truckee Meadows, Nevada

In one of the first examples of automatic calibration of a groundwater-flow model, Cooley (1979) used regression techniques to calibrate a steady-state groundwater-flow model of Truckee Meadows in the Reno-Sparks area of Nevada. Water levels in 59 observation wells were used for model calibration. The model domain was divided into 13 regions that were assumed to have uniform aquifer transmissivity and net recharge rate (recharge minus groundwater evapotranspiration). Optimum values for 22 model parameters (6 head boundaries, 4 transmissivities, 1 streambed conductance, and 11 recharge rates) were determined. Overall model fit was good ($R^2 > 0.99$), but large standard errors of recharge estimates, some in excess of 100%, indicated that the model solution was nonunique. It was concluded that different combinations of parameter values could produce similar simulation results.

Example: Mirror Lake, New Hampshire

A groundwater modeling study was conducted for the Mirror Lake area in central New Hampshire to develop an understanding of the three-dimensional extent of the basin and to quantify components of the groundwater budget, including recharge (Tiedeman *et al.*, 1997). The 10 km² study area is characterized by steep hillsides and relatively flat valleys. The groundwater system is contained in fractured crystalline bedrock and the glacial deposits that overlie the bedrock in most of the basin. Most recharge occurs as direct infiltration from precipitation. Seepage from the lake and streams is a secondary source of recharge.

The three-dimensional model domain was represented by a grid of 89 rows, 85 columns, and 5 layers. Grid spacing ranged from 17 to 150 m on a side. The bottom of the grid was at a depth of 150 m. No-flow boundaries were applied on all sides of the model domain. Streams were simulated with the Streamflow Routing Package of MODFLOW. It was assumed that steady-state conditions existed and that recharge from precipitation was spatially uniform. Recharge and hydraulic conductivity of three zones (glacial deposits, lower bedrock, and upper bedrock) were estimated by using the inverse modeling capability in the MODFLOWP program (Hill, 1992). Four separate calibrations with different hypothetical zonations of hydraulic conductivity were run. Calibration data included long-term average water levels from 89 wells and piezometers and average base flow in three streams that flow into Mirror Lake. Base flow was assumed to be 40% of the long-term average streamflow, on the basis of work by Mau and Winter (1997).

Simulated groundwater levels and base flows were similar to measured values in all calibrations. The estimated recharge rates of 260 to 280 mm/yr (Figure 3.12) were equivalent to 19 to 20% of average precipitation. The approximate, individual, 95% confidence intervals shown in Figure 3.12 were calculated under assumptions

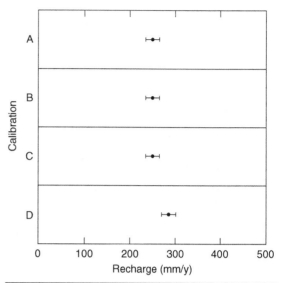

Figure 3.12 Estimated recharge rates (dots) and approximate, individual, 95% confidence intervals (bars) derived from four calibrations of the Mirror Lake groundwater-flow model (Tiedeman *et al.*, 1997).

that the model behaved linearly with respect to parameter values in the neighborhood of the optimal values and that residuals (differences between observed and simulated quantities) were independent and normally distributed. The validity of these assumptions was discussed by Tiedeman *et al.* (1997). Base-flow data allowed model parameters to be uniquely determined. If water-level data alone were used for calibration, only the best-fit ratios of recharge to hydraulic conductivity could have been determined because of high correlation between these variables. An important finding of the modeling analysis was that the groundwater basin within which Mirror Lake lies is much larger than the surface-water basin. If the groundwater model had been constructed with boundaries at the surface-water basin boundaries, the estimated recharge would have been about 50% higher.

Example: Middle Rio Grande Basin

Groundwater in the Middle Rio Grande Basin of central New Mexico serves as the primary water source for the city of Albuquerque and the surrounding area. Sanford *et al.* (2004) describe a groundwater-flow model within the basin that was calibrated with groundwater levels and groundwater ages, as determined by carbon-14 analysis (Section 7.3.2) for samples obtained from 200 observation wells. MODFLOW was used to simulate groundwater flow, and the particle-tracking model MODPATH (Pollock, 1994) was used to simulate the transport of groundwater age (i.e. age was used as a tracer). Groundwater in the basin occurs primarily within unconsolidated materials that range in thickness up to 4000 m. The model grid consisted of 80 columns and 156 rows of 1-km^2 cells (Figure 3.13) and 9 vertical layers of variable thickness.

Focused recharge from the Rio Grande, Jemez River, Rio Puerco, and Rio Salado was simulated with head-dependent boundary conditions (River Package in MODFLOW). Focused recharge also occurred from smaller arroyos that flow from the mountains to the east of the simulated domain and was simulated with the Recharge Package in MODFLOW. Diffuse recharge was assumed to be negligible. Interaquifer flow into the domain occurred along western and northern boundaries and was represented by constant flux boundary conditions. A total of 21 geologic zones were identified; hydraulic conductivity was assumed uniform within each zone. Predevelopment water levels were simulated under the assumption of steady-state conditions.

Model calibration was accomplished with the parameter estimation program, UCODE, but the nonlinear least-squares regression method in UCODE was not entirely compatible with the particle-tracking algorithm used in MODPATH. Small changes in some model parameters could cause dramatic shifts in particle trajectories. As a result, some individual parameter values were manually adjusted from UCODE-determined values to obtain a final fit. UCODE was useful, however, for determining parameter sensitivities (Sanford *et al.*, 2004).

Total flow into the simulated domain was estimated to be 2.14 m^3/s. Approximately 65% is recharge, and the remainder is interaquifer flow across model boundaries. The total inflow is substantially less than estimates of 5.45 m^3/s (Kernodle *et al.*, 1995) and 4.86 m^3/s (Tiedeman *et al.*, 1998); these estimates were also determined by groundwater-flow modeling, but calibration

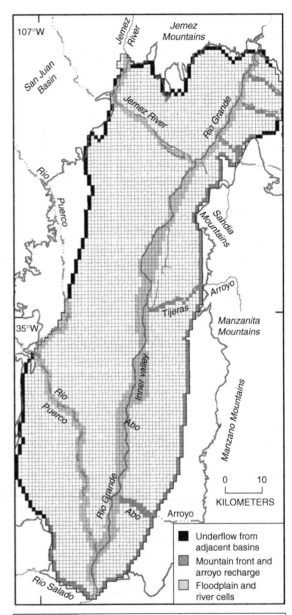

Figure 3.13 Finite-difference grid for the Middle Rio Grande Basin groundwater-flow model showing locations and types of boundary conditions (Sanford *et al.*, 2004).

was based only on groundwater levels. Addition of groundwater-age data should lead to increased accuracy of recharge estimates (Sanford *et al.*, 2004). The availability of data such as base flow or groundwater age for use in model calibration, in addition to water-level data, generally improves model fit (Hill and Tiedeman, 2007).

3.6 | Combined watershed/groundwater-flow models

In recent years, both watershed modelers and groundwater-flow modelers have come to recognize the benefits offered by a linked watershed and groundwater-flow model. A combined model allows watershed modelers to realistically capture the dynamic nature of surface and groundwater exchange as controlled by the head gradient between a stream and an aquifer. Measured groundwater levels provide additional constraints for model calibration. For groundwater-flow modelers, a combined model provides physically based representations of recharge and discharge processes. Both diffuse and focused recharge can vary in space and time, as influenced by climate, geology, topography, vegetation, and other factors. Also, a combined model is able to accommodate settings where watershed boundaries do not exactly coincide with aquifer boundaries. Flow can be simulated in an aquifer, for example, that underlies only a part of a watershed. At present, only a small number of combined watershed/groundwater-flow models are available: SWAT/MODFLOW (Sophocleous and Perkins, 2000), MIKE-SHE (Graham *et al.*, 2006), and GSFLOW (Markstrom *et al.*, 2008), but new and improved models should continue to appear in the future.

To envision a combined watershed/groundwater-flow model, one need only replace the groundwater reservoir shown in Figure 3.6 with a spatially extensive aquifer within which groundwater may move horizontally and vertically. Simulated recharge, depending on model design, can be influenced by water-table height. Similarly, groundwater discharge to a stream can be calculated on the basis of groundwater levels and stream levels. Thus, a combined model allows groundwater recharge and discharge to be represented in a manner more physically realistic than approaches used in stand-alone watershed and groundwater-flow models.

A number of challenges arise in attempting to link a watershed model and a groundwater-flow model. Precipitation, runoff, and streamflow occur at time scales different from those of drainage through the unsaturated zone and groundwater flow (Section 2.2). Time and space discretizations are often quite different between the two types of models. Watershed models invariably operate on a daily time-step basis. Daily time steps can be used in groundwater-flow models, but because groundwater movement is relatively slow, monthly or yearly time steps are more common; use of daily time steps in a multiyear simulation may require substantial computation time. Steady-state groundwater-flow models that assume no change in water levels over time have been widely used in the past, but the steady-state assumption defeats the purpose of a combined model. Typically, a hydrologic response unit (HRU) would overlie multiple computational cells of a groundwater-flow model. So a procedure is required to map and distribute fluxes between an HRU and the appropriate computational cells.

One of the more difficult challenges in linking a watershed and groundwater-flow model is the simulation of water movement through the unsaturated zone between the soil zone and the water table. This issue is addressed in Section 3.3 for unsaturated zone water-budget models, and much of the discussion in that section is relevant here as well. The three-dimensional nature of a combined model introduces additional complexity, as does the fact that recharge fluxes must be calculated for thousands of computational cells within the groundwater-flow model. Most combined models assume one-dimensional vertical flow through the unsaturated zone and use one of the approaches described in Section 3.3, such as a bucket model, a transfer-function model, or an approach based on the Richards equation. More complex models, such as MIKE-SHE (Graham *et al.*, 2006) and InHM (VanderKwaak and Loague, 2001), allow for horizontal water movement within the unsaturated zone by solving the Richards equation numerically in three dimensions. This approach provides a detailed representation of water movement within the unsaturated zone and can be useful when unsaturated zone hydraulic properties are not uniform, but the approach can be quite demanding in terms of computational resources.

Example: GSFLOW simulation of the Sagehen Creek watershed

GSFLOW (Markstrom *et al.*, 2008) is a coupled watershed/groundwater-flow model created by linking PRMS with MODFLOW-2005 (Harbaugh, 2005). The GSFLOW domain consists of three regions (Figure 3.14). Region 1 includes the plant canopy, snowpack, impervious (surface-depression) storage, and the soil zone. Region 2 includes streams and lakes. Region 3 is the subsurface underlying Regions 1 and 2 and includes both the unsaturated and saturated zones. Region 1 is simulated with PRMS, and Regions 2 and 3 are simulated with MODFLOW-2005. Recharge to Region 3 can occur as gravity drainage from Region 1 and as head-dependent leakage from Region 2. Gravity drainage is simulated with a kinematic wave approximation to the one-dimensional Richards equation (Niswonger *et al.*, 2006). Groundwater discharge to Region 1 or 2 occurs when the water table rises into the soil zone or above stream and lake levels.

The Sagehen Creek watershed lies on the east slope of the Sierra Nevada mountains near Truckee, California. The hydrology of the watershed was simulated with GSFLOW for the period 1981 to 1995 (Markstrom *et al.*, 2008). Application of GSFLOW is complex, and only the most salient aspects of the application are presented here. The watershed was delineated into 128 HRUs on the basis of flow planes, climate, vegetation, and depth to water table. Daily precipitation and minimum and maximum air temperatures were obtained from sites near the watershed. Precipitation and temperatures were adjusted on the basis of elevation, which ranged from 1935 to 2653 m above sea level. Daily streamflow data were available from a gauge located near the mouth of the watershed.

The MODFLOW-2005 model consisted of two model layers, each with 73 rows and 81 columns of cells that were all 90 m in length and width. Five hydraulic conductivity zones were identified on the basis of geology. Hydraulic conductivity within each zone was uniform and was determined through model calibration. Other data required for MODFLOW-2005 included saturated water content, a parameter for determining water content within the unsaturated zone, specific storage, thickness of hydrogeologic units, stream channel dimensions, and initial water content beneath streams.

Daily time steps were used for both the PRMS and MODFLOW-2005 components of GSFLOW,

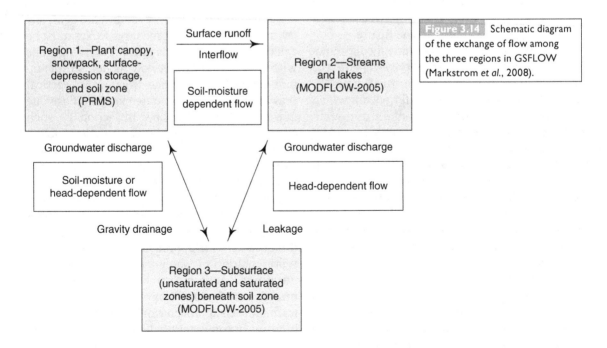

Figure 3.14 Schematic diagram of the exchange of flow among the three regions in GSFLOW (Markstrom *et al.*, 2008).

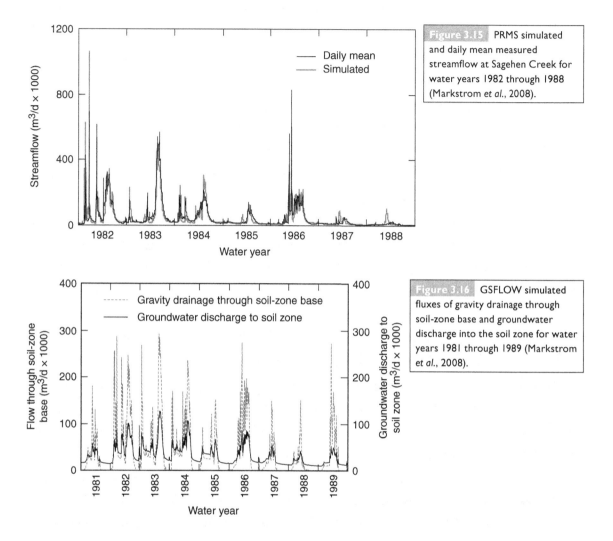

Figure 3.15 PRMS simulated and daily mean measured streamflow at Sagehen Creek for water years 1982 through 1988 (Markstrom et al., 2008).

Figure 3.16 GSFLOW simulated fluxes of gravity drainage through soil-zone base and groundwater discharge into the soil zone for water years 1981 through 1989 (Markstrom et al., 2008).

resulting in more than 6000 time steps for the simulated period of record. The calibration process was fully described by Markstrom *et al.* (2008). Simulated streamflow was generally in good agreement with measured streamflow (Figure 3.15). Drainage from the soil zone was calculated on a daily basis (Figure 3.16) for the entire watershed. Drainage from each HRU was transported to underlying MODFLOW-2005 cells by using the kinematic wave approach. A combined model allows dynamic feedback among the aquifer, soil zone, and surface-water bodies. Water can be transported upward or downward on the basis of calculated water levels. This feature is important for this example because much of the water that recharges the aquifer is returned to the surface as groundwater levels rise, generally in late spring and early summer (Figure 3.16).

Combined watershed/groundwater-flow models such as GSFLOW are useful tools for estimating groundwater recharge. The integrated approach considers the entire hydrology of a watershed/aquifer system; calculated recharge rates may vary in both space and time. These models provide unprecedented opportunities for evaluating the sensitivity of recharge rates to climate, land use, water use, and other natural and anthropogenic factors. The models are also useful for predicting how future changes in any of these factors may affect recharge processes. The major drawback of using these models is that they are difficult to apply. Most model users will require extensive training.

3.7 | Upscaling of recharge estimates

The focus of this section is on upscaling of recharge estimates from one space or time scale to a larger scale, for example, by using point estimates of recharge at a few locations within a watershed to determine average recharge for the entire watershed. Upscaling techniques are also used for developing parameter values for input to hydrologic models. Upscaling is a complex operation that has applications in many areas of science and continues to be a topic of active research. Formal statistical approaches take into account data values, sampling locations, probability distributions, and data uncertainty (e.g. Nœtinger *et al.*, 2005). Such formal approaches are not commonly taken in hydrologic studies, and the level of detail required to apply such approaches is beyond the scope of this text. What will be reviewed are the upscaling approaches that have been used in hydrologic studies, concentrating specifically on recharge.

The simplest procedure for upscaling point measurements is to average all data values. This approach does not account for the spatial distribution of data points. Weighted averaging approaches may be used, with weights calculated on areal distribution of sites or by delineation of subregions of uniform properties. Additional upscaling techniques addressed in this section include regression, geostatistics, and geographical information systems (GIS). Simple empirical equations are frequently used to obtain initial estimates of recharge. They are included in this section because some of the same tools (such as regression analysis) that are used for upscaling can be used to develop those equations and because empirical models are embedded in other techniques for upscaling (e.g. they are often used in conjunction with a GIS).

The nature of the recharge estimates that are extrapolated in space or time is important, regardless of the upscaling approach that is used. Most upscaling procedures are based on estimates of diffuse recharge derived by methods such as streamflow hydrograph analysis (Section 4.5); thus, maps or other products derived from those estimates pertain only to diffuse recharge. Any focused recharge would go unaccounted for. Some models and water-budget methods can account for both diffuse and focused recharge; if recharge estimates generated by these methods are used as the basis for upscaling, then resulting products would represent total recharge.

3.7.1 Simple empirical models

One of the most widely used and easiest approaches for estimating recharge (R) is to assume that it is equal to some fraction, a, of annual precipitation (P):

$$R = aP \qquad (3.9)$$

This is a simple yet useful approach. The ratio of annual recharge to precipitation is reported in countless studies and serves as a parameter that allows comparison of recharge rates from different regions and time spans. Reported ratios for studies cited in this text range from about 0.02 to more than 0.5. There are many factors that influence this ratio, but depth to the water table may be the most influential. Shallow water tables usually receive more recharge (relative to precipitation) than deep water tables. Many variations of Equation (3.9) exist. Sometimes annual precipitation is replaced with precipitation during the nongrowing season (fall through early spring).

Maxey and Eakin (1949) proposed a modified form of Equation (3.9) for estimating total recharge volume to valleys in Nevada where there is a large variation in precipitation between mountains and valley floors. Five zones were delineated on the basis of annual precipitation rates, and a separate coefficient was determined for each zone:

$$R = 0.03P1 + 0.07P2 + 0.15P3 + 0.25P4 \qquad (3.10)$$

where R is now the volume of annual recharge for the watershed; $P1$ is volume of precipitation that fell in zones where precipitation was between 203 and 305 mm; and $P2$, $P3$, and $P4$ are volumes for zones where precipitation was between 305 and 381 mm, 381 and 490 mm, and over 490 mm, respectively. Equation (3.10) was derived under the assumption that recharge was equivalent to groundwater discharge; precipitation data for Nevada were obtained from Hardman (1936). Variations of this method have been extensively used in Nevada over the years since it was first proposed (e.g. Watson *et al.*, 1976; Nichols, 2000; Flint and Flint, 2007).

Keese *et al.* (2005) used a power function model to estimate annual recharge from mean annual precipitation:

$$R = aP^b \qquad (3.11)$$

This equation provided better estimates of recharge across Texas than estimates produced by Equation (3.9). Flint and Flint (2007) used a modified form of Equation (3.11), in which precipitation exceeding a threshold replaced precipitation, to extrapolate recharge estimates in time by using historical precipitation data.

3.7.2 Regression techniques

Regression techniques are commonly employed in hydrologic studies for a multitude of purposes (Helsel and Hirsch, 2002). One such purpose is the extrapolation (or upscaling) of recharge estimates in space or the extension of the estimates in time. Linear regression equations generally take the form:

$$R = aX_1 + bX_2 + c \qquad (3.12)$$

where a, b, and c are coefficients determined by regression analysis and X_1 and X_2 are independent parameters that reflect watershed characteristics, such as soil texture, permeability, elevation, vegetation, and geology, or climate, such as precipitation and temperature. Equation (3.12) is easy to apply; it can be applied at any location where parameter values are known or can be estimated. Nonlinear regression equations, some of the form of Equation (3.11), are also common in recharge studies. Nolan *et al.* (2007) used nonlinear regression to identify factors that influenced recharge estimates for the eastern United States.

Derivation of a regression equation requires a number of known recharge values. Because recharge cannot be measured, but

must be estimated by some technique, there is unmeasurable uncertainty in data used to derive Equation (3.12). Estimates of diffuse recharge derived from analysis of streamflow hydrographs (Section 4.5) have been used to develop regression equations for predicting recharge (e.g. Pérez, 1997; Holtschlag, 1997; Cherkauer and Ansari, 2005; Gebert *et al.*, 2007; and Lorenz and Delin, 2007). Sophocleous (1992) and Nichols and Verry (2001) based their regression equations on estimates of recharge obtained from application of the water-table fluctuation method (Section 6.2). Flint and Flint (2007) and Tan *et al.* (2007) used model-generated estimates of recharge to develop regression equations and subsequently used the regression equations to extrapolate model results.

Example: Minnesota recharge map

Lorenz and Delin (2007) used recharge estimates derived from the streamflow recession-curve displacement method (Section 4.5.2) with data obtained from 38 stream gauges across Minnesota to generate the following equation:

$$R = 14.25 + 0.6459P - 0.022331GDD + 67.63S_y \qquad (3.13)$$

where R is average annual recharge in centimeters (cm), P is annual precipitation in cm, GDD is growing degree days, and S_y is aquifer specific yield. Growing degree days served as a surrogate for evapotranspiration and were calculated as the sum of average daily temperature minus 10°C. Precipitation and growing degree days data were obtained from National Weather Service stations for the period 1971 to 2000. Specific yield served as a surrogate for soil texture and was estimated as the difference between saturated soil-water content and water content at field capacity, as calculated by the method of Rawls *et al.* (1982). Percentages of sand, silt, and clay required for estimating water content were obtained from the STATSGO database.

Regression equations can be conveniently applied manually, in a spreadsheet, and with a GIS. Equation (3.13) was applied with a GIS across all of Minnesota (Figure 3.17). The

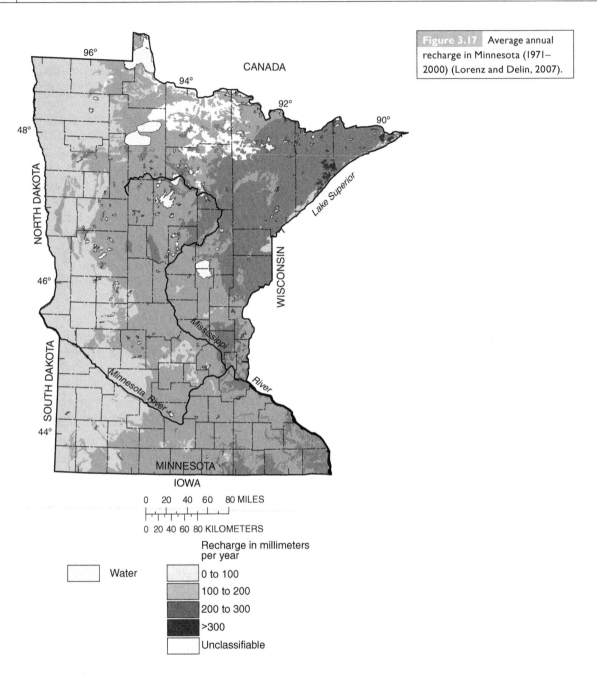

Figure 3.17 Average annual recharge in Minnesota (1971–2000) (Lorenz and Delin, 2007).

recharge map is useful in identifying regional trends, and, as would be expected, the pattern of recharge is similar to that of precipitation, increasing from west to east across the state. Some care should be taken in assessing the absolute accuracy of estimates displayed in the map. Uncertainty in the recharge estimates arises from the fact that Equation (3.13) is derived on the basis of estimates of recharge rather than actual values of recharge; the accuracy of those estimates is unknown. An additional source of uncertainty is the upscaling process whereby Equation (3.13) is applied to areas outside of the 38 gauged watersheds. There is also the question as to whether the recession-curve displacement method provides an estimate of actual recharge or of base flow (Sections 4.1.2 and 4.5.2).

3.7.3 Geostatistical techniques

Kriging is an interpolation method (Journel and Huijbregts, 1978) that can be used to estimate recharge at any location in a watershed if point estimates of recharge are available at other fixed locations within the watershed. By setting up a grid over a watershed and determining kriged estimates of recharge at each grid point, an estimate of recharge integrated over the entire watershed can be obtained. In addition, kriging provides a measure of the uncertainty of the interpolated values. The kriging estimate, R^*, is calculated as a linear combination of all available data values:

$$R^* = \sum_{k=1}^{N_{obs}} \lambda_k R_k \qquad (3.14)$$

where N_{obs} is the number of points where recharge is known, R_k is the recharge value at location k, and λ_k is the kriging weight associated with each data location. The kriging weights are calculated such that the estimation is unbiased and the variance of estimation is minimal. An extension of this method, called cokriging, considers more than one data type in the estimation process (e.g. recharge rate and soil texture). Correlation between these data types can lead to improved estimates of recharge.

Geostatistical tools can be useful when used in conjunction with a GIS. Szilagyi *et al.* (2003; 2005) used streamflow hydrograph separation analysis (Section 4.5) to estimate recharge for selected watersheds in Nebraska; kriging and a GIS were then used to develop recharge maps for the entire state. Lee *et al.* (2006) took a similar approach to develop a recharge map for Taiwan. Kriging and cokriging are used in a variety of other hydrologic applications, such as preparing gridded data for model input and displaying model output. Hevesi *et al.* (1992a, b) cokriged precipitation rates and elevation in the vicinity of Yucca Mountain, NV.

3.7.4 Geographical information systems

A geographic information system (GIS) consists of computers, software, and databases designed to store, manage, analyze, and display information that is spatially referenced to Earth (http://

www.gis.com; accessed September 24, 2009). These systems are widely used in hydrology and other fields of Earth science; many of the examples cited in this book used a GIS. Information relevant to hydrologic studies include elevation, soil properties, land use, population, locations of natural features such as rivers and lakes and human-made features such as roads and buildings, and depth to water table. Regression methods and geostatistics are some of the analytical tools that are usually available in a GIS. In addition, users can create their own tools or models and apply these models at a great many points. For example, simple empirical models have been created and incorporated into a GIS to look at drainage or recharge across large areas (e.g. Keese *et al.*, 2005; Dripps and Bradbury, 2007). Detailed maps of data and of model calculations are easily constructed. A GIS can also be used to upscale (interpolate or extrapolate) point estimates of recharge over large areas.

Keese *et al.* (2005) used a GIS to develop a map of annual recharge rates over Texas for the period 1961 to 1990. Recharge was estimated by using Equation (3.11); coefficients in the equation were determined from a series of simulations of water movement through the unsaturated zone with a Richards equation-based model for 13 regions of the state. Minor *et al.* (2007) used a GIS to upscale point estimates of recharge generated by a chloride mass balance approach to watershed-wide estimates for areas of southwestern Nevada. Dripps and Bradbury (2007) derived recharge estimates that were integrated over an entire watershed by using a GIS to apply a simple soil water-budget model in many locations within the watersheds. Fayer *et al.* (1996) and Rogowski (1996) used a GIS to look at spatial variability of recharge within watersheds.

A GIS is also a useful tool for analyzing and displaying remotely sensed data, climate data, and data on soils and land use contained in databases such as STATSGO2 and SSURGO. The limitations of remote sensing for direct estimation of recharge are discussed in Chapter 2. However, remotely sensed data used in conjunction with a GIS can be quite useful in the preparation of areally distributed parameter values for model input (Brunner *et al.*, 2007). These data could include, for example, land-surface elevation, land use, vegetation type and growth state, surface and air temperature, and soil-water content. Hendricks Franssen *et al.* (2006) demonstrated that uncertainty in estimates of recharge and other parameters of a groundwater-flow model for the Chobe region in Botswana (where data are sparse) could be substantially reduced by inclusion of what the authors referred to as recharge potential. Recharge potential is the difference between precipitation and evapotranspiration, and for that study both of these parameters were estimated solely from satellite data.

Example: Basin and Range carbonate rock aquifer system

The Basin Characterization Model (BCM) is a watershed model that requires as input GIS coverages of geology, soils, vegetation, and elevation and monthly air temperature and precipitation (Flint *et al.*, 2004). Water budgets for 13 hydrographic areas in the Basin and Range carbonate rock system in Nevada and Utah were simulated to generate estimates of drainage through the bottom of the root zone (Flint and Flint, 2007). The BCM was applied on a 270-m grid by using monthly climatic data for the period 1970 to 2004. Precipitation values for the model grid were obtained by interpolation of PRISM (Daly *et al.*, 1994) data that are on a 1.8-km grid. Submodels within the BCM calculated potential evapotranspiration and snow accumulation and melt. Most recharge is predicted to occur in the mountainous areas of the study area. Averaged over the entire study area, drainage for the 35-year period was about 20 mm, or 7% of precipitation and about 47% greater than estimates generated using the Maxey–Eakin method (Flint and Flint, 2007).

Precipitation data for the area are available from PRISM for 1895 to 2006. Flint and Flint (2007) developed power function interpolation equations to extend drainage estimates from the 35-year BCM simulated period to a 112-year period for each of 30 subbasins within the study area. The interpolating equation for one subbasin was:

$$D = 7.7 \times 10^{-5}(P - 27\,100)^{1.65} \text{ for } P > 27\,100 \quad (3.15)$$

where D (drainage) and P (annual precipitation) are in terms of acre-feet for the entire study area, and drainage is assumed to be 0 when P is less than or equal to 27 100. Equation (3.15) was developed by regression analysis using results of the BCM for 1970 to 2004. Average annual recharge for all the watersheds for the 112-year period was estimated at about 18 mm, or 6% of precipitation.

3.8 | Aquifer vulnerability analysis

Many methods have been proposed for determining the vulnerability of groundwater to contamination from surface sources (Gogu and Dassargues, 2000). The underlying assumption of these methods is that contaminants are transported to the water table by recharging water – areas of high recharge rate are more vulnerable than areas of low recharge rate. Two types of vulnerability are usually discussed: intrinsic and specific. Intrinsic vulnerability refers to the susceptibility of contamination based on physical features, such as geology, depth to water table, and climate, and is independent of any particular contaminant. Specific vulnerability refers to susceptibility to one particular contaminant. In addition to physical features of the groundwater system, specific vulnerability analysis must consider the chemical and biological properties of the system and the contaminant. This section provides a brief overview of the link between model estimates of recharge and assessment of intrinsic vulnerability of aquifers.

There is no universally accepted standard by which aquifer vulnerability is measured. Most methods for assessing vulnerability provide overlay maps, or indices, or scores of vulnerability; these methods include models such as DRASTIC (Aller *et al.*, 1985), AVI (van Stempvoort *et al.*, 1993), EPIK (Doerfliger *et al.*, 2000), and PI (Goldscheider, 2005). Although recharge rate is an important factor in predicting aquifer vulnerability, explicit estimates of recharge rates are not required in some assessment methods. Of the four vulnerability models cited above, only DRASTIC requires an estimate of recharge for model input. The other models use parameters such as depth to water table, precipitation, and soil properties as surrogates for recharge.

Schwartz (2006) and Neukem *et al.* (2008) took a different approach for quantifying vulnerability; classes of vulnerability were defined solely on the basis of the estimated travel time of water from land surface to the water table. Many of the models described in this chapter, including unsaturated zone water-budget models, watershed models, and empirical models, could be used to estimate these travel times. Large-scale recharge maps, such as those developed by Holtschlag (1997), Keese *et al.* (2005), and Lorenz and Delin (2007; Figure 3.17), are becoming more available. A first-cut vulnerability assessment might equate areas of relatively high recharge rates on those maps with areas of high vulnerability to contamination.

Another aspect of aquifer vulnerability mapping in which models play an important role is identification of areas that contribute to a water-supply well. Aquifers are susceptible to contamination from recharge that occurs in these areas. Well-head protection rules have been implemented in many areas to safeguard groundwater supplies. Contributing areas are often identified by a particle-tracking routine, such as MODPATH, that is capable of backtracking along the path of a water particle from the point on the well screen where the particle is removed from the aquifer back to the location where that particle first entered the aquifer as recharge (Franke *et al.*, 1998; McMahon *et al.*, 2008; Moutsopoulos *et al.*, 2008). MODPATH requires estimates of groundwater velocity at all points in the aquifer; these velocities are usually determined with a groundwater-flow model such as MODFLOW.

Example: groundwater vulnerability in Namibia

Schwartz (2006) described a study of vulnerability of groundwater to contamination from

Table 3.3. Degrees of vulnerability, simulated travel time through the unsaturated zone, and proportional area of coverage for the country of Namibia (after Schwartz, 2006).

Vulnerability classification	Travel time (yr)	Percent areal coverage
Very low	>25	84.5
Low	10–25	10.6
Medium	3–10	4.3
High	1–3	0.5
Very high	<1	0.1

water percolating downward from land surface in the southwest African country of Namibia. Five classes of vulnerability were identified on the basis of travel time for infiltrating water to reach the water table (Table 3.3). A map of travel time was developed for the entire country. The approach used a combination of numerical modeling, regression analysis, and new and existing GIS data coverages. Flow through the soil zone was simulated with the MACRO4.3 model (Jarvis, 2002), a model based on the Richards equation, for four generic lithological units. Simulations were run for each of six assumed soil-zone thicknesses, ranging from 0 to 1 m. Potential evapotranspiration rates required by the model were estimated on the basis of monthly temperature measurements. Net infiltration (flow out the bottom of the soil zone) was simulated for a 5-year period with measured daily precipitation.

Regression equations were developed to relate net infiltration, I_{net}, to annual precipitation and soil-zone thickness for each lithological unit. Travel time through the unsaturated zone, t, at any location was then calculated as:

$$t = Z\theta / I_{net} \qquad (3.16)$$

where Z is depth to the water table and θ is the average volumetric water content (assumed uniform within each lithological unit). GIS coverages of annual precipitation, lithological unit, and depth to water table were then used to determine and map travel times (Figure 3.18). Results showed that 95% of the country falls

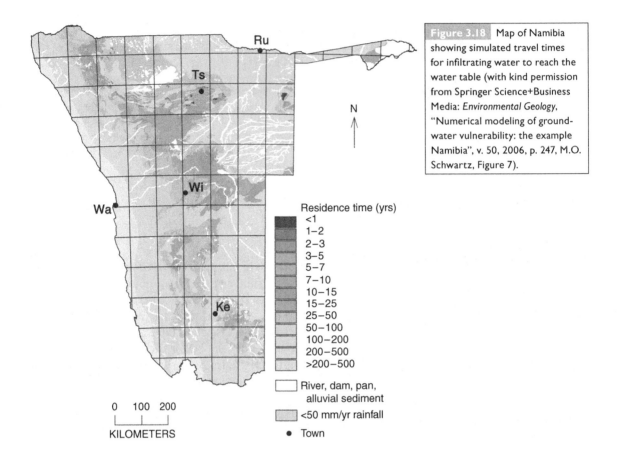

Figure 3.18 Map of Namibia showing simulated travel times for infiltrating water to reach the water table (with kind permission from Springer Science+Business Media: *Environmental Geology*, "Numerical modeling of groundwater vulnerability: the example Namibia", v. 50, 2006, p. 247, M.O. Schwartz, Figure 7).

N

Residence time (yrs)
<1
1–2
2–3
3–5
5–7
7–10
10–15
15–25
25–50
50–100
100–200
200–500
>200–500

River, dam, pan, alluvial sediment

<50 mm/yr rainfall

• Town

0   100   200
KILOMETERS

into the categories of low or very low vulnerability (Table 3.3).

3.9 | Discussion

Hydrologic models are useful tools for estimating groundwater recharge. In addition to providing information on recharge rates, models can provide insight into features of the hydrologic system that influence those rates. Models can be used to predict effects of changes in climate, land use, and water management strategies on recharge rates. Inverse modeling can be used to quantify the uncertainty in model-predicted recharge rates if the model accurately represents the hydrologic system. There is a wide range of complexity in models, and there are tradeoffs in terms of benefits between parsimony and complexity (Hunt *et al.*, 2007). Simple

models are easy to apply and often require only readily available data. They also are amenable to application with a GIS. Complex models can more accurately represent the physics of water flow and can therefore shed additional light on the parameters that most affect recharge rates, but there can be considerable expense, in terms of data collection and compilation, in applying complex models.

As with any method for estimating recharge, assumptions inherent in model design and application should be fully scrutinized before, during, and after application. Assumptions on processes occurring within a hydrologic system vary from model to model. Throughout this chapter important assumptions have been noted, but readers considering the use of any of the modeling approaches discussed here are urged to consult individual model documentation and references for more complete discussion of the underlying assumptions.

Ever more powerful models and graphical display tools enable users to build colorful maps of recharge rates or other water-budget components on basin and regional scales. Such maps can be very useful in a qualitative sense for identifying areas of relatively low or high recharge rates. Some caution is in order, however, when viewing these maps in a quantitative sense. Such maps can give a deceptive impression of accuracy and completeness of areal coverage, when in actuality only a small amount of data was used to construct the map.

Analysis of model assumptions is an important aspect of determining whether a model is appropriate for a particular application, but there are additional factors to consider: availability of required data input, the possible need for training on model use, and financial and time constraints of the study. Because of the difficulties of setting up a complex watershed or groundwater-flow model, one should conduct an evaluation a priori to determine whether the benefits obtained from a model justify the costs that will be incurred.

Methods based on surface-water data

4.1 | Introduction

Streamflow data are commonly used to estimate recharge rates in humid and subhumid regions, in part because of the abundance of streamflow data and the availability of computer programs for analyzing those data. Most of the methods described in this chapter are easy to use, but application of any of the methods should be accompanied by a careful analysis of the underlying assumptions. The methods estimate exchange rates between groundwater and surface-water bodies. That exchange can represent focused recharge from a losing stream, or, as in the case of groundwater discharge to a stream, the exchange can reflect diffuse recharge that occurs over widespread areas. Some of these methods may be unfamiliar to groundwater hydrologists because they were not developed specifically for the study of groundwater recharge; instead, they were developed for purposes such as sizing of culverts and bridge openings, predicting low-flow rates in streams, or developing an understanding of stream-water quality and the ability of a stream to assimilate solutes and contaminants. The fact that base-flow or recharge estimates are generated as byproducts of these methods does not diminish the usefulness or applicability of the methods in recharge studies.

Techniques presented herein include the stream water-budget method, seepage meters, Darcy methods, streamflow duration curves,

traditional streamflow hydrograph analyses (including hydrograph separation and recession-curve displacement), and chemical and isotopic hydrograph separation techniques. Some of these methods are designed specifically for estimating focused recharge; others are for estimating diffuse recharge. Discussions are centered on groundwater movement to or from streams, but the principles discussed and the methods described are equally applicable for groundwater exchange with other surface-water bodies, such as lakes, reservoirs, and wetlands. Proper application of any method requires a good conceptual model of the hydrologic system and a solid understanding of underlying assumptions. Prior to presentation of individual methods, background discussions are given on the exchange of groundwater and surface water and on the relationship between base flow and recharge. These discussions illustrate assumptions inherent to the methods and provide some guidelines for assessing the validity of those assumptions.

4.1.1 Groundwater/surface-water exchange

Exchange of groundwater and surface water occurs in most watersheds and is governed by the difference between water-table and surface-water elevations (Winter et al., 1998). If the water table is higher than stream stage, groundwater discharges to the stream and the stream is referred to as a *gaining stream* (Figure 4.1a). If stream stage is higher than the water table,

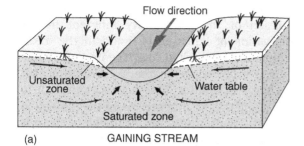

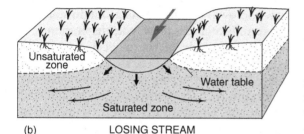

Figure 4.1 Schematics showing (a) gaining stream, stream stage is below water table; (b) losing stream connected to aquifer, stream stage is above water table; and (c) losing stream disconnected from aquifer. (Winter *et al.*, 1998).

the stream is a *losing stream* and surface water flows into the subsurface (Figures 4.1b and c). Paradoxically, both stream losses and stream gains are associated with recharge. A losing stream is a source of focused recharge to the aquifer. Groundwater discharge (base flow) to a gaining stream is water that recharged the aquifer, usually as diffuse recharge, at some point upgradient from the discharge point.

In arid and semiarid regions, focused recharge from ephemeral streams is often the dominant recharge mechanism. These streams usually have water only after rainfalls or snowmelts. Such streams generally have steep slopes and lose water over most reaches. Streams in humid regions are generally gaining streams. However, streams in

any climatic setting can have reaches that gain and reaches that lose; some reaches can gain flow at certain times of the year and lose flow at other times. Hydrologic processes are variable in space and time; the very water that becomes recharge at one point in a stream may be discharged to the same stream at a point 100 m down channel. Hence, the location and timing at which a method is applied can affect estimated recharge rates, a point that underscores the importance of developing a sound conceptual model of recharge processes in any study.

If the water table intersects the stream channel, then stream loss represents actual recharge (Figure 4.1b). However, if an unsaturated zone exists between the channel and the water table (Figure 4.1c), stream loss represents potential recharge because all of the water leaving the stream may not travel to the water table; some may be diverted to evapotranspiration, for example. Streams and rivers tend to be more dynamic than lakes and aquifers in the sense that residence times for water in streams and rivers are typically much less than those for lakes and the subsurface (Healy *et al.*, 2007). Recharge can be slow and steady over time or highly episodic. A stream can quickly change from gaining to losing as a result of a sudden rise in stream stage associated with a large precipitation event. As the stage rises, water flows into stream banks. If the water table is connected to the stream, this inflow from the stream is called *bank storage*. This focused recharge is an important source of recharge for some aquifer systems, but in many other systems the stored water resides in the subsurface for only short periods. As stream stage recedes to levels less than that of groundwater, water flows back from the stream banks to the stream. Release of bank storage can sustain streamflow for extended periods following a runoff event depending on the duration of high stage and the hydraulic properties of the stream banks. If the water table is not connected to the stream, little of the water lost from the stream may return to the stream. This lost water is regarded as potential recharge – it may percolate down to the water table, or it may be evapotranspired to the atmosphere.

If a stream rises above its banks in the course of a runoff event, flood waters could extend over large areas and provide a significant source of infiltration (potential recharge). Recharge from a losing stream is typically viewed as focused recharge. The widespread nature of infiltration from flood waters imparts more of a diffuse nature to the potential recharge. Flood waters may eventually flow back through the subsurface to the stream, but this process can take a long time, especially in flat, swampy areas. Climate, geology, soils, land use, land surface topography, and water use all play a role in the interaction between groundwater and surface water.

4.1.2 Base flow

Base flow is defined in Chapter 2 as groundwater discharge to a stream or other surface-water body. For clarity, it should be noted that this term has been used by others to represent all streamflow that is not derived directly from surface runoff or interflow and can include, in addition to groundwater discharge, release of water from snowpacks, bank storage, and other features (Meyboom, 1961; Linsley et al., 1982; Halford and Mayer, 2000). Some of the methods described in this chapter provide estimates of base flow (groundwater discharge to streams). Base flow is usually associated with diffuse recharge that occurs somewhat uniformly in space as infiltration from precipitation or irrigation. That infiltrating water travels through the subsurface to the water table and eventually discharges to a stream. To determine whether base flow is a good approximation to recharge, two questions must be addressed: (1) Does all water that recharges an aquifer ultimately discharge to a stream? (2) Does streamflow (at certain low flows) consist entirely of groundwater discharge?

The water-budget equation for an aquifer presented in Chapter 2 provides a basis for analysis for the first question:

$$R = \Delta S^{gw} + Q^{bf} + ET^{gw} + (Q^{gw}_{off} - Q^{gw}_{on}) \qquad (4.1)$$

where R is recharge, ΔS^{gw} is change in groundwater storage, Q^{bf} is base flow, ET^{gw} is evapotranspiration of groundwater, and $(Q^{gw}_{off} - Q^{gw}_{on})$ is net groundwater flow out of the aquifer and includes pumping and interaquifer flow. (A discussion of the uncertainty inherent in the derivation of Equation (4.1) and in the measurement of individual components of the equation is provided in Chapter 2.) Assuming that recharge is equal to base flow is equivalent to assuming that all other terms on the right-hand side of Equation (4.1) are negligible – there is no change in aquifer storage, no exchange of water with underlying aquifers, no underflow, no withdrawal or injection of groundwater, and no groundwater loss to evapotranspiration.

The magnitude of the neglected terms in Equation (4.1) should be assessed with any available data; if any terms are determined to be important, recharge estimates can be adjusted accordingly. Groundwater levels, if measured with sufficient frequency, can be used to estimate change in aquifer storage (Section 6.2). In undisturbed natural hydrologic systems, weather patterns result in seasonal fluctuations in groundwater levels and, therefore, aquifer storage. But over the course of years, in the absence of significant climate change, aquifer storage tends to remain constant. The magnitude of interaquifer flow can be assessed by using Darcy/Hantush methods (Section 6.3), but these methods require water-level measurements in both aquifers and an estimate of hydraulic conductivity. Direct measurements of groundwater withdrawal and injection rates can be difficult to obtain, but rates can sometimes be estimated on the basis of permits issued by regulatory agencies, irrigation demands determined by agricultural agencies, and records of utilities and municipal water suppliers. Direct evapotranspiration of groundwater occurs in areas with shallow water tables and is generally more prevalent in humid regions, such as the eastern United States, than in arid regions. However, phreatophytes within riparian zones in arid and semiarid regions can extract significant quantities of groundwater (White, 1932). For many watersheds, the difference between base flow and actual recharge may be within the margin of measurement uncertainty for base flow.

As to the question of whether streamflow at low-flow levels (i.e. in the absence of surface runoff) is derived entirely from groundwater discharge, several other potential sources require consideration. These sources include: stream bank storage; snow and ice packs; slowly draining natural features, such as wetlands, which serve as surface-water reservoirs; and human-developed diversions such as dams, flood-storage basins, and water withdrawal and return structures. The effect on streamflow of slow, steady release of water from these sources may be indistinguishable from that of ground-water discharge (Halford and Mayer, 2000). Hence, methods of estimating base flow could incorrectly interpret these releases as ground-water discharge. It is difficult to assess the magnitude of these release rates from stream-flow record alone. Comparison of stream stage and groundwater elevations over time and use of local-scale groundwater flow models (Section 3.5) can aid in quantifying the rate at which water is released from bank storage. Tracers, such as specific conductance or temperature, could possibly be used to identify the different sources of streamflow (Section 4.6). But per-haps the safest course of action is to identify periods when the contribution to streamflow from these storage reservoirs may be import-ant and to simply remove those periods from the analysis.

Measured or estimated base flow is often divided by the surface drainage area at the measurement point to derive an average rate in units of length per time, such as millimeters per year (mm/yr). This procedure assumes that aquifer boundaries coincide with watershed boundaries and that the area of the aquifer that contributes to groundwater discharge is iden-tical to the surface drainage area (Kuniansky, 1989; Rutledge, 1998, 2007). In fact, ground-water basin and watershed boundaries can dif-fer (Tiedeman et al., 1997). Miscalculation of the aquifer contributing area will lead to a propor-tional error in recharge estimates.

The base-flow index, BFI, is the ratio of mean annual base flow to mean annual streamflow. The BFI offers a convenient means for compar-ing the relative contribution of groundwater to streamflow at different locales. Typically, the BFI ranges between 0.4 and 0.8 (Stricker,

1983; Kuniansky, 1989). Streams that are not connected to an aquifer could have a BFI of 0. Winter et al. (1998) determined BFIs in excess of 0.9 for streams in parts of Michigan and Nebraska.

4.2 | Stream water-budget methods

The water-budget equation for a reach of stream can be written:

$$Q^{sw}_{out} = Q^{sw}_{in} + P - ET^{sw} - \Delta S^{sw} + R_{off} + Q_{trib} + Q_{inter} + Q_{seep} \tag{4.2}$$

where Q^{sw}_{out} is streamflow (discharge) at the downstream end of the reach, Q^{sw}_{in} is streamflow at the upstream end of the reach, P is precipita-tion falling on the stream, ET^{sw} is evaporation from the stream, ΔS^{sw} is change in stream stor-age, R_{off} is surface runoff to the stream, Q_{trib} is flow from tributaries (or to diversions) within the reach, Q_{inter} is interflow (flow from the unsat-urated zone; this term is sometimes lumped in with R_{off}), and Q_{seep} represents exchange of water between the stream and the subsurface. Q_{seep} may be positive, indicating groundwater dis-charge or base flow (Q^{bf}) to the stream, or nega-tive, indicating stream loss to the subsurface. In either case, Q_{seep} may represent recharge. Base flow is indicative of diffuse recharge (as described in the previous section) and can be expressed as a volumetric flow rate (L^3/T) or as a volumetric flow rate per contributing unit area (L/T). Stream loss, on the other hand, indicates focused recharge, recharge from a surface-wa-ter body; such recharge can be expressed as a volumetric rate or as a rate per unit length of stream (L^2/T).

Q_{seep} is often determined as the residual in Equation (4.2); all terms in the equation (except Q_{seep}) are independently measured or estimated. Generally, the surface-flow terms dominate Equation (4.2), and measurement times are often selected so that precipitation, runoff, and inter-flow are zero. ET^{sw} and ΔS^{sw} can be estimated independently, but for practical applications on naturally flowing streams the magnitude of these terms is generally quite small relative to that of the surface-flow terms and often insignificant relative to measurement errors of surface flow. Rantz (1982) provides a simple formula for estimating ΔS^{sw} for the case where stream stage changes over the course of a discharge measurement. Hence, application of the surface-water budget methods usually is based on a simplified form of Equation (4.2):

$$Q_{seep} = Q^{sw}_{out} - Q_{trib} - Q^{sw}_{in} \qquad (4.3)$$

The procedure is to make simultaneous discharge measurements at the upstream and downstream ends of a stream reach and on any tributaries that empty into that reach. This approach is sometimes referred to as a seepage run and is often used in hydrologic studies (e.g. Donato, 1998; Dumouchelle, 2001; Simonds et al., 2004; Turco et al., 2007). Discharge measurement can be made directly with a current or velocity meter or by installing a flume or weir and monitoring stream stage (Section 2.3.4). Oftentimes measurements are made at multiple points on a stream to identify losing and gaining subreaches within the reach of interest.

For a losing stream, the stream water-budget method provides a direct measurement of focused recharge that is integrated over the length of the reach; within that reach there can be gaining and losing subreaches. Data obtained from seepage runs can be useful for calibration of groundwater flow and watershed models (Chapter 3). Because uncertainties in discharge measurements are fairly well understood, error bounds in recharge estimates can be calculated. Rantz et al. (1982), following the analyses of Carter and Anderson (1963) and Carter et al. (1963), determined that the standard error for an average discharge measurement with a conventional current meter was 2.2%. Rantz et al. (1982) also implied that discharge determined from a stage-discharge rating curve is, at best, within 5% of measured discharge. The US Geological Survey rates a streamflow gauging station's record as excellent if 95% of average daily flows are considered to be within 5% of true values. Oberg et al. (2005) suggested that acoustic velocity current profilers be calibrated to within 5% of known discharge at a test location annually. If the smaller error term (2.2%)

is accurate, then Q_{seep} in Equation (4.3) must be substantially larger in magnitude than 2.2% of the sum of the magnitudes of all other terms in the equation in order to provide some confidence in its value. As such, application of the method is limited to relatively small streams. For large streams, errors in discharge measurements may mask any gains or losses. For example, consider a 1-km reach of stream that loses 0.2 m³/s (Q_{seep} = −0.2 m³/s) and has no tributaries. If measured Q^{sw}_{in} = 1 m³/s and Q^{sw}_{out} = 0.8 m³/s, then the estimated value of Q_{seep} would be 0.2 ± (1.8 × 0.022 = 0.04) m³/s or 0.2 ± 0.04 m³/s. If, on the other hand, Q^{sw}_{in} = 100 m³/s and Q^{sw}_{out} = 99.8 m³/s, the estimated value of Q_{seep} is still 0.2, but the error bounds are ± 199.8 × 0.022 = ± 4.4 m³/s. Obviously, little reliability can be placed on the estimate in the latter case. Turco et al. (2007) deemed seepage rates to be acceptable only if they exceeded 10% of the sum of upstream and downstream measured discharges on the Brazos River and tributaries in Texas.

The stream water-budget method is usually applied to estimate focused recharge from losing streams. However, if a stream reach is found to be gaining, the gain in base flow can be attributed to diffuse recharge occurring in the areas of the aquifer that contribute to base flow. Accurate delineation of these contributing areas is not always a straightforward endeavor, but it can be accomplished in some situations, the simplest case being that of the upper end of a watershed to which there is no surface inflow and all measured flow at a gauging site can be attributed to diffuse recharge over the surface area draining to that site.

Ideally, flows in Equation (4.3) are measured simultaneously; however, this is often difficult to accomplish in practice. A long lapse between measurement times can result in additional uncertainty in seepage estimates. Among other limitations of the stream water-budget method, identifying and accessing locations suitable for discharge measurements can be difficult. Substantial resources can be required, especially if manual discharge measurements with conventional velocity meters are required; these measurements are labor intensive. The seepage estimates determined apply only to the period of time over which measurements are made. If streamflow is monitored with recording gauges, then seepage estimates can be obtained throughout the year. Careful planning of a seepage run may be required to ensure that the neglected terms in Equation (4.2) are small in magnitude.

Example: Lemhi River, Idaho
Donato (1998) described a set of seepage runs that were made on the Lemhi River in east-central Idaho to provide information to local and federal agencies to aid in watershed management. Seepage was measured in 14 reaches extending over approximately 100 km of river length (Figure 4.2). Discharge measurements were made with current meters using the standard method described by Rantz *et al.* (1982). Because of the large number of tributaries, diversions for irrigation, and return flows of irrigation, discharge measurements were made at 117 locations. Five days were required to conduct all of the measurements. Two seepage runs were made: August 4–8 and October 27–31, 1997.

August was a period of high groundwater levels due to irrigation return flow; hence, groundwater discharged to almost all river reaches during the first seepage run (Figure 4.3). Groundwater levels were lower in October, after irrigation had ceased for the year; during the October seepage run, stream loss (focused recharge) occurred in 6 of the 14 reaches (Figure 4.3).

The Lemhi River study illustrates several aspects of the stream water-budget method for determining Q_{seep}. Rates of exchange between groundwater and the river were relatively high. Of the 28 measurements over the two periods, only five of the calculated values of Q_{seep} were less in magnitude than the stated acceptability criterion of 5% of measured streamflow at the downstream end of the reach. The large number of discharge measurements shows the intense amount of data collection required when applying the method in a highly managed watershed. Finally, the results demonstrate the dynamic nature and complexity of hydrologic systems in space and time; in some reaches the river gains during part of the year and loses during other parts of the year. Within any particular reach, subreaches could be gaining or losing.

4.3 | Streambed seepage determination

4.3.1 Seepage meters
Exchange of water between streams, or other surface-water bodies such as lakes, and groundwater can be directly measured with seepage meters (Lee, 1977; Lee and Cherry, 1978; Rosenberry *et al.*, 2008). These devices consist of a measurement chamber that isolates an area of the surface-water bed and a variable-volume reservoir that is used to quantify the volume of flow across the measurement area over a specific time interval. Chambers range from the size of a coffee can to a cattle-watering tank but are typically cylinders about 0.2 m in height and 0.5 m in diameter, open only at the bottom (Figure 4.4). Seepage meters are commonly constructed by cutting the end off of a 208-L (55-gal) steel or plastic storage drum. After careful insertion of the bottom of the chamber into the streambed, a collapsible plastic reservoir (commonly a plastic bag) containing a known amount of water is attached to a fitting so that the chamber is completely sealed from the stream (a tube vented to the atmosphere may be attached to allow dissolved gases to escape). Flow into or out of the chamber is determined by measuring the change in volume of water in the reservoir during the time the reservoir is connected to the chamber. Taniguchi and Fukuo (1993) and Rosenberry and Morin (2004) described automated seepage meters in which the flux to or from the meter is measured electronically. Rosenberry (2008) described a seepage meter designed for use in streams and rivers where fluvial forces would corrupt measurements made by standard meters.

Seepage meters are inexpensive, easy to use, and capable of measuring exchange rates ranging from less than 1 mm/d to more than 2 m/d (Rosenberry *et al.*, 2008). Because of the small measurement area (typically 0.25 m²), seepage meters provide a point measurement

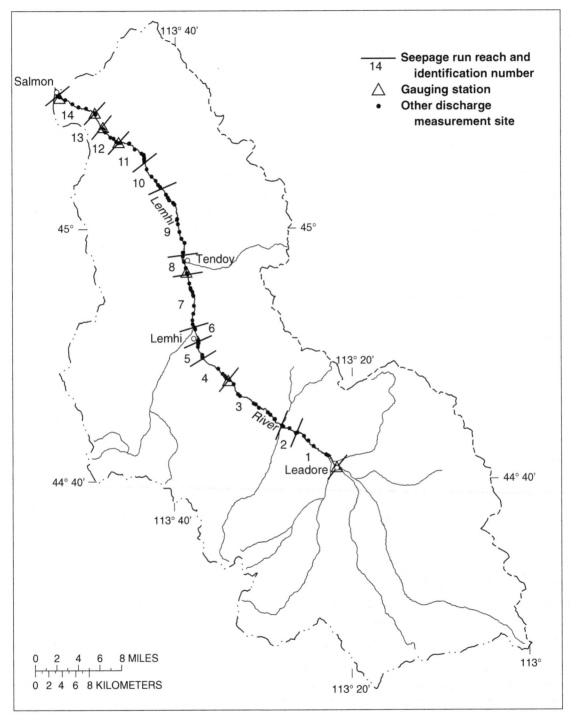

Figure 4.2 Seepage run reaches, gauging stations, and discharge measurement sites in the Lemhi River basin, east-central Idaho, August and October 1997 (Donato, 1998).

of the exchange rates of surface water and groundwater. As such, the meters are useful for examining the spatial and temporal variability of exchange rates. Measurements at multiple locations and times may be required,

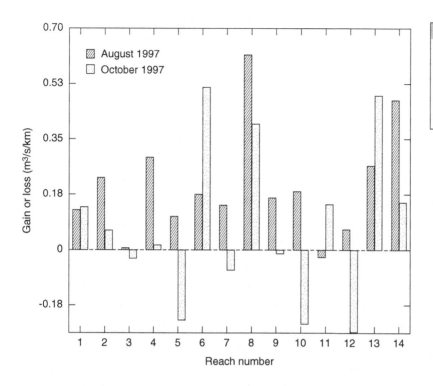

Figure 4.3 Gains (positive) and losses (negative) in streamflow in cubic meters per second per kilometer of channel length over 14 reaches of the Lemhi River, east-central Idaho, August and October 1997 (Donato, 1998).

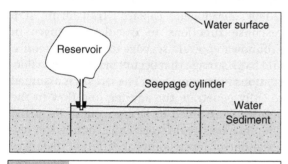

Figure 4.4 Full-section view of seepage meter showing details of placement in the sediment.

however, to obtain representative averages over any stream reach (upscaling of point measurements is addressed in Section 3.7). Installation of the meter at a measurement site requires care; disturbance of bed sediments can alter the natural exchange rate. The meter must be inserted to a depth sufficient to prevent leaks between the meter and the stream. Several studies conducted under controlled laboratory conditions have concluded that seepage meters systematically underestimate actual exchange rates because of resistance to flow within and between the chamber and the measurement

reservoir (Lee, 1977; Belanger and Montgomery, 1992; Murdoch and Kelly, 2003; Rosenberry and Menheer, 2006). Correction factors from 1.05 to almost 2 have been proposed for converting measured fluxes to actual fluxes (Rosenberry et al., 2008). Rosenberry et al. (2008) discussed additional sources of error and provided guidelines for installing and monitoring seepage meters in various stream and lake settings.

While seepage meters are used primarily for quantifying surface-water and groundwater exchange, they also are employed in many water-quality studies. Water samples collected in chamber reservoirs can be removed and analyzed for various chemical and biological constituents. Belanger et al. (1985) made a total of 270 seepage-meter measurements at 25 locations in East Lake Tohopekaliga, Florida, over a 9-month period to study the water and nutrient budget of the lake. Seepage rates ranged from less than 0.2 to more than 7 mm/d, with lowest rates occurring near the center of the 4500-ha lake. It was estimated that groundwater discharge accounted for about 14% of flow into the lake and 9% and 18% of the total phosphorous and nitrogen loads, respectively, to the lake.

4.3.2 Darcy method

The Darcy method is described for flow in the unsaturated zone (Chapter 5) and saturated zone (Chapter 6). Flow from or to a surface-water body can also be estimated by this method (LaBaugh *et al.*, 1995; Wentz *et al.*, 1995; LaBaugh and Rosenberry, 2008). Darcy's equation for vertical flow can be written:

$$q = -K_s \, \partial H / \partial z \tag{4.4}$$

where q is specific flux, K_s is vertical saturated hydraulic conductivity, and $\partial H/\partial z$ is the vertical hydraulic gradient. The gradient is determined from measurements of surface-water stage and groundwater levels in nearby wells or piezometers. The primary limitation in this method, as discussed in previous sections, is the large uncertainty associated with values of hydraulic conductivity. One way to estimate hydraulic conductivity in a streambed or lake bed is to make simultaneous measurements of flux (with a seepage meter) and hydraulic gradient (Murdoch and Kelly, 2003). A hydraulic potentiomanometer (Figure 4.5; Winter *et al.*, 1988; Rosenberry *et al.*, 2008) is a simple device designed specifically for determining vertical head gradient by measuring the difference between surface-water elevation and groundwater level at some depth beneath the stream channel. The potentiomanometer is most often employed to identify areas of seepage into and out of streams and lakes rather than to make direct estimates of exchange. The Darcy method can also be used in two or three dimensions if flow is not strictly vertical. However, multidimensional application requires many additional points for measurement of groundwater levels.

Flow net analysis (Section 6.3.3), which is based on the Darcy equation, also has been used to estimate flow from and to surface-water bodies (Cedergren, 1988; Rosenberry *et al.*, 2008). Flow nets produce a recharge estimate that is integrated over the study region, whereas application of Equation (4.4) produces a point estimate of recharge. Flow nets can help in formulating conceptual models of a hydrologic system, but they have been replaced by numerical groundwater flow models in most studies (e.g. Trommer *et al.*, 2007).

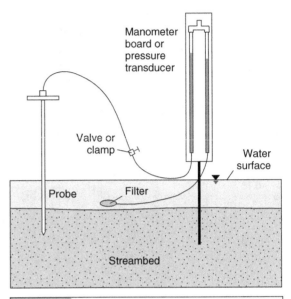

Figure 4.5 Components of the hydraulic potentiomanometer (after Winter *et al.*, 1988).

4.3.3 Analytical step-response function

Moench and Barlow (2000) and Barlow *et al.* (2000) developed Laplace transform step response functions to calculate changes in groundwater levels, seepage to or from streams, and bank storage that occurs in response to fluctuations in stream stage. The stream is assumed to fully penetrate the aquifer, and flow in the aquifer is assumed to be horizontal. Required data include stream stage and groundwater levels at one or more points at different distances from the stream. The method is an economical alternative to a detailed numerical groundwater flow model; Barlow *et al.* (2000) showed that results from the analytical approach are similar to those obtained with a numerical model for a site in eastern Iowa. The approach is particularly useful for predicting recharge occurring in response to sharp rises in stream stage, such as occurs during floods (Ha *et al.*, 2008).

4.4 | Streamflow duration curves

Streamflow duration curves describe the relation between magnitude and frequency of occurrence of discharge in streams (Figure 4.6); they are used in many types of hydrologic studies, such as in the design of bridges and water conveyance structures, flood control planning, and establishment of in-stream flow requirements for aquatic ecosystems (Vogel and Fennessey, 1995). Streamflow duration curves also can be used to approximate base flow to streams. Rutledge and Mesko (1996), for example, determined that a good estimate for base flow was streamflow that was equal to or exceeded 42% of the time for the Valley and Ridge Physiographic Province of the eastern United States.

Flow duration curves for sites with streamflow data are calculated with mean daily flows (weekly or monthly flows also can be used) over a particular period of record. The curves can be easily developed in a spreadsheet program. Flow values are sorted from highest to lowest, irrespective of date. A rank is given to each value, starting with 1 for the highest value and increasing sequentially to N for the lowest, where N is the total number of data points. The probability of exceedance, Pr, for any discharge value can then be calculated assuming a Weibull distribution (Helsel and Hirsch, 2002):

$$Pr = 100 \times [m/(N+1)] \qquad (4.5)$$

where Pr is in percent of time and m is rank of the discharge value of interest. Streamflow is usually plotted against Pr on a log-linear scale to provide detail on the low-flow end of the curve. Q_{Pr} designates streamflow equaled or exceeded Pr percent of the time; it can be read directly from the duration curve (Figure 4.6). Flow duration curves are often normalized by dividing flow rates by drainage area.

Qualitative information can be gleaned from the shape of a streamflow duration curve. Steep slopes in any section of a curve indicate high variability in flow; most likely these streams are dominated by surface flow. Groundwater flow is generally more uniform over time than surface-water flow (although groundwater discharge rates can indeed vary throughout the year). Low slopes in the low-flow section of the curve generally indicate groundwater discharge. As shown in Figure 4.6, flows in Chestuee Creek (drainage area of 38 km²) and South Fork Forked Deer River

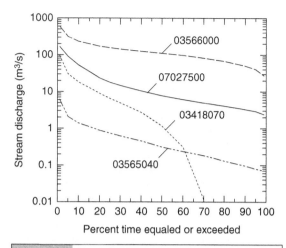

Figure 4.6 Streamflow duration curves for four gauged sites in Tennessee: Hiwassee River at Charleston (USGS station number 03566000, drainage area 5950 km²); South Fork Forked Deer River at Jackson (07027500, 1280 km²); Roaring River at Gainesboro (03418070, 544 km²); and Chestuee Creek above Englewood (03565040, 38 km²).

(1280 km²) appear to be dominated by groundwater discharge; slopes of the curves are low and relatively uniform at exceedance probabilities greater than about 40%. The flow duration curve for the Hiwassee River (5950 km²) has the lowest slope in Figure 4.6, a fact that could be construed as indicating a substantial amount of groundwater discharge; however, flow in this stream is regulated with dams above the gauge; therefore, it is not possible to infer any information on groundwater discharge at this site. The flow duration curve for the Roaring River (544 km²) has a steep slope indicating that flow is dominated by surface runoff, probably in response to storms of short duration. There are many days of zero flow in this stream; thus, little or no groundwater contribution would be expected.

Direct association of streamflow duration with base flow requires an independent estimate of base flow. Nelms *et al.* (1997) applied a streamflow hydrograph separation technique (Section 4.5.1) to estimate base flow for 217 continuous-record streamflow gauging sites within the state of Virginia. Mean annual base flow was found to be well represented by Q_{50} for five different regions in the state (Figure 4.7). Studies

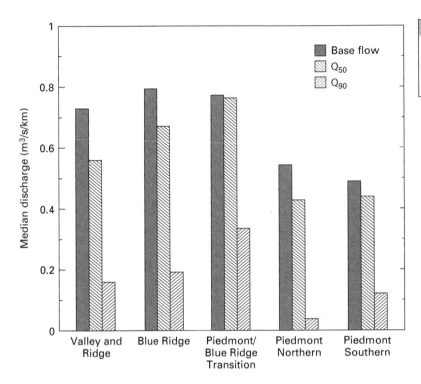

Relation between mean annual base flow, Q_{50}, and Q_{90} for five physiographic regions within the state of Virginia (after Nelms *et al.*, 1997).

in other areas suggest that mean annual base flows can be estimated from flow exceedance percentages of 40 to 90 (Table 4.1).

For estimating base flow, the flow duration curve is usually determined for the entire period of record. However, duration curves can be constructed for individual months (such as for all Januarys) or seasons (such as winter) over the period of record. For gauges with a long history of flow data, duration curves constructed over fixed time intervals (such as decades) can be used to examine trends in base flow that may be attributable to changes in climate, land and water use, or watershed characteristics such as vegetation (Buytaert *et al.*, 2007; Callow and Smettem, 2007; Tuteja *et al.*, 2007). The streamflow duration curve reflects the effects of climate and watershed characteristics over the entire drainage area. Streamflow duration curves are available for many USGS stream gauging sites within the United States (http://streamstats.usgs.gov; accessed February 11, 2009).

The streamflow duration curves described above were constructed with data from gauged locations where stage is continually recorded. Prediction of various low-flow frequency variables at ungauged sites has been a topic of considerable research (Smakhtin, 2001). Regression methods, based on basin and climate characteristics, are the most commonly used approach. Singh (1971) developed equations for predicting Q_{Pr} on the basis of drainage area for 13 different regions in Illinois. Dingman (1978) developed similar equations using drainage area and elevation for streams in New Hampshire. Santhi *et al.* (2008) developed regression equations to predict base-flow index across the entire conterminous United States.

Various low-flow indices have been proposed to represent the variability of base flow, for example $[Q_{25}/Q_{75}]^{0.5}$ (Stricker, 1983) and $[Q_{20}/Q_{90}]$ (Arihood and Glatfelter, 1991). Nelms *et al.* (1997) used $\log(Q_{50}/Q_{90})$ as an index and found a median value of about 0.6 for the sites studied in Virginia. Higher index values indicate higher base flow variability in time at a site. These indices cannot be used to estimate base flow by themselves, but they can be useful for comparing characteristics among sites.

Streamflow duration curves also are useful for the analysis of ephemeral streams that receive little or no groundwater discharge.

Table 4.1. Published values of *Pr* such that Q_{Pr}, daily streamflow that is exceeded *Pr* percent of days, is equivalent to base flow.

Pr, in percent	Region
42	Valley Ridge Province, eastern US (Rutledge and Mesko, 1996)
50	Virginia (Nelms *et al.*, 1997) (Cushing *et al.*, 1973)
60–65	Southeast US Coastal Plain (Stricker, 1983; Kuniansky, 1989)
55	Long Island, nonurbanized (Reynolds, 1982)
60–90	Ohio (Pettyjohn and Henning, 1979)

Ephemeral streams are an important source of focused recharge in many arid and semi-arid regions. Duration curves indicate the percentage of days in which flow is expected. The number of days in which flow occurs is an important factor in calculating the total amount of recharge occurring from an ephemeral stream (Dahan *et al.*, 2007).

4.5 | Physical streamflow hydrograph analysis

Hydrologists have studied streamflow hydrographs since at least the mid 1800s in efforts to identify portions of streamflow that can be attributed to surface runoff and groundwater discharge (Hall, 1968). Over the course of time, various theoretical and empirical approaches have been invented and reinvented for quantifying base flow or recharge. Several empirical hydrograph separation methods are described here, along with the recession-curve displacement analysis. Both approaches are used to develop estimates of diffuse recharge integrated over the area of the aquifer that contributes to groundwater discharge. With hydrograph separation methods, base flow is calculated by summing groundwater discharge to a stream on a daily basis over the period being analyzed; groundwater discharge

to streams is assumed to be a continuous process. Recession-curve displacement methods are based on the assumption of episodic recharge; recharge and rises in stream stage occur only in response to storms. Recharge is estimated by determining the amount of streamflow that can be attributed to groundwater discharge for each rise in stage and summing over all storm events.

Hydrograph analysis techniques require careful selection of the streamflow gauge and watershed to be analyzed. If too large a watershed is selected, rainfall may not be uniformly distributed over the watershed for all storm events and the watershed may overlie several aquifers. Rutledge (1998) suggested that the techniques be applied at stream sites with drainage areas less than about 1300 km². Hydrograph analysis should not be applied to watersheds with flood-control features or reservoirs. Rutledge (2007) and Halford and Mayer (2000) provided some guidance and illustrated basins where hydrograph analysis is inappropriate.

4.5.1 Empirical hydrograph separation methods

Prior to the advent of computers, a number of arbitrary graphical techniques of varying complexity were used for hydrograph separation (Barnes, 1939; Chow *et al.*, 1988). One such technique consisted of plotting stream discharge vs. time on a log-linear scale, determining the slope of the groundwater recession at the point in time after the peak where the recession became approximately linear (at which time streamflow is assumed to be entirely derived from groundwater discharge, point A in Figure 4.8), and extending the recession curve backward to the point in time of the first inflection point on the hydrograph after the peak (point B in Figure 4.8). A straight line was drawn connecting this point to the point of initial rise in the hydrograph (point C in Figure 4.8), and the area under line segments AB and BC in Figure 4.8 is the calculated base flow (Chow, 1964). The method is subjective, results may not be reproducible, and the method can become complicated when streamflow rises occur in quick succession.

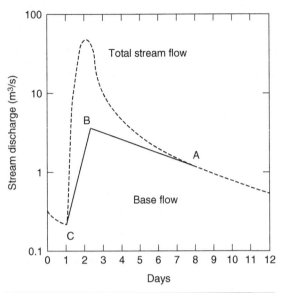

Figure 4.8 Schematic showing steps in manual hydrograph separation; A marks the point in time after which the recession curve becomes approximately linear, B is the time of the first inflection point on the hydrograph after the peak, and C is the point of initial rise in the hydrograph. Base flow for the hydrograph rise is calculated as the area under the line segments AB and BC.

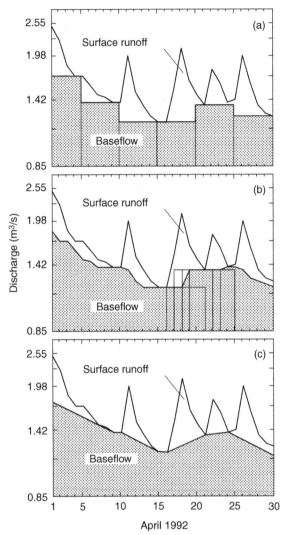

Figure 4.9 Hydrograph separation for French Creek near Phoenixville, Pennsylvania, April 1992, as determined from the HYSEP computer program using (a) fixed-interval, (b) sliding-interval, and (c) local-minimum methods (Sloto and Crouse, 1996).

Recent advances in computer programs for automatically performing hydrograph separation have greatly facilitated the process, removing much of the subjectivity, reducing the amount of time required for analysis, and allowing analyses of complex hydrographs (Pettyjohn and Henning, 1979; Institute of Hydrology, 1980; Wahl and Wahl, 1988; Nathan and McMahon, 1990; Rutledge, 1998; Arnold *et al.*, 1995; Arnold and Allen, 1999; Sloto and Crouse, 1996; Lim *et al.*, 2005). Approaches vary among these models, but they all employ some empirical formula or low-frequency filter for separating base flow from the streamflow hydrograph. Two of the methods are described here.

The HYSEP computer program (Sloto and Crouse, 1996) offers three separate algorithms for hydrograph separation; all three were originally described by Pettyjohn and Henning (1979): local minimum, fixed interval, and sliding interval. Conceptually, these algorithms provide alternative approaches to connecting lines between low points of the streamflow hydrograph (Figure 4.9). The first step is to estimate the length of time following peak discharge over which surface flow continues to contribute to streamflow. A common assumption is that the length of this time interval, N, is constant for all flows and that it can be approximated (Linsley *et al.*, 1982) as:

$$N = 0.83A^{0.2} \qquad (4.6)$$

where N is in days and A is surface drainage area in km^2. The variable $2N^*$, defined by Pettyjohn and Henning (1979) as the odd integer between 3 and 11 nearest to $2N$, determines the time interval used by HYSEP for hydrograph separation. For the fixed-interval method, the period of record is broken down into a contiguous series of time intervals of length $2N^*$. Within each time interval, base flow for all days is set equal to the lowest daily discharge rate within the interval (Figure 4.9a). With the sliding-interval method, base flow for each day of the period, I, is defined as the lowest daily discharge that occurs within the interval extending from $0.5(2N^* - 1)$ days before day I to $0.5(2N^* - 1)$ days after day I (Figure 4.9b). The local-minimum method uses the same time intervals, centered on each day, as the sliding-interval method. Local minima occur on days for which daily discharge is less than that of every other day within the interval. Daily base flow values are determined by linearly interpolating in time between adjacent local minima (Figure 4.9c). Risser et al. (2005b) applied the methods to three sites underlain by fractured bedrock in eastern Pennsylvania. The local-minimum method produced the lowest estimate of base flow (229 mm), followed by the sliding-interval method (292 mm) and the fixed-interval method (295 mm).

Another approach to hydrograph separation uses digital filtering, a technique originally used in signal processing (Nathan and McMahon, 1990; Chapman, 1999; Arnold et al., 1995; Eckhardt, 2005). Although digital filtering is a purely empirical approach, it removes much of the subjectivity from manual separation, providing consistent, reproducible results. A single parameter filtering equation is given by:

$$Q^{fil}_i = \alpha Q^{fil}_{i-1} + (1 + \alpha)(Q_i - Q_{i-1})/2 \quad (4.7)$$

where Q^{fil}_i is filtered direct runoff at time step i, α is the filter parameter, and Q_i is total streamflow at time step i. Base flow at time step i (Q^{bf}_i) is equal to $Q_i - Q^{fil}_i$. Nathan and McMahon (1990) and Arnold et al. (1995) found that a value of $\alpha = 0.925$ produced reasonable results relative to those of manual separation methods. Eckhardt (2008) proposed a recession curve analysis technique for estimating α. Arnold and Allen

(1999) compared results from Equation (4.7) with independent estimates of base flow for six watersheds with historic data; R^2 values for comparisons of monthly values ranged between 0.62 and 0.98. Many variations of Equation (4.7) exist (Chapman, 1999). Eckhardt (2005) proposed a two-parameter model for determining base flow:

$$Q^{bf}_i = [(1 - BFI_{max})\alpha + Q^{bf}_{i-1} + Q_i \\ (1 - \alpha)BFI_{max}]/(1 - \alpha \, BFI_{max}) \quad (4.8)$$

where Q^{bf}_i is the filtered base flow at time step i and BFI_{max} is the maximum value of base-flow index. Eckhardt (2005) suggested values for BFI_{max} of 0.8 for perennial streams, 0.5 for ephemeral streams, and 0.25 for perennial streams connected with hard-rock aquifers. Lim et al. (2005) described a web-based automatic streamflow hydrograph analysis tool (WHAT; http://cobweb. ecn.purdue.edu/~what/; accessed November 10, 2008) capable of using Equations (4.7) and (4.8) as well as the local-minimum approach of HYSEP. Lim et al. (2005) analyzed hydrographs for 50 watersheds in Indiana and found good agreement in results from Equations (4.7) and (4.8). Eckhardt (2008) compared results for Equations (4.7) and (4.8) and the three HYSEP methods and the PART (Rutledge, 1998) and UKIH (Piggott et al., 2005) models for 65 watersheds in North America (Neff et al., 2005). Although there were no measured base flow values to assess the accuracy of the methods, pair-wise correlation coefficients ranged from 0.85 to 1.00, indicating similarity among all methods.

4.5.2 Recession-curve displacement analysis

The recession-curve displacement method (Rorabaugh, 1964) is based on the assumption that an aquifer can be described by one-dimensional flow from a distant no-flow boundary at the edge of the aquifer to a stream (Figure 4.10). The groundwater flow equation under those conditions takes the form:

$$T\partial^2 H/\partial x^2 = S_y \, \partial H/\partial t \quad (4.9)$$

where T is transmissivity, H is hydraulic head, S_y is specific yield, and t is time. Groundwater is assumed to move in a direction, x, perpendicular

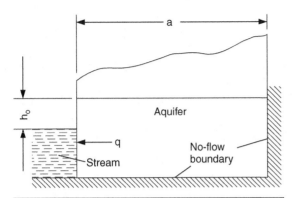

Figure 4.10 Schematic showing idealized, one-dimensional stream-aquifer system on which the Rorabaugh (1964) model is based; a is distance from the stream to the aquifer no-flow boundary, and q is groundwater discharge to the stream in response to an instantaneous and uniform rise in water-table height of h_0.

to the stream. Recharge, transmissivity, and specific yield are assumed to be uniform. Distance from the stream to the aquifer boundary, a, is also assumed to be uniform. Water-table height in the aquifer was initially equal to stream stage. Under these conditions, Rorabaugh (1964) developed an analytical expression for discharge to the stream per unit stream length, q, for the case of an instantaneous, uniform rise in water-table height of h_0:

$$q(t) = (2Th_0/a)e^{-\pi^2 Tt/4a^2 S_y} \qquad (4.10)$$

Equation (4.10) is actually the first term in an infinite series analytical solution. Rorabaugh (1964) suggested application of Equation (4.10) for times greater than critical time, T_c, after the instantaneous rise in groundwater levels, where critical time is a function of aquifer properties:

$$T_c = 0.196\,S_y a^2 / T = 0.21\,K_{RI} \qquad (4.11)$$

where K_{RI} is the recession index, the time required for streamflow to decline through one $\log_{10}$ cycle. Critical time can be determined from aquifer properties, but it is more convenient to determine T_c by estimating the recession index. The recession index can be estimated manually (Mau and Winter, 1997) or with computer programs (Rutledge, 1998; Heppner and Nimmo, 2005) by identifying periods when groundwater discharge accounts

for all streamflow. Following the approach described in Section 4.5.1, the number of days that surface flow continues after peak discharge, N, is usually estimated with Equation (4.6). Discharge is plotted as a function of time on a log-linear scale for these periods. The longest, and often the most useful, recession curves typically occur in late fall and winter when evapotranspiration rates are low. The recession index is set equal to the average slope of the portion of the plotted curves that approaches a straight line.

The total volume of groundwater, V, remaining in storage (and that presumably will eventually discharge to the stream) from a single recharge event can be determined for any time, t, greater than critical time by integration of Equation (4.10) from t to infinity (Meyboom, 1961; Rorabaugh, 1964):

$$V = Q^{bf} K_{RI} / 2.3026 \qquad (4.12)$$

where Q^{bf} is groundwater discharge at time, t, and a log-linear decrease in discharge rate over time is assumed. Glover (1964) and Rorabaugh (1964) estimated that one half of the total volume of groundwater discharge from an instantaneous rise in the water table occurs prior to the critical time. Therefore, total recharge for a water-table rise can be calculated as:

$$R = 2(Q^{bf}_2 - Q^{bf}_1) K_{RI} / 2.3026 \qquad (4.13)$$

where Q^{bf}_2 and Q^{bf}_1 are groundwater discharge rates at critical time after the peak in surface flow for the postrise and prerise, respectively, recession curves (Figure 4.11).

There are six steps in the application of the recession-curve displacement method (Rutledge and Daniel, 1994; Figure 4.11): (1) compute the recession index; (2) compute critical time from Equation (4.11); (3) use critical time to determine t_1, the time at which Q^{bf}_1 and Q^{bf}_2 are calculated; (4) determine Q^{bf}_1 by extrapolation of the prerise recession curve; (5) determine Q^{bf}_2 by extrapolation of the postrise recession curve; and (6) apply Equation (4.13) to compute recharge. The recession index and critical time are assumed to be constant for the period of record under analysis. Q^{bf}_1 and Q^{bf}_2 and recharge are calculated for each rise in stream stage.

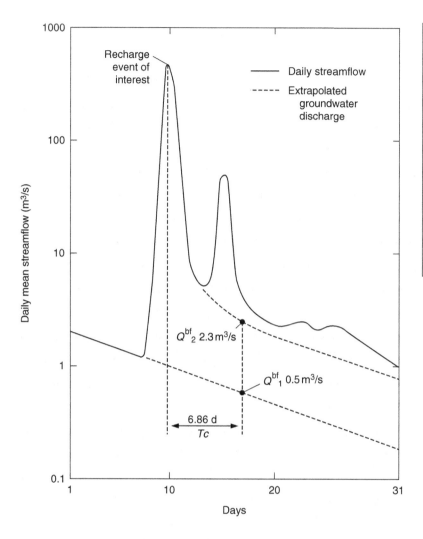

Figure 4.11 Procedure for using the recession-curve displacement method to estimate groundwater recharge in response to a recharge event (after Bevans, 1986; Rutledge, 1998): (1) Compute recession index, K_{RI} (32 d/log cycle); (2) Compute critical time, T_c (0.2144 K_{RI} or 6.86 d); (3) Locate time 6.86 d after peak; (4) Extrapolate pre-event recession to Q^{bf}_1 (0.5 m³/s); (5) Extrapolate post-event recession to Q^{bf}_2 (2.3 m³/s); (6) Compute total recharge, $2 \times (1.8 \text{ m}^3/\text{s}) \times 32 \text{ d}/2.3026 \times 86\,400 \text{ s/1 d} = 4.32 \times 10^6 \text{ m}^3$.

Equation (4.12) estimates the amount of water, from any instantaneous recharge event, that remains stored in the subsurface at any time. On this basis, Meyboom (1961), Rutledge and Daniel (1994), and Rutledge (1998) claimed that the recession-curve displacement method provides an estimate of recharge, as opposed to base flow (i.e. terms in Equation (4.1), such as evapotranspiration of groundwater and interaquifer flow are included in the estimates generated by the method). Results of recession-curve displacement analyses are often reported as recharge. Yet loss of groundwater to evapotranspiration or interaquifer flow violates the underlying assumptions of the method. A tacit assumption in the Rorabaugh approach is that all of the water that arrives at the water table from a recharge event eventually drains to the stream. Rutledge (2000) considered two hypothetical scenarios with the same total recharge. In the first case, all recharge discharges to the stream; in the second, some water is diverted to evapotranspiration. The streamflow recession curves for these two cases differed, as did the estimates of recharge according to Equation (4.13). Rutledge (2000) concluded that the method was estimating net recharge (total recharge minus groundwater evapotranspiration). Nonetheless, several studies (Rutledge and Mesko, 1996; Mau and Winter, 1997; Arnold and Allen, 1999; Chen and Lee, 2003; Risser et al., 2005a) found that the recession-curve displacement method produced estimates that were consistently greater than those from other hydrograph analysis techniques. Daniel (1976) developed a modified form of Equation (4.10) that accounts for interaquifer flow or evapotranspiration of groundwater.

An alternative model proposed by Rorabaugh (1964) assumes that recharge occurs at a constant rate over time as opposed to instantaneously. This approach requires the use of type curves and is not readily adaptable to automatic computer application. Mau and Winter (1997) found good agreement between estimates generated by the two Rorabaugh (1964) methods for a small watershed in New Hampshire.

Few, if any, aquifers conform to all of the assumptions inherent in the derivation of the recession-curve displacement method. Even if critical assumptions are met, Halford and Mayer (2000) pointed out a number of shortcomings and ambiguities; these include the subjective nature with which the recession index and N are defined. Nonetheless, with the abundance of streamflow data and the ease of automatic computer application, the recession displacement method remains a popular technique.

The recession index also can be determined from groundwater hydrographs obtained from observation wells (Rorabaugh, 1964; Halford and Mayer, 2000), but stream hydrographs tend to integrate over a larger part of the aquifer than that represented by a single well. Recession curve analysis is also used for estimating aquifer transmissivity and specific yield (Rorabaugh, 1960; Dewandel et al., 2003).

Example: streamflow hydrograph analyses of Pennsylvania streams

Risser et al. (2005a) analyzed streamflow hydrographs from 197 streamflow sites across the state of Pennsylvania for the period of 1885 through 2001 to generate estimates of groundwater recharge. Each site had at least 10 years of records and a drainage area not greater than 1400 km². Only sites that were not impacted by upstream regulation, withdrawals, or wastewater return flow were selected. Hydrograph separation (PART; Rutledge, 1998) and recession-curve displacement techniques, the RORA computer program (Rutledge, 1998), were applied with mean daily discharges. The

PART program identifies local minima on the hydrograph with a slightly different algorithm from that in HYSEP. PART produces estimates of base flow; with some carefully worded assumptions, the authors referred to results of both PART and RORA as recharge. For the recession-curve displacement method, critical time was determined from Equation (4.11), and the recession index was determined by compilation of a master recession curve using the RECESS computer program (Rutledge, 1998) for streamflow data in all months except June, July, and August (to avoid possible effects of evapotranspiration).

Average annual recharge determined by both methods ranged approximately from 150 to 740 mm. Recharge estimates generated by RORA results were on average about 50 mm/yr greater than those of PART for individual sites (Figure 4.12). Risser et al. (2005a) also applied RORA on a monthly basis (Figure 4.13). They found a clear seasonal trend with almost 80% of recharge occurring between November and May. A drawback to the RORA method is illustrated by the fact that at one site the estimated recharge exceeded mean annual streamflow.

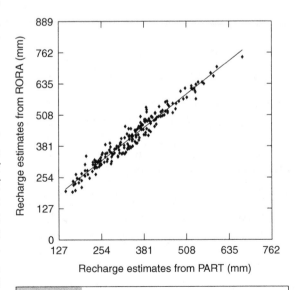

Figure 4.12 Comparison of mean annual estimates calculated by PART and RORA for 197 streamflow sites in Pennsylvania (after Risser et al., 2005a).

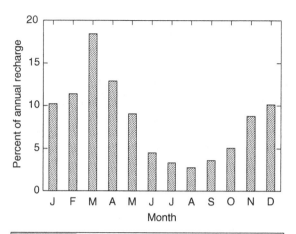

Figure 4.13 Average monthly recharge (as percent of annual recharge) calculated by RORA for 197 streamflow sites in Pennsylvania (after Risser et al., 2005a).

4.6 | Chemical and isotopic streamflow hydrograph analysis

Tracers are used in surface-water studies for a variety of purposes (Kendall and McDonnell, 1998), including determining velocity and residence times in streams (Kimball et al., 1994), identifying flow paths in karst terrains (Katz et al., 1997), and identifying natural and human-induced biogeochemical reactions within watersheds (Soulsby et al., 2004). Tracer methods are also used to quantify base flow over gaining reaches of streams. The most common approach, end-member mixing analysis, consists of constructing a mass balance for the chemical or isotopic tracer. Tracer-injection methods measure tracer concentrations in stream water at multiple distances downstream from the injection point; the rate of groundwater discharge is then inferred from the rate of tracer dilution. The use of tracers in the subsurface to estimate recharge is addressed in Chapter 7.

Tracers should be conservative (i.e. nonreactive) and have no unknown sources in the watershed. Naturally occurring tracers, such as chloride, and tracers such as tritium that have been introduced to the groundwater system through the actions of humans are commonly used in end-member mixing analyses. Stewart

et al. (2007) used specific conductance, a water-quality characteristic that is measured continuously at hundreds of stream-gauging stations across the United States. Radon, an inert gas emanating from source rock, has been used as a naturally occurring tracer for groundwater discharge to streams and coastal waters; virtually all radon measured in surface water comes from groundwater (Gunderson and Wanty, 1991). Rhodamine dye, bromide, chloride, and lithium are often used in tracer-injection studies.

4.6.1 End-member mixing analysis

The first step in developing a tracer mass-budget equation for a stream is to construct a water-budget equation for that stream (Equation (4.2)). The water-budget equation can be simplified by application over a stream reach that has no tributaries and for a time period within which precipitation and evapotranspiration of stream water can be neglected. With the additional assumption that stream stage is constant (and, hence, change in storage is 0), the water-budget equation takes the form:

$$Q^{sw}_{out} = Q^{sw}_{in} + R_{off} + Q_{seep} \qquad (4.14)$$

where Q^{sw}_{out} is stream discharge at the downstream end of the reach, Q^{sw}_{in} is streamflow at the upstream end of the reach, R_{off} is direct surface runoff, and Q_{seep} is base flow. We can then write a mass balance equation for any tracer and denote that tracer's concentration as C, where subscripts refer to one of the four terms in Equation (4.14):

$$C_{out}Q^{sw}_{out} = C_{in}Q^{sw}_{in} + C_{Roff}R_{off} + C_{gw}Q_{seep} \qquad (4.15)$$

For the end-member mixing analysis (EMMA), stream discharge and stream and groundwater concentrations must be measured. C_{Roff} must be measured or estimated; sometimes it is assumed equal to stream concentration at a time when runoff dominates streamflow. At other times C_{Roff} is set equal to tracer concentration in precipitation (the validity of either approach should be examined). Base flow is determined by combining Equations (4.14) and (4.15):

$$Q_{seep} = [(C_{out}Q^{sw}_{out} - C_{in}Q^{sw}_{in}) - C_{Roff} \\ (Q^{sw}_{out} - Q^{sw}_{in})]/(C_{gw} - C_{Roff}) \qquad (4.16)$$

The two-member EMMA described in this section is widely used in hydrologic studies. More complex three-member EMMA applications have been used to study watershed dynamics (Dewalle *et al.*, 1988). Concentrations of end members must be sufficiently different for successful application.

The EMMA method can provide detailed information on watershed hydrology. As with stream water-budget methods, this method is easily adaptable to different hydrologic systems. However, application of the EMMA method may require a large amount of data collection, including continuous streamflow gauging upstream and downstream from the reach of interest (when the method is applied to a small watershed, there may be no surface-water inflow to measure). Tracer concentrations in the stream also must be continuously monitored. Tracer concentration in groundwater (or in base flow) is required, as is concentration in precipitation or in surface runoff. For these reasons, the method is usually applied over small watersheds (1–10 km² drainage area), and even then the method can be expensive and time consuming to apply.

EMMA applications are many and diverse; only a small number are cited in this section. Hooper and Shoemaker (1986) used stable isotopes and major ions to track snowmelt in the Hubbard Brook watershed in New Hampshire. McDonnell (1990) used stable isotopes to identify preferential flow paths through soils. Buttle and Peters (1997) used oxygen-18 and silica to track the pathways of snowmelt in a small basin. Uhlenbrook *et al.* (2002) used oxygen and hydrogen isotopes and dissolved silica to determine groundwater contribution to streamflow in a 40 km² watershed in the Black Forest, Germany. Becker *et al.* (2004) used stream and groundwater temperatures and an energy-budget equation to estimate groundwater discharge to a stream in western New York. Jones *et al.* (2006) used a combined surface/subsurface-water flow and solute transport model to analyze assumptions inherent in the EMMA method and found that due to hydrodynamic mixing of waters, EMMA methods may underestimate actual base flow. Baillie

et al. (2007) used major ions as tracers to distinguish local, monsoon-derived groundwater from waters recharged in distant mountains in base flow in the San Pedro River in southern Arizona and northern Sonora State. Gu *et al.* (2008) used sulfur and oxygen isotopes to determine the contributions of different aquifers to streamflow in southern Arizona. Young *et al.* (2008) used radium isotopes to identify sources of groundwater discharge to a tropical lagoon. Genereux (1998) demonstrated how uncertainty in hydrograph separation can be estimated on the basis of analytical uncertainty in tracer concentrations.

Example: specific conductance as a tracer
Stewart *et al.* (2007) used continuous measurements of streamflow and specific conductance of stream water (Figure 4.14) to estimate base flow. A two-member mixing model was used:

$$Q^{bf} = Q^{sw}[(C_{sw} - C_{Roff})/(C_{gw} - C_{Roff})] \qquad (4.17)$$

where Q^{bf} is base flow, Q^{sw} is stream discharge, C_{sw} is specific conductance of stream water, C_{gw} is specific conductance of groundwater, and C_{Roff} is specific conductance of surface-water runoff. C_{gw} and C_{Roff} were not measured directly in this study. C_{gw} was assumed equal to the stream conductance when streamflow was at its lowest value of the year, i.e. when streamflow is assumed to be entirely derived from groundwater discharge. C_{Roff} was set equal to stream conductance at the highest streamflows of the year; groundwater contribution to streamflow was considered negligible at those times. These assumptions, which were found to be reasonable at an intensively monitored field site (Stewart *et al.*, 2007), greatly facilitated application of the two-member EMMA model. Within the United States, several hundred US Geological Survey streamflow monitoring sites are equipped with continuous specific-conductance meters (http://waterdata. usgs.gov/nwis; accessed October 29, 2009). Conceivably, this model could be applied at many of these sites.

Base flow was estimated by the EMMA model and by hydrograph separation using HYSEP at 10 locations across the southeastern United

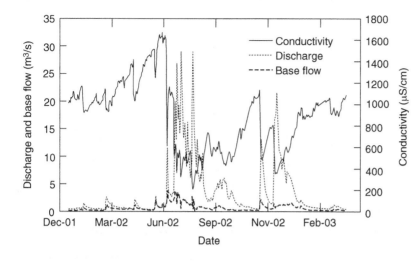

Figure 4.14 Average daily streamflow, conductivity, and base flow as determined from end-member mixing analysis for Joshua Creek, Florida, USGS gauging station 02297100 (Stewart et al., 2007, Figure 3, *Ground Water*, Wiley-Blackwell).

States (Stewart *et al.*, 2007). At six of the sites, the methods produced comparable estimates (differences were not greater than 12%). At the other four sites, the HYSEP estimates of base flow were from 37% to 197% greater than those produced by the EMMA model. These differences were attributed to an inadequate definition of N, the number of days in which streamflow is due solely to surface runoff. A revised version of Equation (4.6) was proposed:

$$N = 0.48A^{0.44} \qquad (4.18)$$

where A is drainage area in km^2. Stewart *et al.* (2007) suggested that the mixing model method may be a convenient approach for calibrating hydrograph separation programs such as HYSEP. Given the simplifying assumptions in both the mixing and hydrograph separation methods, other factors could be contributing to the discrepancy in results. The authors pointed out that overestimation of C_{Roff} will lead to underestimation of base flow.

4.6.2 Tracer-injection method

Streamflow can be measured by injecting a conservative tracer into a stream at a known rate and then measuring the concentration of the tracer at downstream locations (Kilpatrick and Cobb, 1985). The approach is based on a mass-balance equation for the injected tracer in the stream:

$$Q_{pump}C_{inj} = Q_x(C_x - C_{bg}) \qquad (4.19)$$

where Q_{pump} is water injection rate, C_{inj} is tracer concentration in injected water, Q_x is stream discharge at some point x-distance downstream, C_x is measured tracer concentration at that same point, and C_{bg} is background tracer concentration. Q_x is determined by rearranging Equation (4.19):

$$Q_x = Q_{pump}C_{inj}/(C_x - C_{bg}) \qquad (4.20)$$

Estimates of discharge may be obtained in this manner for a number of sampling locations along the course of a stream. A water budget (generally a simplified form of Equation (4.2)) is then constructed for the reach between any two sampling locations. Groundwater discharge and surface-water inflow from tributaries to the reach are estimated from the difference in discharge between the two locations. If a reach between measurement points at distances x1 and x2 downstream from the injection point contains no surface-water inflow, then net groundwater flow to the stream is equal to $Q_{x2}-Q_{x1}$, assuming other terms in Equation (4.2) are negligible. The method is only valid for gaining reaches of streams.

The tracer must be conservative, background concentration in the stream must be uniform, and there should be no significant sources or sinks for the tracer in groundwater or surface water contributing to streamflow. Rhodamine dye, chloride, bromide, and lithium are commonly used tracers. Streamflow and tracer injection rate must be steady throughout the length of the experiment. Ideally, all water inflows to the stream (including injected, surface water, and groundwater) rapidly and uniformly mix in the stream; however, concentrations should be measured at several points in a cross section of the stream over time at selected downstream stations to verify steady conditions. As with the stream water-budget methods discussed in Section 4.2, the tracer injection method is most useful for small streams where groundwater inflow over a reach constitutes a significant percentage of streamflow (in the stream water-budget method, 5% was the minimum suggested percentage).

The tracer injection method can be applied in any stream that satisfies the above requirements. The method has proved particularly useful, however, in mountain streams where steep slopes provide a favorable mixing environment. Traditional current-meter methods of measuring stream discharge are not well suited for these streams because of the steep slopes and the potential for hyporheic flow within cobble-lined streambeds. Hyporheic flow is the transient movement of surface water into and out of the subsurface. The tracer-injection method is usually little affected by hyporheic flow, but large hyporheic zones or surface storage features can complicate efforts to quantify groundwater discharge rates (Harvey et al., 1996).

The tracer injection method has been widely used in studies of acid-mine drainage in mountainous regions of the western United States (Bencala et al., 1987; Kimball et al., 1994, 2004, 2007; Walton-Day et al., 2005). The injected tracer is used to determine groundwater and surface-water inflows along an extended stream reach; simultaneous measurements of stream water chemistry permit calculation of contaminant loads (such as metals) contributed from each subreach. Studies such as these can enhance the effectiveness of remediation efforts by identifying subreaches that contribute the greatest contaminant loads and whose clean-up can best improve stream water quality.

The stream tracer-injection method provides a measure of base flow. Translating a base flow estimate to an estimate of recharge is subject to the caveats discussed in Section 4.1.2. In particular, it may be difficult to determine the area of the aquifer that contributes discharge to a particular stream reach. Recharge estimates from this method are for a single point in time. If hydrologic conditions within a watershed vary throughout the year, it may be desirable to apply the method at different times of the year. Careful planning and intensive sampling over a period of a few days are required, but a large number of locations can be sampled, thus providing very detailed information on stream/aquifer interaction. Walton-Day et al. (2005) obtained samples from 53 locations along and adjacent to a 9.5-km reach of Lake Fork Creek near Leadville, Colorado, during a tracer test that lasted 30 h.

4.7 | Discussion

The methods described in this chapter are similar in that they require data on streamflow, stream stage, or surface-water chemistry. However, there are fundamental differences among the methods, and an understanding of the type of recharge and the time and space scales over which methods are applicable is important. Stream water-budget methods and seepage measurements usually provide estimates of focused recharge or exchange of groundwater and surface water. Streamflow duration curves and hydrograph analysis techniques provide estimates of base flow that is associated with diffuse recharge occurring across a watershed.

Estimates of groundwater and surface-water exchange rates determined from stream water-budget methods are integrated over the reach of the stream where flow measurements are conducted; however, the estimates represent exchange rates only for a particular point in time. Seepage meters provide estimates of exchange rates over a very small area of streambed. Therefore, seepage rates determined with meters represent point estimates in both space and time. Seepage measurements at multiple locations and times can be used to determine variability in exchange rates.

Recharge estimates produced by streamflow duration and hydrograph analysis techniques represent spatially extensive diffuse recharge that is assumed to occur uniformly over some area of the aquifer. Accurate delineation of that contributing area can be problematic; commonly, it is assumed equal to the surface drainage area. Base-flow estimates are integrated over time scales that are largely unknown. Base-flow estimates by themselves provide no insight into the travel time or the travel path that water takes from the time it arrives at the water table as recharge until the time it discharges to a stream. Within any stream reach there can be a wide range of ages in discharging groundwater. Modica et al. (1997, 1998), using a groundwater flow model and chlorofluorocarbon analyses of water, determined that groundwater discharging along a cross section of the Cohansey River in New Jersey ranged in age from about 3 to 29 years. The oldest water was near the center of the channel, and ages decreased with increasing distance from the center.

Flow duration and hydrograph analysis techniques are usually applied with mean daily stream discharge for the entire period of record, but the techniques also can be applied over specific time intervals to examine trends. Callow and Smettem (2007) found that replacement of native vegetation with dry-land crops produced a marked shift in the flow duration curves for streams in southwestern Australia, indicating increased recharge and base flow. Gebert et al. (2007) determined annual base-flow estimates (using hydrograph separation) for streams in Wisconsin and found a trend of increasing base flow with increasing agricultural land use. Risser et al. (2005a) applied the recession-curve displacement method separately to each month for the period of record for 197 stream gauge sites in Pennsylvania (Figure 4.13) and determined that 80% of recharge to these streams occurred between the months of November and May.

Numerous studies have compared results from different streamflow hydrograph analysis methods and found that the methods described herein estimate base flow within a range of about 25%, with the recession curve displacement method providing a high estimate and the local minima method within the computer program HYSEP (Slotto and Crouse, 1996) providing a low estimate (Daniel and Harned, 1998; Risser et al., 2005a). Kinzelbach et al. (2002) determined that hydrograph separation in arid and semiarid regions estimates recharge within a factor of 2 of actual groundwater recharge. Stewart et al. (2007) determined a similar accuracy for hydrograph analysis methods for 10 sites in humid and subhumid regions. Application of the Rorabaugh (1964) method requires more effort than application of most other hydrograph analysis methods, and the results may be no more accurate because real aquifers are often quite different from the hypothetical aquifer described by Rorabaugh (1964). Once the watersheds and stream gauges of interest have been selected and the streamflow data assembled, it is a relatively easy matter to apply several hydrograph analysis methods. Comparisons of results from different methods will provide insight on the uncertainty associated with the methods.

The wealth of available data on streamflow and surface-water quality (Table 2.1) allows methods such as streamflow duration and hydrograph analyses to be applied without the need to collect new data. Streamflow statistics, including duration curves and values of base-flow index, are available for US Geological Survey stream gauging sites (http://streamstats. usgs.gov; accessed February 11, 2009). For methods that require data collection, streams and lakes are much more accessible than aquifers for measuring water levels and flow rates and for collecting water samples.

Computer programs for performing tasks such as surface-water hydrograph analysis and flow-duration analysis provide easy-to-use approaches that permit rapid analysis of many years of streamflow data, but the programs also make it easy to overlook poor data and to ignore important assumptions. Sloto and Crouse (1996) suggested plotting and visually inspecting all data to avoid inclusion of invalid data in base-flow calculations. Assumptions for relating base-flow estimates to estimates of recharge are discussed in Section 4.1.2.

It should be apparent to the reader that a complete evaluation of these assumptions may require substantially more effort than actual application of the methods. Several studies have applied methods described in this chapter to generate estimates of base flow over large areas such as the upper Mississippi River watershed (Arnold *et al.*, 2000) and the conterminous United States (Wolock, 2003). Such studies may provide insight into regional patterns of base flow but not without ignoring many assumptions.

Physical methods: unsaturated zone

5.1 | Introduction

Estimates of recharge can be obtained from measurement of downward water flux or change in water storage within the unsaturated zone. Methods based on physical (as opposed to chemical) data collected within the unsaturated zone are not among the more commonly used techniques for estimating recharge, but they offer some distinct advantages. The methods actually produce estimates of drainage rates below the depth of measurement within the unsaturated zone. The usual assumption is that the draining water will eventually reach the water table, at which time it can properly be called recharge. But there can be a long lag time over which water traverses that depth interval. Estimates generated by these methods are referred to as *drainage* in this chapter. In general, these methods produce point estimates of drainage. The question of how representative a measurement at a single point is of flux through the unsaturated zone as a whole requires careful consideration. These methods can be costly to implement and require intensive instrumentation that is susceptible to measurement inaccuracies. Nonetheless, under certain circumstances, such as rapid movement of a wetting front from land surface to a shallow water table, these methods have a unique ability to provide detailed insight into recharge processes and factors that influence recharge rates.

The methods can be divided into two classes: water-budget methods and methods based on the Darcy equation. Unsaturated-zone water budgets relate changes in the amount of water stored in the unsaturated zone to infiltration, drainage, and evapotranspiration. These methods include the zero-flux plane method and lysimetry. Included in the section on the zero-flux plane method is a discussion on measurement of water storage and change in storage within the unsaturated zone. Darcy methods require measurement or estimation of the hydraulic gradient and hydraulic conductivity at the ambient water content. Natural variability in hydraulic conductivity complicates the application of the Darcy method. Lysimeters can provide precise measurements of drainage rates, but the instruments can be expensive to install and maintain. A brief overview of techniques for measuring water content, pressure head, water-retention characteristics, and hydraulic conductivity of unsaturated-zone sediments, provided in Section 5.2, lays the groundwork for discussions of the zero-flux plane method (Section 5.3), the unsaturated-zone Darcy method (Section 5.4), and the use of lysimeters (Section 5.5).

5.2 | Measurement of unsaturated-zone physical properties

5.2.1 Soil-water content
Soil-water content can be measured gravimetrically or by one of many geophysical techniques, some of which are listed in Table 5.1.

Table 5.1 Summary of some geophysical techniques for determining soil-water content. Application regime refers to location of measurements: land surface (S), subsurface (SS), or borehole logging (BH). Quantitative indicates that quantitative estimates of water content can be obtained (if not checked, technique provides only qualitative information). Measurement scales are from Ferré et al. (2007).

Method	Measured property	Spatial measurement scale (m)	Application regime	Quantitative
Time-domain reflectometry (TDR)	Dielectric permittivity	0.1–1	SS	X
Ground-penetrating radar (GPR)	Dielectric permittivity	1–10	S, SS, BH	Borehole only
Electromagnetic induction (EMI)	Electrical conductivity	1–10s	S, SS, BH	
Temporal gravity	Total mass	10s–100s	S	X
Neutron moderation	Hydrogen density	0.1–1	SS, BH	X

The gravimetric approach consists of collecting a known volume of soil sample at selected depths, weighing the sample, and then drying the sample in an oven at 105°C for 24 to 48 hours and reweighing (Topp and Ferré, 2002). Accurate measurements can be obtained, but the approach is time consuming and, because the method is destructive, measurements cannot be repeated at a specific location. Collection of samples at depths greater than a couple of meters can be expensive. Samples must be weighed quickly after they are obtained; otherwise, evaporation of soil water may affect the accuracy of the measurement. Periodic gravimetric measurements of water content can be used to develop calibration curves for electronic water-content sensors.

Geophysical techniques can provide indirect measurements of soil-water content. The techniques actually measure a certain property (Table 5.1), such as electrical conductivity of the soil-water mixture that varies with water content. A number of techniques have been developed in recent years to relate soil-water content to the dielectric permittivity of a soil/water mixture as determined, for example, with transmission line type electromagnetic (EM) sensors. These sensors measure phenomena such as EM wave travel time, impedance, capacitor charge time, oscillation frequency, and frequency shift (Blonquist, Jr., et al., 2005). The techniques have similarities in terms of probe size, installation methods, and analysis of data, but there are substantial differences in sensor accuracy and cost (Blonquist, Jr., et al., 2005; Robinson et al., 2008b).

Time-domain reflectometry (TDR), first proposed by Topp et al. (1980), is perhaps the most familiar of the EM methods. This paragraph describes some of the features of TDR; many of these features apply to other EM methods as well. TDR probes typically consist of two or three metal electrodes, 150 to 300 mm in length. The probes are installed more or less permanently at selected depths in the unsaturated zone; installation can be through an open borehole, but more commonly a trench is excavated, the probes are installed horizontally into a trench wall, and the trench is backfilled. Wires run from the probes to an analyzer and data logger on land surface for automatic sensing and recording. TDR and other fixed-depth probes offer a fairly high degree of accuracy and the ability to automatically record at high frequencies (in the order of minutes). Disadvantages of the TDR method include the fact that measurements are obtained over a small sample size, a limited number of depths are sampled, and installation techniques can alter natural water movement patterns. Temperature fluctuations, high clay content of soils, and high dissolved solids concentrations in soil water can also complicate determination of soil-water content, although TDR appears to be less affected by these phenomena than other EM methods (Ferré *et al.*, 2007).

Ground-penetrating radar (GPR) is another approach based on measurement of the dielectric permittivity of soils and water. GPR measurements are obtained over a longer distance than TDR measurements (Table 5.1). GPR measurements can be made on land surface; measurements can also be obtained between two or more boreholes in the subsurface (Huisman *et al.*, 2003). Surface measurements are useful for identifying different soil textures, but quantification of soil-water content has not met with much success (Ferré *et al.*, 2007). Borehole GPR can provide accurate estimates of water content (Binley *et al.*, 2001), but instrumentation is expensive, and the requirement for two boreholes limits its use.

The electromagnetic induction (EMI) method measures the electrical conductivity of a bulk soil/water mixture (Ferré *et al.*, 2007). EMI is widely used to determine electric and hydraulic properties of geologic material and to map areas of similar properties. Land-surface-based and airborne EMI surveys are run along transects that can stretch for hundreds of meters. EMI is also an important borehole logging technique. EMI can provide qualitative information on soil texture, but attempts to obtain quantitative estimates of soil-water content with EMI have met with only limited success (Cook and Kilty, 1992; Scanlon *et al.*, 1999).

The neutron moderation method for determining soil-water content has been in use since the early 1950s. Instrumentation has improved over the years, but the basic theory remains the same (Hignett and Evett, 2002). A radioactive source within a probe emits fast-moving neutrons. Hydrogen atoms are very effective at slowing the movement of the neutrons because the atoms are about the same size as the neutrons. A sensor within the same probe detects the number of slowed or thermalized neutrons. Water molecules account for most of the hydrogen atoms in the subsurface, so neutron meters can be calibrated to determine soil-water content, usually by comparing instrument readings with gravimetric analysis of soil samples (Hignett and Evett, 2002). Measurements are obtained by logging a cased borehole or special access tube. Generally, readings must be taken manually because of the radioactive source. Special licensing is usually required to use a neutron probe. In contrast to emplaced sensors (such as TDR probes), which offer high frequency water-content measurements at a limited number of depths, borehole logging tools, such as the neutron probe, allow many depths to be sampled, but generally at a much lower sampling frequency.

The temporal gravity technique described in Section 2.3.3 is included in Table 5.1. As discussed in Section 2.3.3, gravity measurements can be used to estimate total change in subsurface water storage. However, surface measurements cannot distinguish storage changes within specific depth intervals of the unsaturated zone (i.e. intervals above or beneath the zero-flux plane). Therefore, gravity measurements can be used with the zero-flux plane method only if an independent approach is available for estimating storage changes between land surface and the zero-flux plane.

5.2.2 Pressure head

Several devices are available for *in-situ* measurement of pressure head (sometimes referred to as soil-water tension or matric potential), including tensiometers, heat-dissipation probes, and thermocouple psychrometers (Dane and Topp, 2002). Unfortunately, there is no universal method for measuring pressure head that can be applied at any site. Each technique is applicable only within a specific range of pressure heads. Tensiometers are generally regarded as the most accurate instrument for measuring pressure heads in the range of 0 to −10 m of water, and for this reason they are commonly used in applications of the ZFP and Darcy methods (Cooper *et al.*, 1990). Tensiometers consist of a sealed water-filled tube with a porous ceramic cup that is placed in contact with soil. Soil-water pressures are transmitted across the ceramic cup and measured with an attached pressure transducer, manometer, or other pressure gauge. Pressure transducers permit electronic recording of soil-water pressures. Tensiometers are delicate in nature; field installation and operation can be problematic, especially for depths greater than about 2 m, although with recent design advances, specialized tensiometers can be installed through a borehole at any depth within the unsaturated zone (Hubbell and Sisson, 1998). In some studies, duplicate tensiometers are installed at measurement depths to ensure continuous data collection should one of the instruments malfunction.

Other approaches for measuring pressure heads in the unsaturated zone exist, but are not widely used in ZFP or Darcy methods, mostly because the range of pressure heads that they are capable of measuring is in a drier range than that of tensiometers. Heat dissipation probes (HDPs) determine pressure head by measuring the rate at which heat is dissipated within a ceramic probe; that rate is dependent on thermal conductivity of the ceramic, which varies with water content in a predictable manner (Scanlon et al., 2002a). Through calibration, the heat dissipation rate can be converted to determine pressure head. The HDPs are small (about 30 mm in length) and can be installed through boreholes. HDPs typically provide readings in the range of –10 kPa to –1 MPa or –1 to –100 m of water (Scanlon et al., 2002a). The accuracy of each HDP can be assessed through laboratory calibration; McMahon et al. (2003) estimated the accuracy to be ±1 m. Thermocouple psychrometers provide readings in the very dry range, about 0 to –800 m (Andraski and Scanlon, 2002); the probes must be individually calibrated. Other techniques for measuring pressure head are discussed in Dane and Topp (2002). Pressure head can also be inferred from measurements of water content if the water-retention curve is known (Sharma et al., 1991).

5.2.3 Water-retention and hydraulic conductivity curves

The water-retention curve displays the relation between soil-water content and pressure head (Figure 5.1a). Water-retention curves are strongly influenced by soil texture. Coarse-grained sediments, such as sand, tend to drain more easily (i.e. under higher pressure heads) than fine-grained sediments, such as silt loam (Figure 5.1a). Water-retention curves are

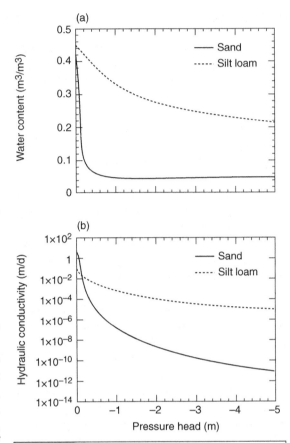

Figure 5.1 Water-retention (a) and hydraulic-conductivity (b) curves for sand and silt loam. Curves are described by the van Genuchten (1980) equations (Equations (5.2) and (5.3)) with parameters as derived by Carsel and Parrish (1988) (Table 3.1).

usually measured in the laboratory on small cylindrical cores (typically 50 to 100 mm in length and diameter) by using Tempe cell or pressure-plate apparatus (Dane and Topp, 2002). The procedure is straightforward: a fixed pressure is applied to the sample, and the sample is allowed to drain. The water content of the sample is determined by weighing the sample once drainage ceases. The equivalent pressure head and the measured water content constitute a single point on the water-retention curve. The procedure usually begins with a fully saturated sample; the fixed pressures are altered incrementally to produce the desired number of points on the curve. A water-retention curve can be measured beginning with a dry sample and allowing the sample to imbibe (rather than drain) water at different pressures. The procedure requires special equipment, and it can take weeks to measure a complete water-retention curve. Water-retention curves also can be determined with simultaneous field measurements of water content and pressure head with methods described in Sections 5.2.1 and 5.2.2.

Unsaturated hydraulic conductivity, $K(h)$, is expressed as the product of saturated hydraulic conductivity, K_s, and relative hydraulic conductivity, $K_r(h)$:

$$K(h) = K_s K_r(h) \qquad (5.1)$$

where h is pressure head. Relative hydraulic conductivity has its maximum value of 1 under saturated conditions; values decrease toward 0 as pressure head decreases. As with water-retention curves, shapes of hydraulic-conductivity curves are influenced by soil texture (Figure 5.1b). A sandy soil may have a much higher hydraulic conductivity than a silt-loam soil at saturation (pressure head of 0 m); however, at a pressure head of –0.5 m, the hydraulic conductivity of the silt-loam soil may be orders of magnitude greater than that of the sandy soil (Figure 5.1b). The capability of the sandy soil to transmit water at a pressure head of –500 mm is greatly reduced relative to that at saturation because of the reduced water content at –500 mm. Methods for measuring or estimating unsaturated hydraulic conductivity are described in Dane and Topp (2002); they include field, laboratory, and empirical approaches. These methods can consume considerable time and require expensive instrumentation. The estimates that they produce may also contain large uncertainties. The instantaneous profile method (Vachaud and Dane, 2002) can encompass a larger sample size than most methods (up to several square meters). It is conducted by ponding water on land surface until a steady infiltration rate is established; the ponded water is allowed to drain, and water content and pressure head are measured at multiple depths beneath the ponded area. A flux is determined for each depth increment from changes in water content (an approach

similar to that of the zero-flux plane method). By relating fluxes to measured head gradients, $K(h)$ can be calculated. The instantaneous profile method requires intensive instrumentation and monitoring, and it can take up to months to complete. In addition, it provides information for only the top part of the unsaturated zone under relatively wet conditions.

The steady-state centrifuge method (SSCM) is a laboratory method for determining unsaturated hydraulic conductivity that relies on centrifugal force to move water through soil core samples (Nimmo et al., 2002a). The SSCM can provide relatively rapid (within a few days) measurements of $K(h)$ for a range of water contents, from very dry to fully saturated conditions. However, the size of the samples that can be analyzed is small (about 50 mm in diameter by 50 mm in length), and centrifuges for running the analyses are available only in a small number of laboratories.

Empirical equations, such as those of van Genuchten (1980) and Brooks and Corey (1964), are often used to represent the water-retention and hydraulic conductivity curves. The van Genuchten equations are given by:

$$\theta(h) = (\theta_s - \theta_r)[1 + (\alpha h)^n]^{-m} + \theta_r \qquad (5.2)$$

$$K_r(h) = \frac{\{1 - (\alpha h)^{n-1}[1 + (\alpha h)^n]^{-m}\}^2}{[1 + (\alpha h)^n]^{m/2}} \qquad (5.3)$$

where θ is volumetric water content, h is pressure head, θ_s is saturated water content, θ_r is residual water content, α and n are referred to as the van Genuchten parameters, and m is taken equal to $1 - 1/n$. The Brooks and Corey equations are given by:

$$\theta(h) = (\theta_s - \theta_r)[h_b / h]^\lambda + \theta_r \qquad (5.4)$$

$$K_r(h) = [(\theta(h) - \theta_r) / (\theta_s - \theta_r)]^{(2+3\lambda)/\lambda} \qquad (5.5)$$

where h_b is the bubbling pressure and λ is the pore-size distribution index. An alternative to laboratory or field measurement of water-retention and hydraulic conductivity curves (which can be time consuming and problematic) is to approximate the curves with pedotransfer functions (e.g. Briggs and Shantz, 1912; Arya and Paris, 1981; Minasny *et al.*, 1999; Minasny and McBratney, 2007). These approximations are based on soil texture and physical properties. The Rosetta model (Schaap *et al.*, 2001) is perhaps the most widely used pedotransfer function. Estimates of van Genuchten parameters (Equations (5.2) and (5.3)) are determined by Rosetta on the basis of percentages of sand, silt, and clay and bulk density of a soil sample. Pedotransfer functions may lack the accuracy of direct measurements, but they are easy and inexpensive to apply. As discussed in Section 3.3.2, Carsel and Parrish (1988) determined average values of saturated water content, residual water content, saturated hydraulic conductivity, and the van Genuchten α and n parameters for 12 soil textural classifications (Table 3.1). Measured unsaturated hydraulic conductivities for a number of different soils are available in the UNSODA database (http://www.ars.usda.gov/Services/docs.htm?docid=8967; accessed September 4, 2008).

5.3 | Zero-flux plane method

The zero-flux plane (ZFP) is the horizontal plane at some depth within the unsaturated zone where the hydraulic gradient in the vertical direction (dH/dz) is 0 (Richards *et al.*, 1956). Above the ZFP, water moves upward in response to evapotranspiration; below the ZFP, water drains downward, possibly becoming recharge. The ZFP is sometimes associated with the maximum depth of rooting (although suction exerted by roots may draw water from beneath their maximum depth). If the depth of the ZFP can be determined, both drainage and evapotranspiration can be estimated by measuring the change in water storage within the unsaturated zone over time. The zero-flux plane method is a water-budget method applied over a control volume that consists of a discrete depth interval of the unsaturated zone. The method was first proposed by Richards *et al.* (1956) and has since been applied in a number of studies (Royer and Vachaud, 1974; Arya *et al.*, 1975; Wellings, 1984; Dreiss and Anderson, 1985; Healy, 1989; Roman *et al.*, 1996;

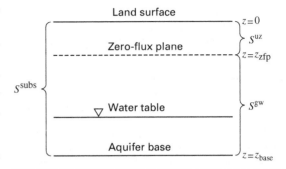

Figure 5.2 Schematic of soil column extending from land surface to the base of the aquifer showing different storage compartments. S^{subs} represents all water stored in column, S^{uz} is water stored between land surface and the zero-flux plane, and S^{gw} is water stored between the zero-flux plane and the base of the aquifer. z_{ZFP} is depth to zero-flux plane; z_{base} is depth to base of aquifer.

Delin *et al.*, 2000, 2007; Delin and Herkelrath, 2005; Schwartz *et al.*, 2008).

Water storage is determined from water-content measurements at multiple depths within the unsaturated zone. Change in water storage is simply the difference in water storage at two different times. Referring to the terminology used in Section 2.3.3 and Figure 5.2, S^{subs} is the amount of water stored within the interval of the subsurface between land surface ($z = 0$) and the base of the underlying aquifer ($z = z_{base}$). S^{subs} is described by the following general equation:

$$S^{subs} = \int_{z_{base}}^{0} \theta dz \qquad (5.6)$$

Equation (5.6) is written for a soil column of unit surface area, so S^{subs} has units of length (e.g. mm). Storage within discrete depth intervals of the subsurface can be represented with Equation (5.6) by adjusting the limits on the integral (Figure 5.2). The zero-flux plane method equates drainage with change in storage in the interval between the zero-flux plane and the water table. Water storage within this interval, S^{gw}, is described by:

$$S^{gw} = \int_{z_{WT}}^{z_{ZFP}} \theta dz \qquad (5.7)$$

where z_{ZFP} is the depth to the zero-flux plane and z_{WT} is depth to the water table. The change in

S^{gw} between successive measurements of water-content profile, ΔS^{gw}, is equivalent to drainage, D, for that period:

$$D_i = -\Delta S^{gw}{}_i = -(S^{gw}{}_i - S^{gw}{}_{i-1})/(t_i - t_{i-1}) \quad (5.8)$$

where i is an index of time, t; $(t_i - t_{i-1})$ is the time interval between the two profile measurements; and D and ΔS^{gw} have dimensions of L/T. In actuality, measurements of water-content profile seldom extend all the way to the base of the underlying aquifer; few studies even attempt measurements down to the water table. Most applications of the ZFP method rely on water-content measurements to some fixed depth within the unsaturated zone; the only requirement regarding that depth, as discussed in the following paragraphs, is that the interval between the ZFP and that depth be of sufficient thickness to fully capture both leading and trailing edges of individual pulses of downward-moving water.

For estimating evapotranspiration rates with the ZFP method, the storage term of interest is S^{uz}, the amount of water stored between land surface and the zero-flux plane:

$$S^{uz} = \int_{z_{ZFP}}^{0} \theta \, dz \quad (5.9)$$

The change in water storage in that depth interval, ΔS^{uz}, is determined as:

$$\Delta S^{uz}{}_i = (S^{uz}{}_i - S^{uz}{}_{i-1})/(t_i - t_{i-1}) \quad (5.10)$$

A simple water-budget equation can then be derived for estimating evapotranspiration, ET, over the measurement time interval:

$$ET = P - R_{off} - \Delta S^{uz} \quad (5.11)$$

where P is precipitation and irrigation, R_{off} is runoff, and it is assumed that all applied water infiltrates, runs off, or is evaporated.

Calculating the amount of water stored in a vertical interval of the unsaturated zone (i.e. evaluating the integrals in Equations (5.6), (5.7), and (5.9)) is straightforward, regardless of the method used to measure water content, if water-content measurements are available at discrete depths. Figure 5.3 shows water contents in a hypothetical 6 m thick unsaturated zone

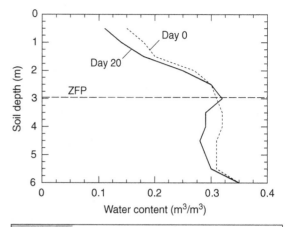

Figure 5.3 Example of water-content profiles obtained over an interval of 20 days. The amount of water stored between the zero-flux plane, ZFP (3 m depth), and the 6 m depth at time t_i is calculated from the formula:

$$S_1 = \sum_{i=0}^{5} 0.5 \times \Delta z_i \times \{\theta_i(t_1) + \theta_{i+1}(t_1)\}$$

where i is the depth index ($i = 0$ at 3 m and $i = 5$ at 6 m), Δz_i is the distance between measurement points (0.5 m), and θ_i is measured water content at depth i. At day 0, 950 mm of water was stored in this interval, and at day 20, 894 mm was stored. So over the 20 day period, 56 mm of drainage is calculated. Similar calculations between land surface and the ZFP indicate that 69 mm of water was lost to evapotranspiration during the same period, assuming no precipitation fell.

profile for measurements made 20 days apart. By assuming that water content varies linearly with depth between sampling points, the equation given in Figure 5.3 can be used to calculate the amount of water stored between the ZFP and the 6 m depth for each measurement date. The drainage rate for the time period between the two measurement dates, 2.8 mm/d, was determined with Equation (5.8). The equation given in Figure 5.3 is derived by applying the trapezoid rule (Anton, 1984) for evaluation of the integral in Equation (5.7). Analogous discrete formulas can be derived for evaluation of Equations (5.6) and (5.9). In the event of a shift in depth of the ZFP between two profile-measurement dates, the water storage change within the interval between the two ZFP depths must somehow be apportioned between drainage and

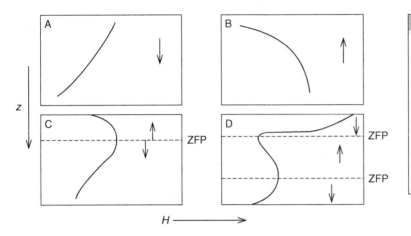

Figure 5.4 Hypothetical vertical profiles of total head (H) within the unsaturated zone indicating (A) downward flux (no zero-flux plane, ZFP); (B) upward flux due to evapotranspiration (no ZFP); (C) a single ZFP with upward flux above and downward flux below; and (D) two ZFPs. Total head is equal to pressure head, h, minus depth, z, and the top of each frame represents land surface.

evapotranspiration. There are no guidelines for that apportionment, so it is left to the discretion of the user.

Central to application of the ZFP method is the assumption that water moves vertically through the unsaturated zone as periodic, distinct "pulses" that can be monitored over time. Steady flow through the unsaturated zone (i.e. flow that occurs with no change in water content) cannot be measured with this method; likewise, the method cannot be applied when water is moving downward throughout the entire extent of the unsaturated zone. The ZFP approach quantifies the amount of water moving in each pulse. In many applications (particularly those in areas of deep water tables), water-content measurements are made only to some finite depth above the water table. Any depth interval for measurements is sufficient as long as that interval is below the ZFP and is of sufficient thickness to include the leading and trailing edge of each pulse of moving water. If measurement depths extend to the water table, the method produces estimates of actual recharge. Otherwise, the estimates should be referred to as drainage. Frequent water-content measurements may be required to adequately capture the movement of a water pulse through a depth interval. Rapidly moving wetting fronts could go undetected with insufficient measurement frequency.

The ZFP moves vertically with time in response to infiltration, evapotranspiration, and drainage. Four vertical profiles of

hydraulic head are depicted in Figure 5.4 to illustrate hypothetical locations of the ZFP. In Profile A, water is moving downward at all depths; there is no ZFP in the subsurface. Profile A would be expected during or immediately following a large precipitation event. The ZFP method cannot be applied under these conditions; an alternative method, such as a Darcy or other water-budget method, must be used (e.g. Hodnett and Bell, 1990; Roman *et al.*, 1996). Profile B indicates that water is moving upward at all depths, most likely in response to evapotranspiration. There is no ZFP and no drainage so the method cannot be used. Profile C contains a single ZFP, and the ZFP method can be applied to estimate drainage and evapotranspiration. Finally, profile D contains a ZFP at two depths. Such an occurrence is not uncommon and could result from precipitation falling when ambient soil conditions were similar to those in profile C. Usually, the two ZFPs would converge within a short time. The ZFP method can be applied under conditions shown in Profile D, but only the change in storage below the lowermost ZFP should be attributed to drainage.

The configuration of hydraulic heads within the unsaturated zone changes over time. Any one of the profiles depicted in Figure 5.4 could exist in the same column of the unsaturated zone at different times of the year. The conditions in profile A must exist from time to time as infiltration occurs; otherwise, no drainage through the column would occur. Frequent

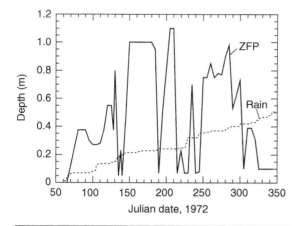

Figure 5.5 Depths of zero-flux plane (ZFP) and cumulative rainfall over a 10 month period for a field near Ambeliet, southwestern France (Royer and Vachaud, 1974).

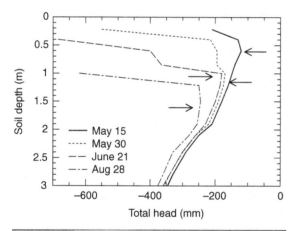

Figure 5.6 Total head profiles and locations of zero-flux plane (arrows) at the Bicton College site in 1989 (Cooper et al., 1990; reprinted with permission of John Wiley & Sons, Inc.).

rates usually exceed precipitation rates in late spring and summer. Hence, water storage decreases during this period and the depth to the ZFP increases. The ZFP usually reaches its maximum depth in late summer, after which time water storage begins to increase and, correspondingly, the depth to the ZFP starts to decrease. The depth to the ZFP can decrease rapidly in response to a large rainfall at any time of the year (Figure 5.5).

Locating the depth of the ZFP requires accurate measurement of hydraulic head at multiple depths within the unsaturated zone. As discussed in Section 5.2.2, tensiometers are commonly used for these measurements (Driess and Anderson, 1985; Cooper et al., 1990; Roman et al., 1996). The limited accuracy of tensiometers and other pressure-head sensors imparts a degree of uncertainty to the estimated depth of the ZFP. Additional uncertainty is introduced if sensors are spaced too widely apart, in which case the hydraulic gradient may be poorly defined.

The ZFP method is best applied in regions that display a wide range of soil-water contents throughout the year. The method is not limited in terms of soil texture, but areas with shallow water tables and coarse-grained sediments can pose a problem. In this type of environment, it is conceivable that a rapidly moving wetting front could escape detection by traveling from land surface to the water table during the interval between water-content measurements. The method provides useful results under the proper conditions, but an alternative method needs to be used for times when it breaks down (e.g. when water movement is downward throughout the entire soil column, as in profile A in Figure 5.4). The method inherently assumes that water in the region below the ZFP moves downward; features that promote horizontal flow in the unsaturated zone, such as layering or variability in texture, can complicate its application.

Example: Chalk and Triassic Sandstone aquifers, England

Seven sites were instrumented for application of the ZFP method to estimate drainage to the Chalk and Triassic Sandstone aquifers

precipitation or irrigation events could be problematic for application of the ZFP method unless head and water-content readings are automatically recorded. For cases where irrigation is applied three or four times per year the method may work well.

Under natural conditions in humid regions, the depth of the ZFP follows a somewhat predictable pattern. In late winter and spring, when evapotranspiration rates are low, water storage in the unsaturated zone increases and the ZFP rises close to land surface. Evapotranspiration

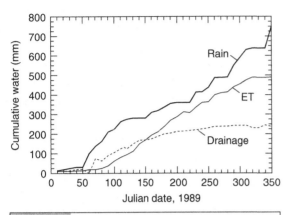

Figure 5.7 Cumulative rainfall, drainage, and evapotranspiration (ET) for 1989 at the Bicton College site (Cooper *et al.*, 1990; reprinted with permission of John Wiley & Sons, Inc.).

in England (Cooper *et al.*, 1990). These aquifers together account for 60% of the country's public groundwater supply. All sites contained at least three neutron-probe access tubes, a vertical profile of tensiometers extending to a depth of 3 m, and a rain gauge. Monitoring frequency of the instruments ranged from one to five times per week. The length of data record was between 2 and 5 years. The ZFP method was used to estimate both recharge and evapotranspiration.

Depth to the zero-flux plane increased over the course of the summer of 1989 at the Bicton College site (Figure 5.6). Most drainage occurred in late winter and spring, and, for 1989, drainage was equivalent to 32% of precipitation (Figure 5.7). Conditions were not appropriate for application of the ZFP method at the Bicton College site for about 45% of the year because of downward drainage throughout the soil column or tensiometer failure. A water-budget method was used to estimate drainage during those periods. Averaging over all years and data-collection sites, drainage was 27% of precipitation. The ratio of drainage to precipitation varied because of differences in hydraulic properties of the subsurface materials at the sites and the year-to-year variability in total amount, timing, duration, and intensity of precipitation (Cooper *et al.*, 1990).

Example: Western Australia

Seasonal variation in drainage through the unsaturated zone at a site on the Gnangara Mound, Western Australia, was studied from 1985 to 1987 by Sharma *et al.* (1991). The depth to the water table was approximately 90 m, and the unsaturated zone consisted mostly of sand with occasional bands of clay. The saturated hydraulic conductivity of the sand was high (approximately 30 m/day); hence, surface runoff was negligible. The subhumid woodland receives about 775 mm of precipitation annually. It was instrumented with rain gauges, observation wells, and eight neutron probe access tubes that penetrated to a depth of 20 m. Water content was measured with a neutron probe at 0.5 m depth intervals every 2 to 4 weeks.

Pressure heads within the unsaturated zone were not directly measured in this study. Water-retention curves (Section 5.2.3) were determined in the lab and parameters of the Brooks-Corey equation (Equation (5.4)) were calculated. Pressure head, h, was then determined from field-measured water contents as:

$$h = h_{\mathrm{b}} / \{ (\theta(h) - \theta_{\mathrm{r}}) / (\theta_{\mathrm{s}} - \theta_{\mathrm{r}}) \}^{1/\lambda} \qquad (5.12)$$

where variables are as defined for Equation (5.4). The location of the ZFP at one access hole varied from about 1 to 10 m in depth (Figure 5.8). Drainage was estimated for two different depth intervals: the ZFP to the 10 m depth and the ZFP to the 18 m depth, referred to as the 10 m and 18 m drainage estimates, respectively. The two estimates should be similar, but several factors can contribute to differences. Heterogeneities, such as a clay lens, can impart a horizontal component to the flow field; there can be a long lag from the time water passes the 10 m depth until it reaches the 18 m depth; and there are inaccuracies in water-content measurements. For periods when the entire profile was draining, Sharma *et al.* (1991) used the Darcy method in conjunction with the assumption of a unit hydraulic gradient (Section 5.4) to estimate drainage.

There was a strong seasonal trend in precipitation, 80% falling between May and October. The 10 m drainage estimates for boreholes 4

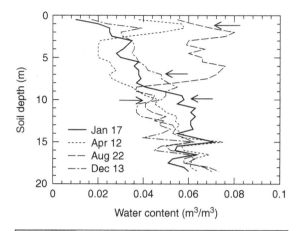

Figure 5.8 Profiles of water content and locations of zero-flux plane (arrows) for 4 days in 1985 at a site in Western Australia (Sharma et al., 1991; reprinted with permission of John Wiley & Sons, Inc.).

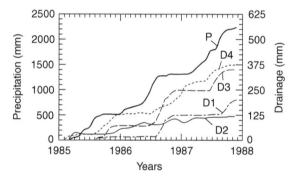

Figure 5.9 Precipitation (P) and drainage for 1985–1987 at a site in Western Australia. Drainage calculated with ZFP method for borehole 4 for 10 m (D1) and 18 m (D2) depths and for borehole 6 for 10 m (D3) and 18 m (D4) depths (Sharma et al., 1991; reprinted with permission of John Wiley & Sons, Inc.).

and 6 showed a similar seasonal trend (Figure 5.9); highest rates occurred during periods when precipitation exceeded evapotranspiration (which was also estimated by the ZFP method). A seasonal trend was less apparent in the 18 m drainage estimates. The two estimates for 3 years of cumulative drainage at borehole 6 are in good agreement. The agreement is not so good for the borehole 4 estimates, perhaps because data collection was terminated at a time when drainage rates at the deeper depth would have been increasing (Sharma et al.,

1991). Drainage at borehole 6 is greater that at borehole 4, and evapotranspiration is less. This is attributed to denser vegetation near borehole 4. Average annual drainage for the period of study was 90 mm (13% of precipitation) at the 10 m depth and 70 mm (10% of precipitation) at the 18 m depth.

The approach used to estimate the depth of the ZFP in this study was viable because of the uniform texture of the sediments, the relatively deep water table, and the measurement of the soil-water retention curves. Roark and Healy (1998) and Schwartz et al. (2008) also used water-content measurements to identify the depth of the ZFP.

5.4 | Darcy methods

The Darcy equation was first introduced in Section 2.3.5 to describe groundwater flow. A variant of the equation may also be used to estimate drainage through the unsaturated zone. For vertical flow in the unsaturated zone, the Darcy equation takes the form:

$$q = -K_s K_r(h) \partial H / \partial z \qquad (5.13)$$

where q is drainage rate; $K_s K_r(h)$ is the vertical hydraulic conductivity at the ambient pressure head, h; H is total head; and z is depth. The unsaturated-zone Darcy method requires measurements or estimates of the vertical total head gradient (Section 5.2.2) and hydraulic conductivity as a function of water content or pressure head (Section 5.2.3). The method has been applied in many studies under various climatic conditions. Enfield et al. (1973), Stephens and Knowlton (1986), and Sammis et al. (1982) used the method in arid or semiarid environments. It has also been applied in humid regions (e.g. Ahuja and El-Swaify, 1979; Steenhuis et al., 1985; Kengni et al., 1994; Normand et al., 1997; and Coes et al., 2007).

Unlike the zero-flux plane method, application of the Darcy method is not limited to certain times of the year. Equation (5.13) can be expanded to account for water movement in two or three dimensions. If head measurements are made in multiple dimensions, fluxes can be estimated for cases where horizontal flow in the unsaturated zone is important (such as with layered or heterogeneous soils; generally, simulation models based on the Richards equation (Section 3.3.2) are used to study two- or three-dimensional flow in the unsaturated zone). The Darcy method allows monitoring of individual recharge events and therefore can provide insight into the mechanics of the recharge process. Concerns associated with the method include the highly variable nature of hydraulic conductivity and the limited accuracy with which unsaturated hydraulic conductivity curves and hydraulic gradients can be measured.

Tensiometers are usually used to measure the hydraulic gradient, as discussed in Section 5.2.2 (Stephens and Knowlton, 1986; Healy, 1989). Because the hydraulic gradient can vary greatly with depth and time, it is desirable to have head measurements at several depths. Steenhuis et al. (1985) installed tensiometers at seven depths between land surface and 2.13 m. Stephens and Knowlton (1986) installed eight tensiometers at 0.3-m depth intervals down to 2.4 m. The frequency with which head measurements are made may also be important. Electronic recording facilitates a high frequency of measurement. As with the zero-flux plane method, if frequent readings are not taken, there is a possibility of a wetting front passing undetected.

Instruments other than tensiometers are occasionally used to measure hydraulic gradients for application of the Darcy method. Flint et al. (2002) measured pressure heads with heat-dissipation probes to apply the Darcy method near Yucca Mountain, Nevada. Also in the vicinity of Yucca Mountain, Kwicklis et al. (1993) applied the Darcy method with pressure-head data measured with thermocouple psychrometers installed at multiple depths within the 500-m thick unsaturated zone. Sophocleous et al. (2002) applied the Darcy method with the hydraulic gradient determined with heat dissipation probes at three sites in southwestern Kansas. Schwartz et al. (2008) measured water content with TDR probes in field plots near Bushland, Texas; water contents were converted to pressure heads by means of computed

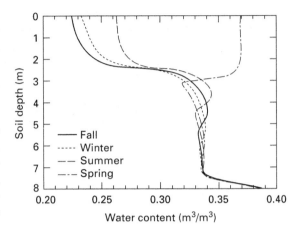

Figure 5.10 Hypothetical water-content profiles at four different times of the year. As depth increases, the seasonal variation in water content decreases (after Healy et al., 2007).

soil-water retention curves, and the drainage was calculated with the Darcy equation.

Precipitation and evapotranspiration patterns usually produce large variations in pressure head and water content near land surface over a period of a year. These fluctuations tend to be dampened with depth (Figure 5.10). If the unsaturated zone is deep enough, fluctuations can be almost completely damped out at some depth and a region of uniform pressure head and water content can exist over some depth interval. Within this depth interval the vertical pressure-head gradient will be 0, flow will be constant and driven strictly by gravity, and the vertical hydraulic gradient will be equal to –1. The concept of the unit hydraulic gradient in natural systems was described more fully by Gardner (1964), Childs (1969), and Nimmo et al. (1994). Black et al. (1969) proposed use of the unit-gradient assumption for estimating drainage rates from soils. The assumption greatly simplifies application of the Darcy method by removing the need to measure the hydraulic gradient. Substituting –1 for the gradient in Equation (5.13) gives:

$$q = K_s K_r(h) \qquad (5.14)$$

The drainage rate is equal to the value of hydraulic conductivity at the ambient pressure head (or water content). Application of the unit-gradient approach requires a measurement of pressure head (or water content) and a measurement or estimate of hydraulic conductivity at that pressure head. Drainage estimates obtained in this manner represent long-term average values. The unit-gradient assumption has been invoked in a number of studies (e.g., Chong *et al.*, 1981; Stephens and Knowlton, 1986; and Nolan *et al.*, 2007).

Initially the unit-gradient assumption was restricted to uniform material. Chong *et al.* (1981) and Sisson (1987) extended the unit-gradient theory for application to nonuniform, layered soils. Layers must be thick enough to permit development of a uniform pressure-head profile. The flux must be nonzero only in the vertical direction and uniform throughout all layers. Data from Roark and Healy (1998; Figure 5.11) support the notion of unit gradients within heterogeneous sediments. The considerable variability in water content with respect to depth at the site near Roswell, New Mexico, is due to heterogeneity in soil properties. At any depth below about 2 m there is little temporal variability in measured water content. Whether lateral flow occurred along interfaces of layers is an important and difficult question to address in this and other studies.

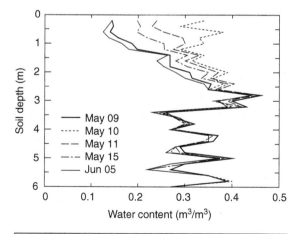

Figure 5.11 Water-content profiles for 5 days in 1996 in a field near Roswell, New Mexico (after Roark and Healy, 1998). The field was irrigated after the May 9 measurement.

Example: Sevilleta Grasslands, New Mexico

Stephens and Knowlton (1986) applied the Darcy method to estimate drainage through the unsaturated zone at a site in the semiarid Sevilleta Grasslands, New Mexico. Data were collected from November 1982 through May 1984 with tensiometers and a neutron probe. Drainage estimates were made with measured hydraulic gradients (Equation (5.13)) and with the assumption of a unit hydraulic gradient (Equation (5.14)). The site is located on an old floodplain of the ephemeral Rio Salado; the soil was a relatively uniform fine sand. Vegetation was sparse, and the rooting depth was visually determined to be 1.5 m. The water table was at a depth of about 6 m. Eight tensiometers were installed at 0.3 m depth intervals to a maximum depth of 2.4 m. A neutron probe access tube was installed to a depth of 6 m, and readings were taken biweekly. Unsaturated hydraulic conductivity was determined in the field with the instantaneous profile method.

Measured hydraulic gradients, based on average monthly pressure heads at depths of 1.53 and 2.44 m, produced annual drainage estimates of 7 and 37 mm, or 4 and 20%, respectively, of annual precipitation of 179 mm. Invoking the unit-gradient assumption and using mean monthly water contents, annual drainage was estimated to be 97 mm at the 1.22-m depth and 37 mm at the 1.53-m depth, or 54 and 20% of average annual precipitation, respectively.

The wide scatter in drainage estimates underscores the previously mentioned concerns with the Darcy method. According to the authors, the most likely source of estimation error is the uncertainty in the unsaturated hydraulic conductivity. The instantaneous profile method generated hydraulic conductivities primarily in the wet range of soil conditions. Most of the calculated drainage occurred under drier conditions; hence, the generated curve had to be extrapolated beyond the measured points. Instrumentation problems were also a concern. Tensiometers did not always operate properly because of leaks. Slight inaccuracies in measured water contents or pressure heads could translate into large uncertainties in hydraulic conductivities.

Example: agricultural field, France

As part of an intensive experiment to determine an optimal fertilizer application scheme for irrigated maize near Grenoble, France, eight plots were instrumented with tensiometers and neutron probe access tubes (Kengni *et al.*, 1994). Detailed water budgets were developed for the period April 1991 to February 1992, with drainage calculated by the Darcy method. Soil was a highly permeable sand. A layer of coarse gravel at a depth of 1 m restricted plant root growth. Two of the sites were bare soil; the other six were planted with maize under different fertilization methods. Each site had tensiometers at depths of 150, 300, 500, 700, and 900 mm, and a neutron probe access tube. Tensiometer readings were taken daily during the growing season and weekly thereafter. Water contents were measured with a neutron probe at 100 mm intervals to a depth of 900 mm on a weekly basis. Unsaturated hydraulic conductivity was determined with a modified instantaneous profile method. When a zero-flux plane was present, the ZFP method was used to determine downward fluxes. These fluxes were used with measured water contents and hydraulic gradients to construct the unsaturated hydraulic conductivity curve.

Drainage rates for the bare and cropped sites were similar in the spring (Figure 5.12), but by Julian date 180, evapotranspiration from the crops was much greater than that from bare soil. As a result, drainage from the cropped sites was essentially 0 for Julian date 180 to 280, whereas drainage from the bare soil sites for that period exceeded 200 mm. After day 280, drainage rates for all sites were again similar. Average drainage was 551 mm for the two bare sites for the study period (which extended beyond the time shown in Figure 5.12) or 63% of the total rainfall and irrigation of 871 mm. For the six cropped sites, average drainage was 364 mm or 42% of rainfall and irrigation. The Darcy method appears to be an appropriate choice for estimating drainage in this rather unique study area. The fact that the rooting depth is restricted to the top 1 m by

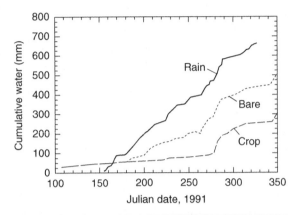

Figure 5.12 Average cumulative rainfall and drainage for bare and cropped soil sites near Grenoble, France, April through December 1991 (reprinted from *Journal of Hydrology*, v. 162, Kengni *et al.* (1994), Figure 5, copyright (1994), with permission from Elsevier).

the underlying gravel layer limits the number of depths that must be instrumented. Because of the permeable nature of the soil and, at times, high application rate of rainfall and irrigation, wetting fronts moved rapidly through the soil. Thus, a high frequency of data collection was required.

5.5 | Lysimetry

Many different types of lysimeters have been used in hydrologic studies. Perhaps the most common types are those whose upper boundary is the land surface; these are containers filled with soil and placed in a field so as to mimic as closely as possible surrounding soils and vegetation. They are used to study atmosphere, soil, water, and plant interactions under natural or artificial conditions. Originally designed to measure leaching and percolation of solutes through soil, these lysimeters can provide direct measurements of evapotranspiration as well as drainage. Brutsaert (1982) traces their use back to the early 1700s. In the last several decades lysimeters have been used primarily to measure evapotranspiration rates, although they also have been used for estimating rates of water drainage through the unsaturated

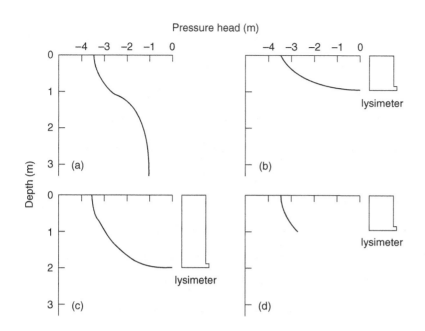

Figure 5.13 Idealized pressure-head depth profiles in: (a) undisturbed soil; (b) shallow lysimeter with free drainage bottom; (c) deep lysimeter with free drainage bottom; and (d) shallow lysimeter with tension controlled bottom drain (after van Bavel, 1961).

zone (Kitching and Shearer, 1982; Gee *et al.*, 1994; Gburek and Folmar, 1999; Brye *et al.*, 2000; Scanlon *et al.*, 2005; Fayer and Gee, 2006; Heppner *et al.*, 2007; Gee *et al.*, 2009). Surface areas of lysimeters can range from less than 0.1 m² (Evett *et al.*, 1995) to more than 100 m² (Tyler *et al.*, 1992). For studies of evapotranspiration, the rate of change in water storage in the lysimeter is required. This is usually determined by weighing the lysimeter at regular time intervals. For recharge studies, the rate of water drainage out the bottom of the lysimeter is the component of interest.

Careful design, construction, and maintenance of a lysimeter are essential for obtaining accurate measurements. Ideally, a surface lysimeter consists of a block of undisturbed soil. Excavating a large soil block is expensive and requires great care. Some large-scale lysimeters are built in place by driving sidewalls downward from land surface and sealing the bottom through an access tunnel. When it is not feasible to obtain an undisturbed block of soil, lysimeters are repacked with soils, layer by layer, each to its original bulk density, so that it closely resembles the natural profile. The top of the lysimeter should be flush with the soil surface. To avoid the introduction of

artificial water-flow patterns, the surface area should be large enough to remove any "edge" effects introduced by the border of the lysimeter. Vegetation in the lysimeter should be maintained in the same manner as that in the adjacent soil. Different vegetation on and around the lysimeter, sidewalks, fences, trees, or nearby structures may introduce serious errors into the lysimeter measurements (Brutsaert, 1982).

The depth of the lysimeter base influences hydraulic conditions within the lysimeter because the base imposes an artificial flow boundary. Most lysimeters have a drainage collection system open to the atmosphere, which creates a boundary condition equivalent to that of a water table or a seepage face. This artifact can produce pressure-head profiles within the lysimeter that differ from those in the adjacent undisturbed area (Figure 5.13). The effect decreases as the depth of the base increases. To minimize the influence of the lower boundary, the base of some lysimeters consists of a porous plate that can be set to a prescribed pressure head (Pruitt and Angus, 1960; Brye *et al.*, 1999). If the ambient pressure head in an area adjacent to the lysimeter can be measured with a tensiometer or other device, then the pressure head in the porous plate can be adjusted by means of a vacuum system to match that ambient value.

Microlysimeters can be constructed from steel or plastic pipe, typically about 80 mm in diameter and 100 to 300 mm in depth (Evett et al., 1995). The pipe is pushed into the soil, extracted, sealed at the base to secure the soil and prevent drainage, and reinserted into the soil. These lysimeters can be weighed daily to obtain an estimate of bare soil evaporation. Because of their small size, these devices are appropriate only for estimating evaporation from bare soils for periods of several days following rainfall or irrigation. After time, the lower boundary will influence the water flow regime. Although microlysimeters are not capable of providing a direct measurement of drainage, they can be useful in water-budget methods for estimating recharge. They also are useful in studying the spatial variability of evaporation rates.

Pruitt and Angus (1960) described an elaborate weighing lysimeter with a 28 m² surface area. This device is temperature controlled and allows for setting a fixed pressure head at the bottom. It is capable of recording weights with an accuracy equivalent to 0.03 mm of water storage. The large surface area should minimize edge effects. Precision weighing lysimeters such as this may be the most accurate instruments for measuring evapotranspiration. However, for recharge studies, the capability to resolve small changes in soil-water storage is not as critical as the capability to accurately monitor drainage from the base of the lysimeter. Hence, weighing lysimeters are not commonly used in recharge studies.

Other types of lysimeters are designed for installation in the subsurface, usually beneath an undisturbed section of soil. A trench or access tunnel is required for installation. These lysimeters are designed to collect water draining through the overlying soil column. Careful installation is needed to ensure that the top of the lysimeter is in good hydraulic connection with the overlying sediments. Failure to obtain a good connection may compromise lysimeter performance. Pan lysimeters (sometimes referred to as zero-tension or gravity-drain lysimeters) have been used for many years (Parizek and Lane, 1970). They consist of a pan or collection vessel and small diameter tubing for withdrawing accumulated water. The surface area of the collector typically is in the range 0.01 to 0.5 m². Originally designed to collect water samples for chemical analyses, pan lysimeters have been used to estimate drainage rates through the unsaturated zone (e.g. Healy and Mills, 1991; Gburek and Folmar, 1999; Heppner et al., 2007). Pan lysimeters are inexpensive to construct and maintain, but a few considerations should be addressed when using them. As discussed above, an unnatural pressure head profile may develop above the lysimeter base. The installation process may also disrupt the natural flow system, and the small size of the collector provides only a point estimate of drainage.

Other subsurface lysimeters rely on an applied pressure head or tension to collect draining water. The equilibrium-tension lysimeter (ETL) (Brye et al., 1999, 2000; Masarik et al., 2004) has as its collection surface a stainless-steel porous plate across which suction equivalent to ambient levels in adjacent soils can be applied by means of a small vacuum pump. Relative to a pan lysimeter, the ETL should cause less disruption of the natural flow field and thus provide a more reliable estimate of drainage. Wick lysimeters rely on the tension provided by a hanging column of fiberglass wick material to collect draining soil water (Holder et al., 1991; Louie et al., 2000; Gee et al., 2002). The collection surface usually consists of a glass or stainless-steel plate across which single strands of the wick are spread and carefully glued. The wick runs from this surface down to a collection vessel. The difference in elevation between the collection surface and the end of the wick that hangs in the collection vessel is equivalent to the effective pressure head exerted by the strands on the collection surface. The water-flux meter described by Gee et al. (2002, 2009) uses a wicking system and includes a small tipping-bucket mechanism to measure drainage. For other wick lysimeters, the drainage rate must be determined by measuring the volume of water in the collection vessel. Wick lysimeters do not require a vacuum pump, but the applied pressure head is constant and cannot be adjusted. A trench is required for installation, and sufficient depth below the collection surface must be provided for the hanging wick and collection vessel. Also, the collection surface must be in good hydraulic connection with the overlying soil.

Carefully installed and maintained lysimeters can provide direct measurements of drainage rates with excellent temporal resolution (Fayer and Gee, 2006). Drainage and precipitation rates can be used to infer travel times of water moving through the unsaturated zone. Depending on design, lysimeters can capture both matrix and preferential flow, but their very presence may disrupt the natural flow system. Lysimeters with large surface areas (a few m² or more) are prohibitively expensive for most studies, but they can provide an integrated drainage estimate that may account for field heterogeneity. Small lysimeters provide point measurement of fluxes; they are inexpensive to fabricate and operate, and they can provide useful information for many studies. For any lysimeter, it is difficult to ensure that conditions inside or above it are similar to those in the adjacent undisturbed soil. Preferential flowpaths can be inadvertently created during lysimeter installation.

Example: Las Cruces, New Mexico

An experiment was initiated in 1983 in south-central New Mexico to determine drainage rates for unvegetated areas under natural precipitation (Gee *et al.*, 1994). A 2.44-m diameter, 6-m deep lysimeter was filled with Berino loamy fine sand packed to a density of 1670 kg/m³. The surface of the lysimeter was kept free of vegetation. There were two access tubes for neutron probe measurements of water contents, and tensiometers were installed at nine depths (Figure 5.14). Drainage at the base of the lysimeter was collected through porous ceramic cups attached to a vacuum pump. Water contents and soil-water pressure heads were measured monthly.

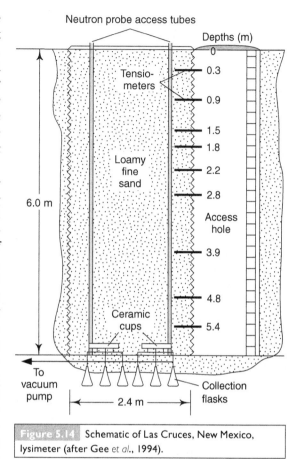

Figure 5.14 Schematic of Las Cruces, New Mexico, lysimeter (after Gee *et al.*, 1994).

Water content of the lysimeter sediments increased throughout the profile from 1983 to 1992 (Figure 5.15). The increase initially appeared only at shallow depths. Over time the wetting front moved slowly downward. A quasi-steady state appears to have been reached in the middle portion of the lysimeter by 1988 indicating that the rate of infiltration to the lysimeter was being balanced by evaporation and drainage. It is interesting to examine water travel times through the lysimeter. For 1989 to 1991, the average drainage rate was 20 mm/yr (Table 5.2). If the average water content was 0.16, then the average vertical velocity should be about 125 mm/yr. By this calculation, however, it would take 48 years for the wetting front to travel from land surface to the lysimeter drain. A possible explanation for this inconsistency is that the net infiltration (infiltration minus evapotranspiration) for

Table 5.2. Precipitation, change in water storage, and drainage for the Las Cruces, New Mexico, lysimeter (after Gee *et al.*, 1994).

Year	Precipitation (P), mm	Change in storage, mm	Drainage (D), mm	D/P, %
1984	385	172	0	0
1985	337	150	0	0
1986	320	145	0	0
1987	323	47	0	0
1988	344	6	0	0
1989	363	55	12	3.3
1990	250	−65	17	6.8
1991	278	124	30	10.8

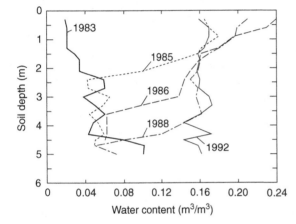

Figure 5.15 Plot of water content with depth within the Las Cruces, New Mexico, lysimeter for Aug. 1983, Sep. 1985, Nov. 1986, Feb. 1988, and Jul. 1992 (after Gee *et al.*, 1994).

the early years of the experiment was greater than 20 mm/yr and that the low precipitation in 1990 and 1991 resulted in a reduction in the rate of water movement through the lysimeter. In a separate study, Wierenga *et al.* (1987) used neutron-probe measurements along a 2.7 km transect near the lysimeter to examine drainage under vegetated conditions. Those measurements indicated that in the presence of native plants (mostly creosote bushes) there was negligible deep drainage.

Example: Chalk aquifer, Fleam Dyke, England

Kitching and Shearer (1982) reported on the construction and operation of a large-scale lysimeter at the Fleam Dyke pumping station near Fulbourn, Cambridgeshire, in eastern England. The lysimeter encompasses a 5 m³ block of undisturbed chalk. The chalk is highly fractured due to jointing and faulting; typical spacing between fractures is in the order of 0.2 m. The water table was at a depth of about 18 m. The lysimeter was constructed by initially driving steel piling for the four walls, then excavating to a depth of 6 m along the outside of the walls to allow driving of additional steel pilings that served as the base of the lysimeter. The base was installed as two panels driven in from opposite sides at an angle of 10°. Finally, a tunnel was dug beneath the centerline of the base, and a sloping trough was installed to collect freely draining water and transport it to a measuring tank. During construction in 1977, the surface was not disturbed and retained its original grass cover. After construction, the excavations were backfilled, and grass was replanted to provide a uniform cover over the entire area.

Most drainage occurred between the months of January and May (Figure 5.16). Winter and early spring rainfall appears to be responsible for most of the drainage. Annual totals for drainage were 201 mm for 1978 and 197 mm for 1979. Average annual rainfall for these years was 642 mm; the long-term average for the site is 543 mm. Over the 2-year period, drainage accounted for 31% of total rainfall.

The large size of this lysimeter is well suited for measuring drainage through the chalk

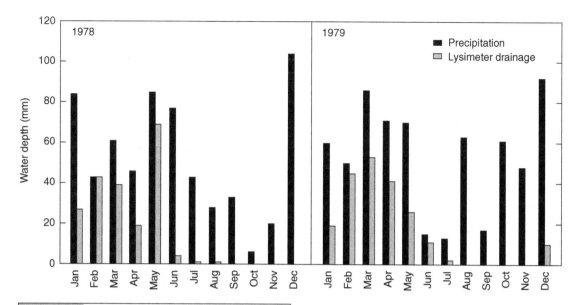

Figure 5.16 Monthly rainfall and lysimeter drainage at Fleam Dyke (reprinted from *Journal of Hydrology*, v. 58, Kitching and Shearer (1982), Figure 2, copyright (1982) with permission from Elsevier).

where water flow is primarily in fractures. The lysimeter surface area is large enough to minimize edge effects. Drainage rates can be measured at a high frequency. The lysimeter has some drawbacks, however. Installation and maintenance of such a facility are expensive. It is unlikely that many studies could afford such an undertaking. Installation also could disrupt natural flow patterns within the unsaturated zone, although Kitching and Shearer (1982) examined this issue and found minimal disruption. An additional consideration was the condensation of water in the underground tunnels. Precautions were taken to ensure that condensed water did not pass through the piling walls into the lysimeter.

Example: Columbia County, Wisconsin

Brye *et al.* (2000) determined water budgets for a 132-week period between 1995 and 1998 at three sites in Columbia County, southern Wisconsin: a restored natural prairie, maize under no tillage, and maize with chisel-plow tillage. Instrumentation included rain

and snow gauges, a neutron probe to monitor soil-water content, and equilibrium tension lysimeters (ETLs) to measure water drainage below the root zone. The lysimeters have a porous stainless steel surface that allows collection and measurement of drainage. Tensiometers and a vacuum system allowed the pressure head at the surface of an ETL to be set slightly less than that recorded in the bulk soil surrounding the lysimeter. The flow into an ETL should be similar to natural drainage. The top surface of each ETL, which was 0.75 m long and 0.25 m wide, was set at a depth of 1.4 m beneath undisturbed soil through the wall of a 2 m deep trench.

Precipitation rates were similar at all sites (Figure 5.17). Evapotranspiration rates for the prairie site were slightly greater than those for the no-tillage site, which were slightly greater than those from the chisel-plow site. Drainage occurred from late January to mid-June at all sites. There were substantial differences among sites in terms of total drainage for the study period: 199 mm for the prairie, 563 mm for the no-tillage, and 793 for the chisel-plow site. Drainage increased with increasing disturbance of the land surface. The ETLs are difficult to install and maintain, but they provided excellent temporal and spatial resolution of drainage rates for this study.

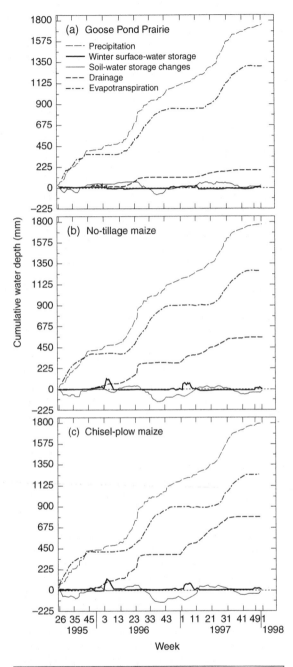

Figure 5.17 Water budget for (a) prairie site, (b) no-tillage maize site, and (c) chisel-plow maize site in central Wisconsin (after Brye et al., 2000). Imbalance in the water budgets is attributed to runoff for the prairie and no-tillage site and to runon from melting snow at the chisel-plow site.

5.6 | Discussion

The ZFP and Darcy methods both require measurements or estimates of the hydraulic gradient; therefore, the methods are often applied in tandem. Hydraulic gradients in the unsaturated zone can be measured with clusters of instruments such as tensiometers or heat-dissipation probes. However, if the gradient is small in magnitude, it can be difficult to accurately quantify because of instrument error and natural heterogeneity in unsaturated-zone sediments. Neither method is limited in terms of the time scales over which they can be applied. They are particularly useful for estimating drainage from a fast-moving wetting front propagated by an individual storm (assuming automatic recording of data). Both methods can provide useful insights on recharge processes, although they only produce point estimates of drainage. The ZFP method cannot be applied when drainage occurs as steady flow, whereas the Darcy method can measure drainage occurring as steady as well as transient flow. When the unit hydraulic gradient assumption is invoked with the Darcy method, steady drainage is assumed. Both ZFP and Darcy methods are relatively expensive to apply. Values of hydraulic conductivity are required for the Darcy method; this parameter is difficult to accurately determine, and it can vary substantially in natural systems.

A wide variety of lysimeters have been applied in hydrologic studies, ranging from simple collection devices, such as pan lysimeters, to very expensive computer-monitored systems. Carefully constructed and maintained lysimeters can provide precise measurements of drainage, but by their mere presence, lysimeters can disrupt the natural flow system. Lysimeters can be used in almost any environment, and if their surface areas are large enough they can capture both preferential and matrix flow.

Physical methods: saturated zone

6.1 | Introduction

Among the most widely used techniques for
estimating recharge are those based on measurement of groundwater levels over time and
space. The abundance of available groundwater-level data and the simplicity of these
methods facilitate straightforward application. The water-table fluctuation method uses
fluctuations in groundwater levels over time
to estimate recharge for unconfined aquifers; it is the focus of most of this chapter.
Included in the discussion of the method are
an analysis of mechanisms that can cause
water-table fluctuations and a review of methods for estimating specific yield. Other methods addressed in this chapter are based on
the Darcy equation and include an approach
developed by Theis (1937), the Hantush (1956)
method for estimating interaquifer flow, and
the application of flow nets. The chapter also
includes a discussion of approaches based on
time-series analyses of measured groundwater
levels. The content of this chapter draws from
and expands upon the material presented in
Healy and Cook (2002).

6.1.1 Groundwater-level data
Many local, state, and federal agencies maintain
databases of measured groundwater levels in
individual countries. Within the United States,
the US Geological Survey maintains the largest
database on real-time and historic groundwater

levels (Table 2.1). Groundwater levels can be
measured manually by using a graduated measurement tape to determine the depth to water
in a well from a reference point at the top of
the well casing. Historically, groundwater levels
in some observation wells were automatically
recorded by using a float that was attached by a
steel tape or wire to a wheel sensor; a strip-chart
or paper-punch device was used to record movement of the wheel (Rasmussen and Andreasen,
1959). Submersible pressure transducers have
come into widespread use for monitoring
groundwater levels since the 1990s (Freeman
et al., 2004). These electronic devices can automatically sense and record groundwater levels
at user-selected frequencies. The depth (relative
to the reference point on the well casing) at
which a transducer is placed in a well must be
carefully measured. If groundwater elevation
is desired, the elevation of the reference point
needs to be determined.

Total head within an aquifer varies by location and depth. A measured groundwater level
represents a head value that is averaged across
the depth interval over which an observation
well is screened. So the length and depth of the
screened interval affects measured water levels
(Taylor and Alley, 2001). A water level measured
in a well screened across the entire thickness of
an aquifer represents a depth-averaged head for
the aquifer at the well location. Multiple wells
installed at different depths and with narrow
screened intervals are required to determine
vertical head gradients within an aquifer.

6.2 | Water-table fluctuation method

A simple water budget for some part of a groundwater system can be written as (Schicht and Walton, 1961):

$$\Delta S^{gw} = R - Q^{bf} - ET^{gw} - Q^{gw}_{off} + Q^{gw}_{on} \qquad (6.1)$$

where ΔS^{gw} is change in storage in the saturated zone (this includes all storage changes that occur at depths greater than that of the zero-flux plane), R is recharge, Q^{bf} is base flow, ET^{gw} is evapotranspiration from groundwater, and Q^{gw}_{off} and Q^{gw}_{on} are subsurface flow away from or into the area of interest, including pumping. All terms are expressed as volumetric rates per unit surface area (e.g. mm/year). Equation (6.1) states that changes in subsurface water storage are attributed to recharge and groundwater flow into the basin minus base flow (groundwater discharge to streams or springs), evapotranspiration from groundwater, and groundwater flow out of the basin.

The water-table fluctuation (WTF) method is based on the premise that rises in groundwater levels in unconfined aquifers are due to recharge water arriving at the water table. With the assumption that the amount of available water in a column of unit surface area is equal to specific yield times the height of water in the column, recharge can be calculated from:

$$\Delta S^{gw} = R = S_y\, \Delta H / \Delta t \qquad (6.2)$$

where S_y is specific yield and ΔH is change in water-table height over time interval Δt. Derivation of Equation (6.2) assumes that water arriving at the water table goes immediately into storage and that all other components of Equation (6.1) are zero during the period of recharge. A time lag occurs between arrival of water during a recharge event and redistribution of that water to the other components of Equation (6.1). If the method is applied during that time lag, all of the water going into recharge can be accounted for. This assumption is most valid over short periods of time (hours or a few days), although the method has been successfully applied over periods of years and decades. The method may not be appropriate if water is transported away from the water table at a rate that is not substantially slower than the rate at which recharge water arrives at the water table.

Application of Equation (6.2) for each individual water-level rise will generate an estimate of total or *gross* recharge. To determine total recharge, ΔH is set equal to the difference between the peak of the rise and low point of the extrapolated antecedent recession curve at the time of the peak (Figure 6.1). Equation (6.2) also can be used to calculate net change in saturated-zone storage over any time interval (e.g. days, months, or years). This change in storage is determined by replacing ΔH in Equation (6.2) with ΔH_n, which is the difference in head between the end and the beginning of the time interval (Figure 6.1). The difference between recharge and net change in subsurface storage is equal to the sum of evapotranspiration from groundwater, base flow, and net subsurface flow from the site. With some additional assumptions, the WTF method can be used to estimate any of these parameters.

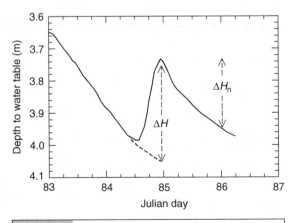

Figure 6.1 Hypothetical groundwater hydrograph. Recharge for the water-level rise, in terms of depth of water, is calculated as S_y times ΔH, where ΔH is the difference between the peak of the rise and low point of the extrapolated antecedent recession curve (dotted line) at the time of the peak. The change in subsurface storage between days 85 and 86 is calculated as S_y times ΔH_n, where ΔH_n is the difference in water level at midnight on days 85 and 86.

The antecedent recession curve is the trace that the well hydrograph would have followed in the absence of the rise-producing precipitation (Figure 6.1). That trace can be drawn manually, a practice that is time consuming and prone to subjectivity but one that accommodates hydrologic intuition. Automated techniques for generating recession curves have also been developed (Crosbie et al., 2005; Heppner and Nimmo, 2005; Delin et al., 2007). These techniques generally use long-term hydrograph records to fit linear or exponential equations that define the rate of water-table decline as a function of water-table height. At any particular well, the rate of water-table decline decreases as the height of the water table decreases. Automated techniques are effective for processing long-term water-level records and removing subjectivity from the recharge estimation process. Approaches for representing water-table recession curves are similar to those employed for base flow in streams (Section 4.5).

The WTF method for estimating groundwater recharge was applied as early as the 1920s (Meinzer, 1923; Meinzer and Stearns, 1929) and since then has been used in numerous studies (e.g. Rasmussen and Andreasen, 1959; Gerhart, 1986; Hall and Risser, 1993; Crosbie et al., 2005; Lee et al., 2005; Coes et al., 2007; and Delin et al., 2007). White (1932) used diurnal fluctuations in groundwater levels to estimate evapotranspiration in the Escalante Valley of Utah. Weeks and Sorey (1973), Loheide et al. (2005), Schilling and Kiniry (2007), and others have used water-table fluctuations to estimate ET^{gw}.

The WTF method is simple and easy to use. Because no assumptions are made on the mechanisms of water movement through the unsaturated zone, the presence of preferential flow paths does not restrict its application. Recharge estimates by the WTF method are representative of an area of several to perhaps thousands of square meters. So the WTF method can be viewed as less of a point measurement than those methods that are based on data in the unsaturated zone (Chapter 5).

The WTF method has some limitations. Recharge rates may vary substantially within a watershed because of differences in elevation,

geology, land-surface slope, vegetation, and other factors (Lee et al., 2005). Data from multiple wells should be used to ensure that recharge estimates are representative of the catchment as a whole. In the WTF method, recharge is assumed to occur as discrete events in time, in direct contrast to methods, such as the unit hydraulic gradient method (Section 5.4), in which a steady recharge rate is assumed. If the recharge rate to an aquifer was constant and equal to the drainage rate away from the aquifer, the groundwater levels would not change, and the WTF method would estimate a recharge rate of zero. Therefore, the method is not applicable under these conditions. As discussed in the following section, not all fluctuations in groundwater levels indicate recharge. Difficulties in estimating specific yield also contribute to the overall uncertainty of the method. In addition, the frequency with which water levels are measured can affect recharge estimates. Morgan and Stolt (2004) found that magnitudes of water-table fluctuations determined on the basis of weekly measurements were 33% less than those determined on the basis of measurements obtained at half-hour intervals for the same observation well and time period. Delin et al. (2007) also found that decreasing the frequency of water-level measurements led to a decrease in estimated recharge rates; those authors recommended that water levels be measured on a weekly or higher frequency for application of the WTF method. In spite of its limitations, the simplicity of the approach and the widespread availability of groundwater-level data combine to make the water-table fluctuation method one of the most widely used approaches for estimating recharge.

6.2.1 Causes of water-table fluctuations in unconfined aquifers

Groundwater levels fluctuate over multiple time scales. Long-term changes, over periods of perhaps decades, may be related to changes in recharge rates due to natural variations in climate or to anthropogenic effects. Allison et al. (1990) attributed rising groundwater levels and increased recharge in the Murray Basin of Australia to replacement of native eucalyptus trees with agricultural crops; likewise, Leduc *et al.* (2001) traced the continuous rise in water tables since the 1960s in southwest Niger to removal of native vegetation. Groundwater levels in Memphis, Tennessee declined from 1935 to 1975 (Figure 6.2a) as rates of pumping increased; after 1975, pumping rates and water levels stabilized (Taylor and Alley, 2001). Seasonal fluctuations in groundwater levels are common in many areas due to the seasonality of evapotranspiration, precipitation, groundwater pumping, and irrigation (Figure 6.2b). Short-term water-table fluctuations occur in response to rainfall, pumping, barometric pressure fluctuations, evapotranspiration, and other phenomena (Figure 6.2c).

The WTF method is most often applied for short-term water-level rises that occur in response to individual storms, conditions that typically exist in humid regions with shallow water tables. Wetting fronts moving downward within the unsaturated zone tend to disperse with increasing depth, so water tables in deep aquifers may not react to individual storms but will instead display seasonal rises and falls. The WTF method, however, can also be applied using these seasonal or long-term water-table fluctuations.

Groundwater levels rise and fall in response to several phenomena. Application of the WTF method for estimating recharge requires identification of the water-level rises that are attributable to recharge from precipitation or a surface-water body. It can be a difficult task. The following sections give some details on mechanisms other than recharge that can induce short-term fluctuations in the water table.

Evapotranspiration

Shallow water tables may exhibit diurnal fluctuations, declining during daylight hours in response to evapotranspiration and rising through the night when ET^{gw} is essentially zero. Figure 6.2c shows diurnal fluctuations in depth to the water table beneath a field of alfalfa in the Escalante Valley of Utah before and after cutting (White, 1932). White (1932) developed a formula similar to Equation (6.2) for estimating ET^{gw} based on such

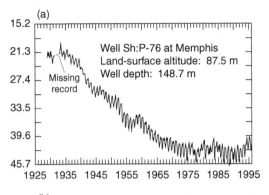

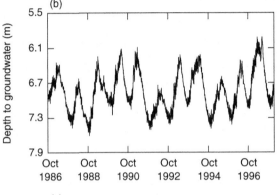

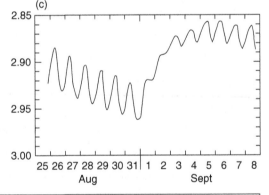

Figure 6.2 (a) Water level trends in an observation well in Memphis, Tennessee between 1928 and 1995 (Taylor and Alley, 2001); (b) Hydrograph of daily water-level measurements for a well in Vanderburgh County, Indiana (Taylor and Alley, 2001); (c) Diurnal fluctuations in response to evapotranspiration by alfalfa in the Escalante Valley, Utah, August 25 to September 8, 1926. Alfalfa was cut on August 3 (White, 1932).

fluctuations. He assumed that ET^{gw} was zero between midnight and 4:00 a.m. and defined h' as the hourly rate of water-table rise during those hours (Figure 6.3). The total amount of

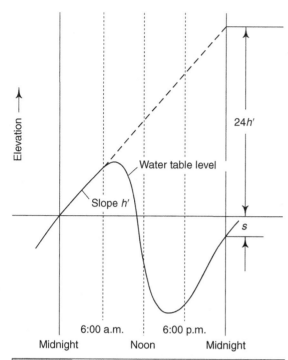

Figure 6.3 Diagram of water-level fluctuations, showing variables needed for the White (1932) method for estimating evapotranspiration (s is change in water level over 24 hours, h' is hourly rate of water table rise between midnight and 4:00 a.m.) (after Troxell, 1936).

well than through the sediments overlying the aquifer. These fluctuations are observational effects rather than effects on the aquifer itself. A time lag exists over which the pressure change at land surface is propagated through the unsaturated zone to the water table. Therefore, an imbalance exists between the pressure on the water in the well and the water in the aquifer until the pressure front arrives at the water table. This imbalance produces a change in the observed water level in the well. The length of the time lag increases with increasing depth to the water table and with decreasing vertical air diffusivity of the unsaturated-zone sediments. Techniques for identifying and removing the effects of atmospheric-pressure changes from observed water levels are described in Weeks (1979), Rojstaczer (1988), and Rasmussen and Crawford (1997). Pressure transducers, which are commonly used to measure water levels, can also be affected by changes in atmospheric pressure (Rasmussen and Crawford, 1997).

Entrapped air
Water levels in unconfined aquifers also can be affected by the presence of entrapped air between the water table and a wetting front advancing downward from land surface. This phenomenon is particularly difficult to identify because it occurs in response to precipitation and, thus, is easily mistaken for recharge. The phenomenon takes place when surface soils become saturated, and therefore impermeable to air, and is most prevalent in fine-textured soils. Figure 6.4 illustrates the phenomenon, which has been termed the *Lisse* effect (Krul and Liefrinck, 1946) after the village in Holland where it was first identified. The entrapped air restricts infiltration at land surface; therefore, it not only gives the false impression of recharge, but it also reduces the amount of recharge that would be expected in its absence. Meyboom (1967) attributed the immediate increased runoff in the Qu'Appelle River after a light rainfall to the Lisse effect. In most field settings the effect is probably not important because irregularities in land surface and the presence of macropores in soils tend to inhibit widespread air entrapment (Weeks, 2002).

groundwater discharged during one day, V_{ET}, was then calculated as:

$$V_{ET} = S_y(24h' + s) \qquad (6.3)$$

where s is the water-level elevation at midnight at the beginning of the 24 hour period minus the water-level elevation at the end of the period. Using a specific yield of 0.073, White (1932) estimated seasonal ET^{gw} of the alfalfa to be 700 mm. Loheide *et al.* (2005) analyzed White's method using a groundwater flow model and found that the greatest source of uncertainty with the method was in estimating S_y.

Atmospheric pressure
Changes in atmospheric pressure can cause fluctuations of tens of millimeters in water levels measured in observation wells. The fluctuations occur because pressure changes are transmitted more rapidly through the open

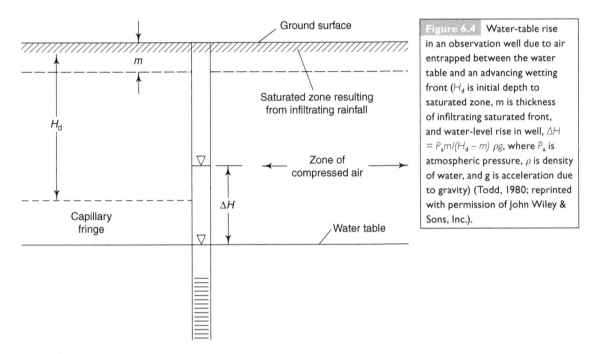

Figure 6.4 Water-table rise in an observation well due to air entrapped between the water table and an advancing wetting front (H_d is initial depth to saturated zone, m is thickness of infiltrating saturated front, and water-level rise in well, $\Delta H = P_a m/(H_d - m)\,\rho g$, where P_a is atmospheric pressure, ρ is density of water, and g is acceleration due to gravity) (Todd, 1980; reprinted with permission of John Wiley & Sons, Inc.).

Other mechanisms

Other mechanisms can induce fluctuations in water levels in unconfined aquifers. In general these mechanisms are more easily identified or are less frequently encountered than those already discussed. Temperature variations affect groundwater levels due to freeze-thaw action and the temperature dependency of surface tension, air solubility, and air density. Pumping of wells and natural or induced changes in surface-water elevations can greatly affect groundwater levels. Earthquakes and ocean tides can also affect groundwater levels. Changes in groundwater flow into or out of the study area, due to processes occurring in adjacent areas, can produce water-level fluctuations (Dickinson et al., 2004). Additional information on these processes can be found in Freeze and Cherry (1979) and Todd and Mays (2005).

Correcting water-level data

Tamura et al. (1991) and Toll and Rasumssen (2007) described computer programs for removing the effects of barometric fluctuations and Earth tides from water-level records. Crosbie et al. (2005) presented an algorithm for applying the WTF method to eliminate rises attributed to air entrapment and check that there was

sufficient recent rainfall to induce a water-level rise. Also, von Asmuth et al. (2008) developed a time series approach for decomposing a series of water-level fluctuations into partial series that represent the influence of individual stresses, such as evapotranspiration, atmospheric pressure fluctuations, river- and lake-level fluctuations, groundwater pumping, and precipitation. The approach allows the effects of all stresses that do not contribute to recharge to be removed from water-level records. In the absence of an automated approach, manual inspection and correction of water-level data benefit from side-by-side comparison with data on precipitation, barometric pressure, temperature (to check for snowmelt), and nearby streamflow.

6.2.2 Specific yield

Meinzer (1923) defined specific yield of a rock or soils as the ratio of (1) the volume of water which, after being saturated, it will yield by gravity to (2) its own volume. Specific yield has traditionally been represented by the formula:

$$S_y = \varphi - S_r \qquad (6.4)$$

where φ is porosity and S_r is specific retention (the volume of water retained by the rock per

Table 6.1 Statistics on specific yield from 17 studies compiled by Johnson (1967).

Texture	Average specific yield	Coefficient of variation (%)	Minimum	Maximum	Number of determinations
Clay	0.02	59	0.0	0.05	15
Silt	0.08	60	0.03	0.19	16
Sandy clay	0.07	44	0.03	0.12	12
Fine sand	0.21	32	0.10	0.28	17
Medium sand	0.26	18	0.15	0.32	17
Coarse sand	0.27	18	0.20	0.35	17
Gravelly sand	0.25	21	0.20	0.35	15
Fine gravel	0.25	18	0.21	0.35	17
Medium gravel	0.23	14	0.13	0.26	14
Coarse gravel	0.22	20	0.12	0.26	13

unit volume of rock). Specific yield is treated as a storage term, independent of time, that accounts for the instantaneous change in water storage upon a change in total head. In reality, the release of water is not instantaneous. Rather, the release can take an exceptionally long time, especially for fine-grained sediments. King (1899) determined S_y to be 0.20 for a fine sand; however, it took two and a half years of drainage to obtain that value. The wide range of values that has been reported for S_y in the literature is attributed to natural heterogeneity in geologic materials, different methods used for determining S_y, and, in large part, to the amount of time allotted to the determination (Prill *et al.*, 1965). Johnson (1967) compiled values of S_y from 17 studies (Table 6.1). As shown in the table, S_y tends to increase with increasing sediment grain size. The variability within each textural class (based on the coefficient of variation) tends to decrease with increasing grain size. Coarser sediments drain more quickly and, hence, S_y for these sediments shows less time dependency.

Specific yield varies with depth to the water table, decreasing as that depth decreases. As the water table approaches land surface, the sediments above the water table are unable to fully drain to S_r. The interested reader is referred to the analysis of Childs (1960) for further insight into this phenomenon. The extreme case occurs when depth to the water table is less than the height of the capillary fringe; here, no water

is released when water levels change. This phenomenon is sometimes referred to as the reverse Wieringermeer effect and is reflected by a nearly instantaneous rise in water level in response to only a small amount of infiltration (Gillham, 1984). Dos Santos Jr. and Youngs (1969) and Duke (1972) suggested the following expression for specific yield:

$$S_y = \varphi - \theta(H_d) \tag{6.5}$$

where H_d is depth to water table and θ is volumetric water content. $\theta(H_d)$ can be determined from measured water-retention curves or estimated on the basis of texture from published tables of soil properties (Section 5.2.3). The effect of depth to water table on S_y, as described in Equation (6.5), is shown in Figure 6.5 for two soils described by Duke (1972).

Specific yield also varies with time following a change in water-table height. Nachabe (2002) developed a formula for determining S_y as a function of depth to water table and time following an instantaneous change in water-table elevation. It was assumed, as in the analysis of Childs (1960), that the material was uniform and a static equilibrium pressure head profile initially existed above the water table. Water-retention and hydraulic conductivity curves were represented by slight rearrangements of the Brooks and Corey (1964) formulas given in Section 5.2.3:

$$\Theta = (\theta(h) - \theta_r)/(\theta_s - \theta_r) = (h_b/h)^\lambda \tag{6.6}$$

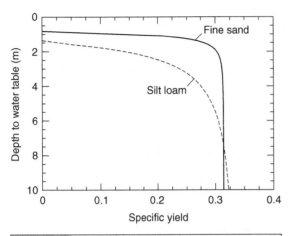

Figure 6.5 Specific yield as a function of depth to water table for a fine sand and a silt loam using Equation (6.5).

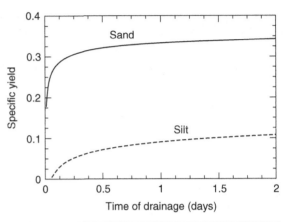

Figure 6.6 Specific yield as a function of drainage time as described by Equation (6.8) (Nachabe 2002) with $\Delta h = 0.25$ m for the sand and silt in Table 6.2.

$$K(\Theta) = K_s \Theta^n \qquad (6.7)$$

where Θ is effective saturation, h is pressure head, θ_s is saturated water content, θ_r is residual water content, h_b is the bubbling pressure head, λ is pore-size distribution index, K is hydraulic conductivity, K_s is saturated hydraulic conductivity, and $n = (2 + 3\lambda)/\lambda$. Applying the method of characteristics to solve the one-dimensional flow equation under conditions of gravity drainage, Nachabe (2002) developed the following closed-form approximate solution for S_y:

$$S_y = K_s[\Theta^n_b - \Theta^n_{sura}]\Delta t / \Delta H + (\theta_s - \theta_r)/(1 - \Theta_b) \qquad (6.8)$$

where

$$\Theta_b = [\Delta H(\theta_s - \theta_r)/(nK_s\Delta t)]^{1/(n-1)} \qquad (6.9)$$

$\Theta_{sura} = (h_b/H_d)^\lambda$ is the effective saturation at land surface for a water-table depth of H_d, and Δt is time since rise or fall of the water table. For Equation (6.8), a negligible change in saturation at land surface is assumed over the time interval of interest; for cases where this assumption is not valid, a more detailed equation was given by Nachabe (2002). Figure 6.6 shows how S_y changes with time, according to Equations (6.8) and (6.9), following an instantaneous decline in water table for two sediments. Table 6.2 (modified from Loheide *et al.* (2005)) contains

estimates of specific yield as calculated from Equations (6.8) and (6.9) for generic soils.

The foregoing discussion illustrates how, in theory, S_y is affected by the depth to the water table and time since water level rise or fall, but the assumptions used in these analyses represent ideal conditions that may not be typical of actual field conditions. Sediment properties are seldom uniform. True steady-state conditions probably never exist in the real world and many years may be required for some soils to fully drain. Water tables in most aquifers exhibit daily, seasonal, or annual fluctuations, so complete drainage of the sediments is unlikely. Similarly, water-table rises and declines typically do not occur instantaneously. The following sections describe several other methods that have been used for estimating specific yield.

Laboratory methods for estimating S_y

In the laboratory, specific yield is usually determined by measurement of porosity and specific retention and application of Equation (6.4). Johnson *et al.* (1963) described a column-drainage approach, whereby a column, filled with undisturbed or repacked sediments, is saturated with water and in turn allowed to drain. As previously mentioned, the time allowed for drainage can have a large effect on the calculated values of specific retention

Table 6.2 Hydraulic properties of generic sediments from Carsel and Parrish (1988) and Loheide et al. (2005). θ_s, θ_r, n, and α are parameters of the van Genuchten (1980) equation (Equation 5.2). K_s is saturated hydraulic conductivity. Estimates of S_y are from: A) Equation (6.5) after Duke (1972) with depth to water table, H_d, of 1 m; B) Equation (6.5) with H_d = 2 m; C) Equation (6.8) after Nachabe (2002) with H_d = 1 m and drainage time, Δt = 12 hours; D) Equation (6.8) with H_d = 2 m and Δt = 12 hours; and E) Loheide et al. (2005).

Sediment				α	K_s		S_y				
Texture	θ_s	θ_r	n	m^{-1}	m/d	$\theta_s-\theta_r$	A	B	C	D	E
Sand	0.43	0.045	2.68	14.5	7.1	0.39	0.38	0.38	0.35	0.35	0.32
Loamy sand	0.41	0.057	2.28	12.4	3.5	0.35	0.34	0.35	0.31	0.31	0.26
Sandy loam	0.41	0.065	1.89	7.5	1.1	0.35	0.29	0.31	0.27	0.27	0.17
Loam	0.43	0.078	1.56	3.6	0.25	0.35	0.19	0.23	0.19	0.20	0.08
Silt	0.46	0.034	1.37	1.6	0.06	0.43	0.11	0.15	0.11	0.13	0.03
Silt loam	0.45	0.067	1.41	2	0.11	0.38	0.12	0.17	0.12	0.15	0.04
Sandy clay loam	0.39	0.1	1.48	5.9	0.31	0.29	0.17	0.20	0.16	0.17	0.07
Clay loam	0.41	0.095	1.31	1.9	0.06	0.32	0.08	0.11	0.08	0.10	0.02
Silty clay loam	0.43	0.089	1.23	1	0.02	0.34	0.04	0.06	0.04	0.06	0.01
Sandy clay loam	0.38	0.1	1.23	2.7	0.03	0.28	0.07	0.09	0.07	0.07	0.02

and yield. The height of the column is also important. A column that is shorter than the height of the capillary fringe is likely to produce a value of S_y that is much less than that obtained from a column that is greater in height than that of the capillary fringe (Prill et al., 1965). Romano and Santini (2002) described similar laboratory techniques for approximating field capacity. For practical purposes, S_r and field capacity are often considered equivalent.

Many variants of Equation (6.5) exist, whereby S_r or field capacity is estimated from information on water-retention curves. Jamison and Kroth (1958) proposed setting field capacity equal to volumetric water content at a pressure head of –33 kPa; this approach has gained widespread acceptance within the agricultural community. Cassel and Nielsen (1986) pointed out that field capacity for coarse-textured soils is usually better represented by water content at –10 kPa, whereas water content at a pressure head of much less than –33 kPa is appropriate for fine-textured soils. If the water-retention curve is measured (or

approximated by pedotransfer functions as described in Section 5.2.3), water content at any prescribed pressure head can be used to obtain an estimate of S_r.

Field methods for determining S_y

AQUIFER TESTS

Traditional aquifer tests provide in-situ measurements of S_y and other aquifer properties that are integrated over fairly large areas. Observation wells are located at various distances from the pumping well. Water-level drawdown vs. time data from observation wells are matched with theoretical type curves developed by Boulton (1963), Prickett (1965), Neuman (1972), and Moench (1994, 1995, 1996) to generate estimates of S_y and transmissivity. The methods have some drawbacks. Interpretation of results is nonunique (Freeze and Cherry, 1979), and the validity of assumptions inherent in the techniques is difficult to verify. Expense is also a key concern; installation of pumping and observation wells may be cost prohibitive for many recharge studies. Aquifer tests require careful planning; useful guidelines are provided by Walton (1970), Stallman (1971), and Batu (1998).

The volume-balance method combines a traditional aquifer test with a water budget of the cone of depression created by the withdrawal of water from the pumping well. Specific yield is defined as:

$$S_y = V_w / V_c \quad (6.10)$$

where V_w is the volume of water pumped out of the system at some point in time and V_c is the volume of the cone of depression (the region between the initial and the final water table) at that same time (Clark, 1917; Wenzel, 1942; Remson and Lang, 1955). Nwankwor et al. (1984) used this method to analyze results from an aquifer test at the Borden field site in Ontario, Canada. Their results (Table 6.3) show a trend of increasing S_y with increasing time: the longer the pump test, the larger the calculated value of S_y became. These results were attributed to delayed drainage from the unsaturated zone. Wenzel (1942) observed a similar trend in a sand-gravel aquifer. Difficulties in accurately measuring the volume of the cone of depression limit the applicability of this method, but advances have been made with ground-penetrating radar for these measurements (Endres et al., 2000; Bevan et al., 2003; Ferré et al., 2003).

WATER-BUDGET METHODS

Walton (1970) proposed using a water budget in conjunction with Equation (6.2) to estimate S_y during winter periods when evapotranspiration is low and soil is near saturation (so change in storage in the unsaturated zone is low). A simple water budget for a basin can be written as:

$$P = Q_{on} = ET + \Delta S + Q_{off} \quad (6.11)$$

where P is precipitation plus irrigation, Q_{on} and Q_{off} are surface and subsurface water flow into and out of the basin, ET is evapotranspiration, and ΔS is change in water storage. ΔS can be written as the sum of change in storage of

Table 6.3 Values of specific yield after Nwankwor et al. (1984).

Method	S_y	Elapsed time (min)
Neuman (1972)	0.07	
Boulton (1963)	0.08	
Volume balance	0.02	15
	0.05	40
	0.12	600
	0.20	1560
	0.23	2690
	0.25	3870
Laboratory ($\theta_s - \theta_r$)	0.30	

surface reservoirs, ΔS^{sw} (including water bodies as well as ice and snowpacks), the unsaturated zone, ΔS^{uz}, and the saturated zone, ΔS^{gw}. Substituting into Equation (6.11) and recalling that Equation (6.2) can be used to determine ΔS^{gw}, we can write:

$$\Delta S^{gw} = P + Q_{on} - ET - \Delta S^{sw} - \Delta S^{uz} - Q_{off}$$
$$= S_y \Delta H / \Delta t \quad (6.12)$$

Rearranging this equation produces an estimate for S_y:

$$S_y = (P + Q_{on} - Q_{off} - ET - \Delta S^{sw} - \Delta S^{uz}) / \Delta H / \Delta t \quad (6.13)$$

Gerhart (1986) and Hall and Risser (1993) applied this method during winter periods (usually of week-long duration) with the assumption that ET, ΔS^{uz}, and net subsurface flow were zero. Crosbie et al. (2005) and Saha and Agrawal (2006) used similar approaches for estimating S_y. Best results should be expected in areas where surface and subsurface flows are easily determined.

GEOPHYSICAL METHODS

The temporal gravity technique of Pool and Eychaner (1995) for estimating change in subsurface water storage (Section 2.3.3) can also be used to estimate specific yield. Pool and Eychaner (1995) made microgravity measurements over transects that were kilometers in length. Changes in gravity over time were attributed to changes in subsurface water storage (combined unsaturated and saturated zones). Estimated changes in storage and measured changes in groundwater levels in several monitoring wells were inserted into Equation (6.2) and values of S_y in the range of 0.16 to 0.21 were calculated for an alluvial aquifer in the Pinal Creek watershed of central Arizona (Pool and Eychaner, 1995).

Meyer (1962), Weeks and Sorey (1973), Loeltz and Leake (1983), and Sophocleous (1991) used neutron meters to determine change in subsurface water storage due to changes in groundwater levels. The method requires installation of a neutron-probe access tube adjacent to an observation well. The total amount of stored water at any point in time is determined by measuring water contents between the water table and the zero-flux plane (the horizontal plane in the unsaturated zone that separates downward moving water from water that moves upward in response to evapotranspiration). The difference in stored water between any two measurement times, ΔS^{gw}, is the amount of water that has been added or subtracted from storage (Section 5.3). Specific yield can be determined from Equation 6.2:

$$S_y = \Delta S^{gw} / \Delta H / \Delta t \qquad (6.14)$$

MODELING METHODS

Numerical models that simulate water movement through variably saturated porous media can be used to estimate S_y. Loheide et al. (2005) generated estimates of S_y by simulating plant transpiration with a model of variably saturated subsurface flow. The objective of that work was to evaluate the White (1932) method for estimating transpiration. A series of simulations were run for flow through a two-dimensional vertical cross section for various hypothetical flow-system geometries and sediment textures. The model allowed for precise monitoring of transpiration rates and water-table depth. Readily available specific yield was calculated as the transpiration rate divided by the change in water-table depth. Halford (1997), Moench (2003), and El-Kadi (2005) used the VS2DI software package (Hsieh et al., 1999) to assess the importance of flow in the unsaturated zone

on aquifer test analysis. Results of these simulations could be used to determine specific yield in a manner similar to that employed by Loheide et al. (2005).

Concluding thoughts on S_y

Selection of appropriate values of S_y for use in the WTF method remains something of a puzzle. A large variability exists in both laboratory and field determinations of S_y, but laboratory values are generally greater than those obtained from field tests simply because laboratory tests are usually run for longer periods of time. Meyboom (1967) chose to multiply laboratory-determined values of S_y by 0.5 to arrive at a value of readily available specific yield that he then used with Equation (6.2) to estimate recharge in a prairie setting. Loheide et al. (2005) found, in general, that specific yields of coarse-grained sediments were similar to $\theta_s - \theta_r$; for fine-grained sediments, though, specific yield was much less than $\theta_s - \theta_r$.

If the water-retention curve and saturated hydraulic conductivity of the geologic materials are known, the method of Nachabe (2002) can be used to estimate S_y. For cases where these data are not available, Table 6.2 (modified from Loheide et al., 2005) contains estimates of specific yield by different methods for generic soils. The parameters of the van Genuchten equation listed in Table 6.2 were determined by Carsel and Parrish (1988) on the basis of thousands of soil samples; values represent the averages of all samples in each texture class. Loheide et al. (2005) comment that for drainage times of 12 hours or less, the depth dependency of specific yield is minimal for depths greater than 1 m. Table 6.2 indicates good agreement among estimates of S_y (including $\theta_s - \theta_r$) for the sand and loamy sand classes. Estimates of S_y for the finer-textured sediments show less agreement with each other and with $\theta_s - \theta_r$; these estimates are also more sensitive to depth to the water table, a predictable trend given that coarser soils are more easily drained.

6.2.3 Fractured-rock systems

Interpretation of water-table fluctuations in fractured-rock aquifers requires some special attention (Healy and Cook, 2002). Because of the complexities of such systems, the validity of applying the WTF method demands diligent analysis. Porosities of igneous and metamorphic rocks are often less than 1%, and even some sandstones and limestones have total porosities of less than a few percent (de Marsily, 1986). Recharge to low porosity, fractured-rock aquifers is often characterized by large variations in groundwater levels. Bidaux and Drouge (1993), for example, measured water-level rises of approximately 15 m in response to rainfall events of approximately 50 mm over 24 hours in fractured Cretaceous limestones and marly limestones in southern France.

In fractured-rock systems, the permeability of the matrix is usually very low, so that a long time is required for pores to fill and drain. The fracturing of low-porosity formations creates a total porosity distribution that is essentially bimodal. Most recharge through the unsaturated zone occurs rapidly along discrete, permeable fractures, which may become saturated during rain events, even though surrounding pores remain unsaturated. Thus, water levels in fractures may rise while most of the formation remains unsaturated (Figure 6.7a). In this case, the specific yield would be equal to the fracture porosity. This situation is most likely to occur in response to large rainfall events where the matrix permeability is very low, so the rate of water-level rise would be very rapid, as would the subsequent water-level decline (Figure 6.7b).

Some fractured-rock systems have deep water tables that display only seasonal fluctuations; water levels in fractures and matrix tend to rise and fall together. In the case of a declining water table, the time for the matrix to drain may be extremely long. The aquifer matrix may supply water as base flow to streams for many months after the water table (as reflected in fractures and piezometers) has declined (Price *et al.*, 2000).

Hydrograph variations for piezometers installed in fractured rocks often provide a poor record of water-level variations within the aquifer itself. This counterintuitive scenario occurs where the permeability of the aquifer is low and the storativity of the aquifer is very low relative to the storativity of

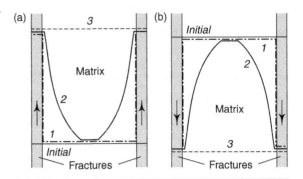

Figure 6.7 Diagrams of saturation of a fractured-rock matrix under (a) rising and (b) falling water tables. Where the rate of water-table rise is rapid relative to the matrix permeability, the matrix remains unsaturated as the water level in the fracture rises (1). The specific yield is equal to the fracture porosity. Where the rate of water-level rise is slow and the matrix permeability is high, the water table rises evenly in both the fracture and matrix (3). More usually, the matrix partially fills as the water table rises, and the specific yield is between these two limiting values (2). Congruent behavior occurs for declining water tables. (With kind permission from Springer Science+Business Media: *Hydrogeology Journal*, "Using groundwater levels to estimate recharge", v. 19, 2002, p. 101, R. W. Healy and P. G. Cook, Figure 9.)

the piezometer or well. Simmons *et al.* (1999) noted that short-term variations in aquifer water level are substantially attenuated within the well, particularly where the well radius is large. The degree of attenuation increases as the storativity of the aquifer decreases. Longer-term variations in aquifer water level (such as annual cycles) are less attenuated. This effect sometimes manifests itself as very smooth hydrographs in low-porosity aquifers, which do not show responses to daily rainfall events, except in extreme cases. Figure 6.8 shows a well hydrograph from a piezometer completed in the Mintaro Shale near Clare, South Australia. At this site, the fracture porosity is estimated to be about 10^{-3} on the basis of outcrop mapping, and the matrix porosity is estimated to be 0.01 to 0.05, from helium porosimetry. The hydraulic conductivity of the matrix is extremely low (<10^{-12} m/s). The water table at this site varies smoothly throughout the year; most rainfall events do not produce measurable changes in water level in the piezometer.

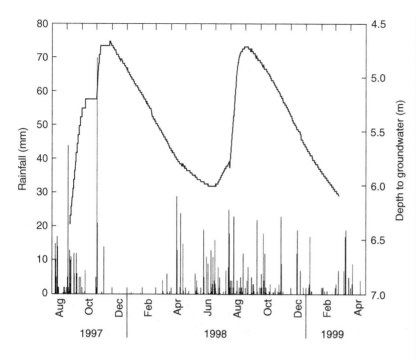

Figure 6.8 Hydrograph of water levels in a piezometer screened in the Mintaro Shale, and bar chart of daily rainfall at Clare, South Australia. The magnitude of the seasonal water-level fluctuations and the independently estimated recharge rate are consistent with a value of specific yield close to the total porosity. Short-term fluctuations in response to daily rainfall events are generally absent due to attenuation of short wavelength variations by the large storage capacity of the well. (With kind permission from Springer Science+Business Media: *Hydrogeology Journal*, "Using groundwater levels to estimate recharge", v. 19, 2002, p. 102, R. W. Healy and P. G. Cook, Figure 11.)

This is not attributed to a lack of recharge, but rather attenuation of these short-term signals by the large storage capacity of the piezometer (50 mm inside diameter) relative to that of the fractures. The annual cycle in water level is approximately 2 m, which is consistent with a recharge rate of approximately 40 mm/year and a specific yield (0.02) that is closer to the matrix porosity than to the fracture porosity.

Estimating specific yield in fractured rocks is problematic. Aquifer tests in fractured-rock systems are usually unreliable for determining S_y (Bardenhagen, 2000). The water-budget method is perhaps the most widely used technique for estimating specific yield in fractured-rock systems because it does not require any assumptions concerning flow processes. Gburek and Folmar (1999) and Heppner *et al.* (2007) used a water-budget approach to determine S_y for a fractured sandstone, siltstone, and shale system in east-central Pennsylvania. The depth to the water table was approximately 7 m. Pan lysimeters were installed at 1 to 2 m depths beneath undisturbed soil columns. These lysimeters were designed to capture all downward-moving water. The rate of water percolation measured by the lysimeters was assumed equal to the recharge rate. Water levels were measured in wells near the lysimeters, and S_y was calculated from Equation (6.2). An average value for S_y of 0.009 was obtained by Gburek and Folmar (1999) for eight events between 1993 and 1995. At that same study site, Gburek *et al.* (1999) compared the recession of well hydrographs with the base-flow recession curve for a stream draining the aquifer over a 40-day period. Through calibration of a groundwater flow model, specific yield was estimated to be 0.01 in the overburden and 0.005 in the highly fractured rocks within which most wells were screened. Moore (1992) compared stream-flow hydrographs with slopes of groundwater recession curves for shale and limestone aquifers in Tennessee and estimated a specific yield of 0.002. In spite of the efforts that have been devoted to evaluating S_y in fractured-rock systems, it remains unclear whether estimates of S_y generated for these systems are of sufficient accuracy to permit their use in estimating recharge.

Example: Beaverdam Creek Basin
Rasmussen and Andreasen (1959) studied the water budget of the Beaverdam Creek watershed on the Delmarva Peninsula of Maryland.

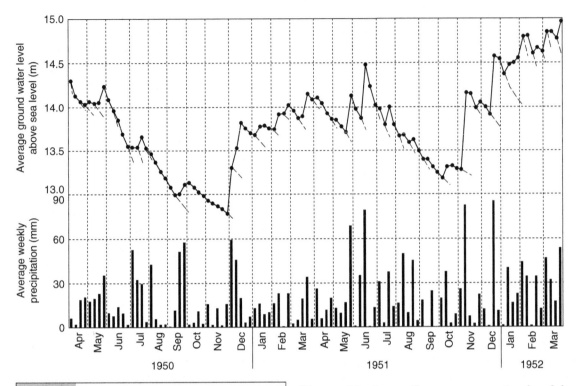

Figure 6.9 Hydrograph of average groundwater level and bar graph of weekly average precipitation for Beaverdam Creek Basin, Maryland (after Rasmussen and Andreasen, 1959). Dashed lines represent the expected level to which the water table would have receded in the absence of precipitation.

The 51 km² drainage basin ranges in elevation from 4 to 26 m above sea level and receives on average 1.09 m of precipitation annually. Beneath the basin, which is located on the Atlantic Coastal Plain, the unconsolidated sedimentary rocks consist mostly of sand, silt, clay, greensand, and shell marl. The water table is generally within a few meters of land surface. Quaternary-age surficial sands and silts of up to 22 m in thickness overlie Tertiary-age sand aquifers. Twenty-five observation wells, ranging in depth from 3 to 8 m, were distributed fairly uniformly (subject to constraints of accessibility) across the watershed. Groundwater levels were manually measured on a weekly basis. Stream discharge was monitored at the outlet of the basin. Precipitation was measured weekly at 12 sites within the basin.

Figure 6.9 shows the average water level in the observation wells and precipitation on a weekly basis. Water levels were highest in late winter and early spring. Precipitation was fairly evenly distributed throughout the year. Recharge was calculated with Equation (6.2) on a monthly basis. A water-budget approach (Equation (6.13)) was used to develop a basin-wide estimate of 0.11 for S_y. The authors noted that this value is less than the average value of 0.21 determined in the laboratory on core samples obtained from the basin and attribute this difference to inadequate time for the sediments to fully drain between successive water-level rises. ΔH was taken as the cumulative rise in water level for the month (i.e. the sum of all rises that occurred). Water levels prior to rises were extrapolated to their expected positions had there been no precipitation (dashed lines in Figure 6.9). The rise was then determined as the difference between the peak level and the extrapolated antecedent level at the time of the peak. Table 6.4 shows monthly estimates of recharge for April 1950 through March 1952. Average annual precipitation for the 2-year period

Table **6.4** Monthly groundwater recharge in millimeters for Beaverdam Creek Basin, Maryland, for April 1950 to March 1952 (Rasmussen and Andreasen, 1959).

	1950	1951	1952
January		27	82
February		42	54
March		44	107
April	23	15	
May	50	55	
June	7	74	
July	37	49	
August	7	27	
September	30	10	
October	0	23	
November	50	119	
December	72	79	

was 1.05 m; average annual recharge was estimated at 0.54 m.

Base flow, Q^{bf}, change in storage in the saturated zone, ΔS^{gw}, and evapotranspiration from groundwater, ET^{gw}, were quantified by using a modified form of Equation (6.1):

$$R = \Delta S^{gw} + Q^{bf} + ET^{gw} \qquad (6.15)$$

ΔS^{gw} was calculated as $S_y \, \Delta H_n \,/\, \Delta t$, where ΔH_n is the net change in head over each month (i.e. the difference in head between the end and the beginning of the month). Base flow was determined by streamflow hydrograph separation, and ET^{gw} was calculated as the residual of Equation (6.15). Average annual base flow was estimated to be 0.27 m; annual estimates of ΔS^{gw} and ET^{gw} were 0.02 and 0.25, respectively.

Example: Tomago sand beds

Crosbie *et al.* (2005) applied the water-table fluctuation method to estimate recharge to the Tomago sand beds, an unconfined aquifer consisting of unconsolidated aeolian deposits of fine sands, near Newcastle, Australia. The aquifer, which supplies about 25% of the potable water used in the Newcastle region, covers an area of 152 km² and has an average thickness of 18 m. Average depth to the water table is about 2 m. The aquifer is recharged by rainfall; discharge is to local surface-water bodies and to regional pumping centers. Six observation wells were instrumented with pressure transducers for monitoring water levels and rainfall gauges. Readings were made at 5-minute intervals over a period of 3 years beginning in August, 2000.

An automated time-series procedure was used to estimate recharge from the water-level and rainfall data. Exact details of the procedure are given in Crosbie *et al.* (2005); a general description is presented here. The rate of water-level recession at each well was determined by analyzing the record for water-level declines in the absence of rainfall. The recession rate was determined to be a linear function of water-table height, with higher rates for higher water-table heights. The procedure checks that rainfall precedes a water-table rise; if a rise occurs when there is no rainfall, the rise is not included in the recharge calculation. It also checks for instances of water levels being influenced by entrapped air by analyzing the rate of water-level decline after each rise. If that decline is too rapid, the height of the water-table rise is adjusted to remove the effect of entrapped air.

Three methods were used to calculate specific yield: Equation (6.4) with S_r set to residual water content as determined from laboratory measurements, analysis of aquifer pumping tests, and a water-budget approach (Equation (6.13)). This third option was determined to be the most appropriate. For calculating recharge, this value of S_y was adjusted to account for depth to the water table. Figure 6.10 shows water levels measured in one observation well along with calculated values of S_y and estimated recharge rates. The procedure in this study is attractive because it eliminates much of the subjectivity in applying the water-table fluctuation method, although estimated recharge may be sensitive to parameters for assessing the effects of entrapped air and the maximum time lag between rainfall and water-table rise.

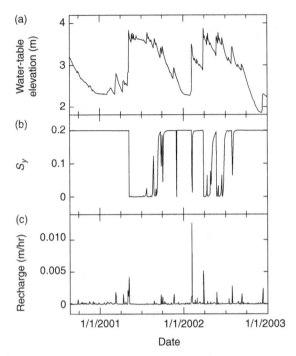

Figure 6.10 Data from observation well tapping the Tomago sand beds: (a) water-table elevation; (b) apparent specific yield; and (c) estimated recharge rate (Crosbie et al., 2005).

6.3 | Methods based on the Darcy equation

6.3.1 Theis (1937)

Theis (1937) used Darcy's equation to estimate flow through a cross section of the Southern High Plains aquifer in New Mexico. A simple water budget was constructed assuming steady conditions and no water extraction. The calculated flow rate through the cross section was divided by the contributing upgradient area to give an estimate of recharge. The approach requires estimates of saturated hydraulic conductivity (K_s), hydraulic gradient, cross sectional area, and the area up gradient from the cross section over which recharge occurs. Using a range of values of K_s determined from laboratory and aquifer tests and water levels from a network of observation wells, Theis (1937) estimated the annual recharge rate to be 3 to 7 mm. The method is easy to apply if information on K_s and water levels is available. However, determination of K_s

and the hydraulic gradient can be problematic and costly (Section 2.3.5). K_s is highly variable in space, so the method is limited by the ability to determine representative values of K_s.

6.3.2 Hantush (1956)

Leakage through aquitards is an important source of recharge for regionally extensive, deep, confined aquifers. Leakage does not conform to the definition of recharge given in Chapter 1; instead, leakage is referred to as interaquifer flow. Estimates of interaquifer flow usually rely on knowledge of groundwater levels. As pumping lowers water levels in the confined aquifer, water is drawn out of the confining units or adjoining aquifers, in accordance with Darcy's law. For steady flow of water through confining beds, Hantush (1956) expressed Darcy's law as:

$$R = Q_c / A_c = (k'/m')\Delta H \qquad (6.16)$$

where Q_c is the volumetric rate of leakage through the confining bed; A_c is the area over which leakage is occurring; (k'/m') is the leakage coefficient or leakance, where k' is the vertical hydraulic conductivity and m' is the thickness of the confining bed; and ΔH is the difference in total head between the aquifer and the top of the confining bed. The method requires measurements of water levels (head) in the aquifer and at the top or above the confining bed as well as measurements or estimates of k'. Hydraulic conductivity can be measured in the laboratory if core samples of the confining bed are available (Section 2.3.5). Piezometers in the confining bed can also be used to conduct single-borehole slug tests for estimating k' (Neuzil, 1986). Alternatively, (k'/m') can be determined through analysis of a constant-rate pumping test on the confined aquifer or from drawdown vs. distance or drawdown vs. time data in a manner similar to that described for type-curve matching (Walton, 1970; Neuman and Witherspoon, 1972).

6.3.3 Flow nets

Estimates of groundwater recharge can also be obtained by graphical analysis of flow nets for both confined and unconfined aquifers.

Although flow-net analyses have been replaced by groundwater flow models in most applications, they nonetheless deserve discussion because they have been widely used in the past. The technique was first proposed by Forchheimer (1930). A brief description is given here; details are provided in Cedergren (1988). In the approach, steady flow is assumed in a two-dimensional section (either vertical or horizontal). A flow net for a homogeneous, isotropic system consists of two sets of orthogonal lines: flow lines and equipotential lines. A flow line represents a path over which a water molecule travels from a recharge zone to a discharge zone. The region between two adjacent flow lines is termed a streamtube. An equipotential line represents a line of equal head or water level. The two sets of lines are drawn to form a series of rectangles. The flow through a rectangle, Q, can be written according to Darcy's law as:

$$Q = AK_s \, \Delta H \, / \, \Delta l \qquad (6.17)$$

where A is the cross-sectional area of one of the rectangles, K_s is hydraulic conductivity, and $\Delta H / \Delta l$ is the hydraulic gradient across the rectangle. In a simple case of a flow net for a vertical section of aquifer, recharge is inflow to the uppermost rectangle in a streamtube and is calculated as Q/A.

Example: Central North Dakota

Rehm *et al.* (1982) conducted a study on recharge to the Hagel Lignite Bed aquifer in central North Dakota. Three methods were used for estimating recharge: the water-table fluctuation method, the Hantush method, and a flow-net analysis. The 150-km^2 study area contained 175 piezometers and water-table wells. Thirty-eight observation wells were used to estimate recharge to the water-table aquifer by the water-table fluctuation method. S_y was determined from soil cores by using Equation (6.4) with S_r assumed to be equal to water content at a pressure head of −30 kPa. An average value for S_y of 0.16 was obtained. Estimates of recharge were in the range of 1 to 80 mm/year.

Ten groups of nested piezometers were used to determine vertical hydraulic gradients for

the Hantush method. Hydraulic conductivity was measured in the field at each piezometer by using single-hole slug tests. The measured hydraulic conductivity was assumed to be equal to the vertical hydraulic conductivity. Sites were located in sandy and fine-textured bedrock, as well as below two of the many sloughs that were present in the study area. Measured values of K_s ranged from 3×10^{-9} m/s for clayey bedrock to 6×10^{-6} m/s for sands. Vertical gradients ranged from 0.006 in the sands to 1.2 in the fine-textured material. Estimates of recharge ranged from 80 to 730 mm/year, rates considerably greater than those estimated from the water-table fluctuation method. The differences were attributed to natural heterogeneities within the system and inherent uncertainties in the methods.

An areal average recharge rate for the study area was obtained by averaging the water-table fluctuation method estimates and the Darcy method estimates and weighting the estimates on the basis of areal coverage of different hydrogeological settings. The average recharge rate was estimated to be between 10 and 35 mm/year.

A flow net analysis was conducted for the aquifer under the assumption that water-level data for April 1977 represented steady-state conditions. The potentiometric surface was contoured at 2 m intervals and a series of flow lines were drawn perpendicular to the contours, thereby forming rectangles of approximately equal area. The flow through each rectangle is described by Equation (6.17). Average calculated recharge rates were between 10 and 33 mm/year.

6.4 | Time-series analyses

Soil water-budget models have been developed to relate infiltration patterns to patterns of recharge and fluctuations in groundwater levels (e.g. Besbes and de Marsily, 1984; Morel-Seytoux, 1984; Bierkens, 1998; O'Reilly, 2004; Berendrecht *et al.*, 2006). The approach is to construct a soil-water budget; the calculated excess drainage from the soil zone is transported to the water table at a rate determined by a transfer-function model, a Richards equation-based model, or some similar approach. These models can be calibrated by matching model-calculated groundwater levels to measured values. These models are discussed in some detail in Sections 3.3.1 and 3.3.2.

Su (1994) developed a method for estimating recharge from knowledge of the time series of water levels in an observation well. His approach assumes one-dimensional groundwater flow in an unconfined aquifer of uniform properties overlying a sloping impervious base with spatially uniform diffuse recharge. An analytical solution was developed for the Boussinesq equation and from it an equation was derived for estimating recharge as a function of head, the derivative of head with respect to time, specific yield, and aquifer thickness, slope, and hydraulic conductivity. For the field example described by Su (1994), most terms in the analytical solution were negligible, and the solution took the form:

$$R = S_y H (\Delta H / \Delta t) / b \qquad (6.18)$$

where b is average aquifer thickness, and H is measured relative to the base of the aquifer. When there is little variation in aquifer thickness, Equation (6.18) reverts to Equation (6.2).

Ostendorf et al. (2004), using assumptions similar to those of Su (1994), developed a method for predicting steady, monthly, and annual recharge rates by using analytical solutions to the one-dimensional groundwater-flow equation. Dickinson et al. (2004) describe a method for predicting focused recharge on the basis of groundwater level fluctuations at different distances from the point of recharge; assumptions included one-dimensional flow from the point of focused recharge, uniform aquifer properties, and no additional recharge to the aquifer along the flow line. Dickinson et al. (2004) used an analytical solution to the groundwater-flow equation.

Predictions of depth to water table and groundwater-level fluctuations have also been made with empirical time series models on the basis of daily or monthly records of water level, hydraulic gradient, precipitation, or precipitation intensity (Viswanathan, 1984; Tankersley et al., 1993; Coulibaly et al., 2001; Jan et al., 2007). These time series models make no attempt to represent the physics of water storage and movement in the subsurface. Although the models were not developed specifically for estimating recharge, the water-table fluctuation method can be applied with any time series of predicted groundwater levels. These empirical models can be useful in management of groundwater resources in regions where little aquifer information is available (Coulibaly et al., 2001).

6.5 | Other methods

Other approaches based on groundwater level data have been used to study aquifer recharge and discharge. Streamflow hydrograph recession analysis (Chapter 4) is a popular approach for determining base flow, but the analysis of groundwater hydrographs has received much less attention. Shevenell (1996) analyzed water-level recession curves to determine values of specific yield for different elements of a karst aquifer. Salama et al. (1994a) used groundwater hydrograph separation techniques to estimate evapotranspiration rates. Ketchum et al. (2000) used spring discharge measurements to separate a groundwater hydrograph and estimate recharge in a small watershed. Schicht and Walton (1961) combined stream discharge measurements and measured groundwater levels to develop a rating curve for estimating base flow as a function of groundwater level. Moench and Barlow (2000) and Barlow et al. (2000) describe a method for using groundwater level and stage data for a fully penetrating stream to estimate bank storage and discharge and basinwide recharge rates (Section 4.3.3).

In the work of van der Kamp and Schmidt (1997), water levels in fine-grained geologic units were found to respond to changes in soil-water storage much like weighing lysimeters. Water-level measurements in fine-grained units, or in confined aquifers (Sophocleous et al., 2006; Bardsley and Campbell, 2007; Rasmussen and Mote, 2007), can be used to estimate changes in water storage. The method, sometimes referred to as a geological weighing lysimeter, relies on the principle that changes in mechanical load (attributed to changes in water storage) are reflected in changes in groundwater pressures. Accurate measurements of groundwater levels are required, and the measurements must be corrected for the effects of atmospheric pressure and Earth tides. Barr *et al.* (2000) indicated that results obtained from measurements at a depth of 34.6 m in a single well were representative of an area of about 10 ha, a much larger area than that represented by a typical weighing lysimeter. Geological weighing lysimeters do not provide direct estimates of recharge; instead, they provide estimates of changes in surface and subsurface water storage, much like gravity measurements (Sections 2.3.3 and 2.5.3). The use of geological weighing lysimeters for calibration of gravity measurements was suggested by Bardsley and Campbell (2007). Estimated water storage changes could be a useful part of a water-budget approach for estimating recharge and evapotranspiration.

6.6 | Discussion

Recharge estimation methods that are based on measurements of groundwater levels are widely used because of the ease with which they can be applied and the abundance of available data in local, state, and federal databases. Recent technological advances have improved the accuracy, reduced the expense, and eased the complexity of making automatic measurements and recordings of groundwater levels.

Within any watershed, recharge rates vary with location because of differences in precipitation, vegetation, soil properties, land use, geology, and other factors. Obtaining an average recharge rate requires multiple wells that are distributed across the watershed in a manner that accounts, as much as is practical, for this variability. Groundwater levels can be affected by phenomena other than recharge, including atmospheric pressure, pumping, evapotranspiration, tides, and surface loads. Measured water levels should be adjusted to account for these phenomena. Difficulties in estimating specific yield contribute to the overall uncertainty of the water-table fluctuation method, yet it remains one of the most widely used methods for estimating recharge. In light of the uncertainty inherent in any method for estimating recharge, the desirability of applying multiple estimation methods, and the favorable features cited above, any hydrologic study would likely benefit from careful analyses of available groundwater-level data.

Chemical tracer methods

7.1 | Introduction

Tracers have a wide variety of uses in hydrologic studies: providing quantitative or qualitative estimates of recharge, identifying sources of recharge, providing information on velocities and travel times of water movement, assessing the importance of preferential flow paths, providing information on hydrodynamic dispersion, and providing data for calibration of water flow and solute-transport models (Walker, 1998; Cook and Herczeg, 2000; Scanlon et al., 2002b). Tracers generally are ions, isotopes, or gases that move with water and that can be detected in the atmosphere, in surface waters, and in the subsurface. Heat also is transported by water; therefore, temperatures can be used to trace water movement. This chapter focuses on the use of chemical and isotopic tracers in the subsurface to estimate recharge. Tracer use in surface-water studies to determine groundwater discharge to streams is addressed in Chapter 4; the use of temperature as a tracer is described in Chapter 8.

Following the nomenclature of Scanlon et al. (2002b), tracers are grouped into three categories: natural environmental tracers, historical tracers, and applied tracers. Natural environmental tracers are those that are transported to or created within the atmosphere under natural processes; these tracers are carried to the Earth's surface as wet or dry atmospheric deposition. The most commonly used natural environmental tracer is chloride (Cl) (Allison and Hughes, 1978). Ocean water, through the process of evaporation, is the primary source of atmospheric Cl. Other tracers in this category include chlorine-36 (^{36}Cl) and tritium (^{3}H); these two isotopes are produced naturally in the Earth's atmosphere; however, there are additional anthropogenic sources of them.

Historical tracers refer to tracers that have been widely introduced to the atmosphere by human activity. These are sometimes referred to as anthropogenic tracers (Cook and Böhlke, 2000). Included in this category are ^{3}H and ^{36}Cl (Figure 7.1), which were produced by nuclear weapons testing in the 1950s and early 1960s (Phillips et al., 1988; Scanlon, 1992; Cook et al., 1994); iodine-129 (^{129}I) from nuclear reprocessing and the Chernobyl accident (Englund et al., 2008); and gases such as chlorofluorocarbons (CFCs) and sulfur hexafluoride (SF$_6$), which are generated by industrial activities (Plummer and Busenberg, 2000; Plummer, 2005). As with environmental tracers, non-gaseous historical tracers arrive at the land surface as wet or dry atmospheric deposition and enter the subsurface in a dissolved phase. Gaseous tracers, on the other hand, are assumed to diffuse from the atmosphere through the unsaturated zone and to dissolve in groundwater at the water table.

Applied tracers are those that are applied directly to land surface (or in some instances in the shallow subsurface) as a result of human activity. The application can be specifically for purposes of conducting a tracer experiment

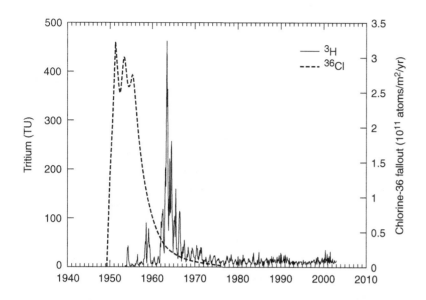

Figure 7.1 Chlorine-36 fallout based on Zerle *et al.* (1997) and atmospheric tritium concentrations from Ottawa, Canada (International Atomic Energy Agency, http://www-naweb.iaea.org/napc/ih/; accessed April 1, 2009), decay corrected to 2008.

(Rangarajan and Athavale, 2000; Wang *et al.*, 2008), in which case bromide (Br), Cl, and fluorescent dyes are commonly used. In other cases, use of applied chemicals as tracers may be secondary to other purposes (e.g. fertilizers and pesticides can be used to track water movement in the subsurface).

Ideal chemical and isotopic tracers are highly soluble in water, behave conservatively in the environment (i.e. they are chemically inert with no subsurface sources or sinks), and can be measured accurately and inexpensively. In addition, in the case of applied tracers, background concentrations in the system should be low. There is no *perfect* tracer that can be used in all studies. Theoretically, however, any chemical or isotope could be used as a tracer. ^{3}H is a useful tracer because it is part of the water molecule. Inorganic anions, such as Cl and Br, often are used as tracers because they are usually not affected by adsorption or other geochemical or biochemical processes.

The nature of tracer application is important and in part determines the manner in which subsurface tracer concentrations are interpreted. Mass application rates can be relatively constant over time. Many studies have relied on the assumption that the atmospheric deposition rate of Cl has been constant over the last several thousand years (Scanlon, 1991; Phillips, 1994). Tracer application can occur at a single point in time, as is usually the case with applied tracers; this is referred to as a pulse application. Tracer application can also vary in time, such as with gases like CFCs, SF$_6$, and krypton-85 (^{85}Kr) whose atmospheric concentrations continually change. CFC concentrations increased from the 1930s through the mid-to-late 1990s, after which gradual declines have been observed (Figure 7.2). Whether a tracer application can be considered a single pulse is a somewhat relative question depending on the time lag between application and sampling. As previously mentioned, ^{3}H and ^{36}Cl were released to the atmosphere in the 1950s and early 1960s. More than 40 years later, these isotopes are commonly thought of as occurring as a pulse input (the term *bomb-pulse tritium* frequently appears in the hydrologic literature), even though the application period extended for more than a decade. It is conceivable that by 2050, atmospheric concentrations of CFCs will have diminished to a point whereby hydrologists will treat their introduction to a hydrologic system as a pulse in time.

Tracer methods have some important attributes that need to be considered in light of the goal of a recharge study. To clarify terminology, estimates of vertical flux within the unsaturated zone are referred to as *potential recharge* or *drainage* because the term *recharge*, by definition, is reserved only for water that

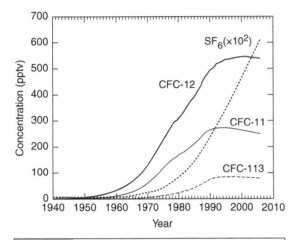

Figure 7.2 Concentrations of CFC-11, CFC-12, CFC-113, and SF$_6$ in dry air in the northern hemisphere (International Atomic Energy Agency, 2006), in parts per trillion by volume (pptv).

rates (or tracer input) of environmental and historical tracers must be estimated, often on the basis of data collected for relatively short time periods at a small number of sites.

It is not possible to discuss every tracer that has been used in hydrologic studies or all aspects of tracer sampling, chemical analysis, and data interpretation. Instead, generic approaches, which are applicable to many different tracers, are described in some detail in this chapter, first as they pertain to tracers within the unsaturated zone and then to tracers in groundwater. Characteristics of some of the more commonly used tracers are discussed, and examples are used to illustrate approaches.

7.2 | Tracers in the unsaturated zone

Most tracer methods are based on one or more assumptions on how water moves through the unsaturated zone; the validity of these assumptions should be closely scrutinized. At depths greater than the zero-flux plane, water generally moves vertically downward. (The zero-flux plane, ZFP, as described in Chapter 5, is the horizontal plane in the subsurface that marks the divide between water moving upward in response to evapotranspiration and water draining downward; the ZFP is somewhat synonymous with the bottom of the root zone.) Variability in soil properties and layering of geologic units, however, can induce some horizontal movement. Water flow within the unsaturated zone is often assumed to be governed by the Richards equation (Section 3.3.2). Tracers are transported by moving water and by mixing due to hydrodynamic dispersion and diffusion, as described by the advection–dispersion equation. Figure 7.3 shows a concentration-depth profile that would be predicted by these equations for steady infiltration of tracer-free water following a pulse application of tracer at land surface. The smooth theoretical curves are not always supported by field data. Concentration-depth profiles determined from field data typically resemble theoretical profiles, but there

actually arrives at the water table (Chapter 1). The term *drainage* is used throughout this chapter to denote vertical flux in the unsaturated zone. Most tracer methods are designed to determine the time required for a tracer to move a certain distance in the subsurface, resulting in an effective tracer velocity that typically is equated with the water velocity. To calculate a flux rate (drainage or recharge), the velocity is multiplied by volumetric water content (unsaturated zone) or porosity (saturated zone); therefore, water content or porosity must be determined along with tracer concentrations to estimate flux. Flux rates determined from tracer tests represent values that are integrated over the time interval between tracer application and sampling and the distance between the point of application and the point of sampling. Environmental tracers such as Cl can be used to determine average drainage rates over centuries to millennia in arid regions (Phillips, 1994; Tyler *et al.*, 1996). In humid regions, applied tracer tests can provide estimates of drainage rates over periods of days to months to years. Analysis of unsaturated-zone tracers usually provides a point drainage estimate in space, whereas analysis of tracers in groundwater provides an estimate that is integrated over a larger, but sometimes ill-defined, area. Application

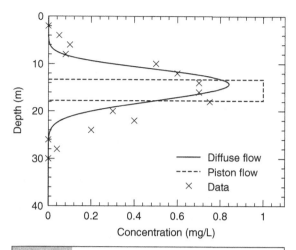

Figure 7.3 Hypothetical tracer concentration-depth profiles in the unsaturated zone after pulse application of a tracer at land surface followed by infiltration of tracer-free water: typical field data (×), theoretical profile (solid line) based on the Richards and advection–dispersion equations, and piston-flow profile (dashed line).

usually is some scatter in the actual data points as a result of microscopic variability in velocities (caused by heterogeneity in soil properties) and uncertainty in analytical methods for determining tracer concentrations (Figure 7.3). The piston-flow assumption provides a simplified explanation of water and tracer movement; it states that water moves vertically downward through the unsaturated zone, pushing existing water and solute to greater depths in the soil column with no mixing or variation in velocity (Figure 7.3). True piston flow probably never occurs in real hydrologic systems, but the assumption is inherent in some methods to be discussed, and the term *piston-like flow* is commonly used to represent uniform flow with little dispersion. In contrast, the term *preferential flow* refers to nonuniform downward movement of water in which some water moves rapidly along preferred pathways, such as decayed roots and fractures, and bypasses much of the matrix. Multiple peaks in tracer concentration-depth profiles have been attributed to the presence of preferential flow paths (Scanlon and Goldsmith, 1997).

There are three general approaches for using tracers in the unsaturated zone to estimate velocities of water movement: the tracer-profile method, the peak-displacement method, and the mass-balance method. Inherent to all of these methods are assumptions that water and tracers move vertically within the unsaturated zone. Profile and mass-balance methods can be applied by obtaining tracer concentration-depth data at a single point in time, but these methods require information on the rate and timing at which tracers are introduced at land surface. Peak-displacement methods do not require information on tracer inputs, but concentration-depth data must be obtained at two different times.

The tracer-profile method can be used to analyze tracers that are introduced as a single pulse at land surface. Thus, this method is applicable for the study of most applied tracers and some historical tracers (such as ^{3}H and [36]Cl). Measurements of tracer concentration throughout a vertical profile of the unsaturated zone at some point in time after the pulse was introduced are used to determine the penetration depth of the tracer (z_T). The equivalent depth of water in the profile between land surface and z_T represents the total water flux since tracer introduction. The idea is to determine the depth of penetration of the tracer, measure soil-water content between land surface and z_T, and determine the drainage rate, D, as:

$$D = \int_0^{z_T} \theta \, dz \, / \, \Delta t \tag{7.1}$$

where θ is volumetric water content, Δt is the time interval between tracer introduction at land surface and subsurface sampling, and the integral represents the mass of water in the unsaturated zone column above z_T. The penetration depth is usually set equal to either the depth of peak tracer concentration or the depth of the center of tracer mass (Walker, 1998):

$$z_T = \int_0^{\infty} z C(z)\theta(z) dz \, / \, M \tag{7.2}$$

$$M = \int_0^{\infty} C(z)\theta(z) dz$$

where $C(z)$ is tracer concentration in pore water and M is total mass of tracer in the subsurface. Figure 7.4 describes a hypothetical application

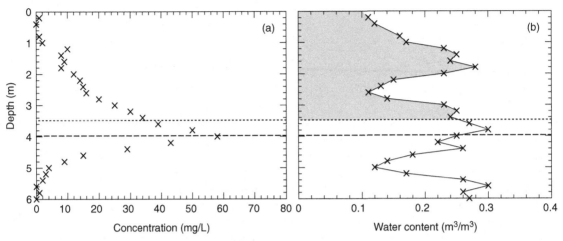

Figure 7.4 Hypothetical tracer concentration-depth profile (a) and water-content profile (b) in the unsaturated zone are used to describe application of the tracer-profile method for estimating drainage rates. Depth of peak concentration is 4 m (dashed line); depth of center of mass, as calculated with Equation (7.2), is 3.46 m (dotted line). The mass of water in the soil column (the integral term in Equation (7.1)) is equivalent to 0.64 m above the 3.46 m depth (represented by shaded area) and 0.79 m above the 4 m depth. Assuming 20 years between tracer-application and sampling dates, drainage rates calculated with Equation (7.1) are 32 mm/yr for z_T = 3.46 m and 40 mm/yr for z_T = 4 m. Evaluation of integrals in these equations is by the trapezoid rule as described in Figure 5.3.

of the profile method. If transport is by piston flow, similar drainage rates can be calculated by using either the depth of the peak or the center of mass. Where preferential flow occurs, drainage rate should be based on the depth of the center of tracer mass (Walker, 1998; Cook *et al.*, 1994). The integrals in Equations (7.1) and (7.2) can be evaluated with the trapezoid rule on the basis of water-content and tracer-concentration measurements at discrete depths, as described in Figure 5.3.

The integral in Equation (7.1) extends from $z = 0$ (land surface) to z_T and so includes the root zone. Soil-water content in the root zone can vary widely over the course of a year, a fact that complicates drainage calculations. Typically, only a small portion of infiltrating water in natural or agricultural settings drains past

the ZFP; most of that water is returned to the atmosphere via evapotranspiration. Ideally, tracers would be introduced at or below the ZFP, and the integral in Equation (7.1) would extend from the depth of the ZFP to z_T. Only applied tracers provide the option of injecting tracers below land surface. If the root zone is thin relative to the depth of tracer movement, variable water contents in the root zone will have little impact on estimated drainage rates. However, when tracer peaks are located above the ZFP, such as has been found for ^3H and ^{36}Cl in arid regions (Norris *et al.*, 1987; Phillips *et al.*, 1988; Scanlon, 1992), it is not possible to accurately quantify long-term average drainage rates (Tyler and Walker, 1994). For situations in between these extremes, Equation (7.1) should be modified to use average annual soil-water contents for depths above the ZFP and water contents measured at the time of sampling for greater depths.

The peak-displacement method is similar to the profile method. Profiles of tracer concentrations and soil-water contents are obtained at two times. Vertical tracer velocity, v, is calculated by dividing the change in penetration depth (Δz_T) by the length of time (Δt) between the two profile samplings:

$$v = \Delta z_T / \Delta t \qquad (7.3a)$$

Drainage rate is equal to velocity multiplied by average volumetric water content:

$$D = v\theta \qquad (7.3b)$$

This approach requires the collection of samples at two times, but no information on tracer input is required. The approach is particularly useful for analyzing a tracer that was accidentally introduced to the environment, such as through a contaminant spill, and for which release dates and amounts are unknown.

Inherent to Equation (7.3b) is the assumption that velocity is constant over time. The displacement method also can be useful in evaluating changes in drainage rates due to changes in land use (Cook et al., 1994; Stonestrom et al., 2004; Scanlon et al., 2007). Following a change in land use there will be a time period before new equilibrium conditions are fully established. Walker et al. (1991) proposed an alternative equation to determine drainage during that period. This equation was originally developed to describe displacement of a front of uniform Cl concentration as a result of removal of native vegetation in a semiarid region of Australia:

$$D = \left[\int_{z_{T0}}^{z_{T1}} \theta(z)dz + \int_{z_{ZFP}}^{z_{T0}} \Delta\theta(z)dz + \frac{C_n}{C_b} \int_0^{z_{ZFP}} \Delta\theta(z)dz \right] / \Delta t \tag{7.4}$$

where z_{T0} is initial depth of the front, z_{T1} is current depth of the front, z_{ZFP} is depth of zero-flux plane, $\Delta\theta$ is the difference between current and initial water content, C_b is the initial concentration in the front, C_n is the concentration of the new equilibrium front, and Δt is the time between land-use change and sampling. The first term on the right side of Equation (7.4) is the integral form of Equation (7.3b). The second term accounts for the change in water content in the depth interval between the zero-flux plane and the initial front, and the third term accounts for the change in chloride mass within the root zone. In practice, Equation (7.4) is not often applied, but it is included here for completeness.

The mass-balance method relates the mass of tracer in the unsaturated zone with the rate at which the tracer arrives at land surface. As with the profile method, tracer concentration and soil-water content depth profiles and tracer source term information are required. The method can be used for constant or varying tracer introduction. For the simplest constant-source case, we can write:

$$D = PC_P / C_{uz} \tag{7.5}$$

where P is precipitation rate, C_p is average concentration in precipitation, and C_{uz} is average concentration in unsaturated zone pore water. Equation (7.5) is quite simplistic. Precipitation is assumed to be the sole source of the tracer. All precipitation infiltrates the surface; there is neither runoff nor runon, and the tracer is conservative. For a variable tracer source, the equation for estimating drainage takes on a more complex form (Heilweil et al., 2006):

$$D = \int_0^\infty C_{uz}(z)\theta(z)dz / \sum_{i=1}^\infty w_i C_{Pi} \Delta t_i \tag{7.6}$$

where the integral represents the total mass of tracer in the unsaturated zone, the summation represents the total mass added to the unsaturated zone (again under the assumption that precipitation is the only source of tracer), w_i is a weighting factor to correct for variations in annual drainage, C_{Pi} is tracer concentration in precipitation i time increments before sampling, and Δt_i is time increment corresponding to the periods of application (typically, 1 year for historical tracers and days for applied tracers). An obvious drawback to this approach is the need to know the time history of application. This method can be used with historical tracers and with tracers applied specifically for tracing water or with those applied as part of some ongoing operation (e.g. agricultural chemicals, such as nitrate). Allison (1987) recommends use of Equation (7.6) because it is not overly influenced by preferential flow and because it often requires fewer sampling depths than profile methods.

7.2.1 Tracer sampling: unsaturated zone

Tracer concentrations can be determined from soil samples collected throughout the depth interval of interest or by installing instruments more or less permanently at fixed depths to allow sampling or analysis of pore water. Each approach has benefits and limitations. Obtaining soil cores allows data points to be collected at many different depths and thus provides detailed spatial information. Installed instruments usually provide limited depth resolution but allow frequent data collection.

Soil samples can be obtained by using core barrels with traditional auger-type or pneumatic-type drill rigs. Samples can also be collected manually with a soil auger. No drilling fluids should be used, and because soil-water content must be determined, the samples should be sealed in sample containers immediately after they are collected to prevent evaporation. An extraction procedure is used to analyze tracer concentrations in soil samples. Deionized water is mixed with dried soil in a mass ratio of generally 1:1 to 10:1. The mixture is placed on a reciprocal shaker for a prescribed period of time, then spun in a centrifuge at 7000 rpm. Finally a filtered sample (usually a 0.45 μm filter is used) is withdrawn from the supernatant for chemical analysis (McMahon *et al.*, 2003).

Installed instruments do not provide the depth resolution that can be obtained by collecting core samples (expense usually determines the maximum number of instrumented depths), but they allow a much higher frequency of sample collection. Soil-suction lysimeters (sometimes called soil-solution samplers) or gravity-drain lysimeters are often used to collect pore water. However, pore water samples can be obtained only if ambient soil matric potentials are greater than –100 kPa. This restriction limits lysimeter use to humid and semihumid regions, in general. Specific conduction probes can be installed in the unsaturated zone to monitor tracer concentrations over time without the need for physically collecting samples. Time domain reflectometry (TDR) probes also have been used to determine specific conduction and anion concentrations (Kachanoski *et al.*, 1992; Mallants *et al*, 1996; Hart and Lowery, 1998). One concern with the use of permanently installed probes is that recalibration of the probes may not be possible.

Whichever approach is used to determine tracer concentrations, soil-water content also must be measured. If soil samples are collected, water content can be determined gravimetrically; soil samples are weighed, dried in an oven at 105°C for 24 hours, and then reweighed.

Alternatively, TDR or resistance probes can be installed in the subsurface to provide continuous measurements of water content. A neutron moderation meter can be used to measure water content at any depth in the unsaturated zone, but a borehole access tube is required, and the meter must be operated manually. Additional information on measuring soil-water content can be found in Section 5.2.1 and in Dane and Topp (2002).

7.2.2 Natural environmental tracers
Chloride
Meteoric chloride is an excellent tracer of water movement in the subsurface. It is conservative because of its anionic form (it does not adsorb onto negatively charged silicates). It is highly soluble in water (the solubility of sodium chloride is about 220 000 mg/L; Holser, 1979), and it generally does not participate in geochemical or biochemical reactions. Chloride is carried to the atmosphere by evaporation from oceans. Atmospheric chloride is deposited onto land surface with precipitation or in the form of dry deposition. Chloride moves with infiltrating water into the subsurface. Chloride concentrations in pore water tend to increase with depth through the root zone as a result of evapotranspiration because plants exclude chloride when they take up water and water returned to the atmosphere through bare-soil evaporation is pure. According to the conceptual model proposed by Gardner (1967), steady-state conditions exist within the unsaturated zone and chloride concentrations are uniform beneath the zero-flux plane (Figure 7.5a). Drainage rates are inversely proportional to this uniform chloride concentration; high drainage rates flush chloride through the system quickly, resulting in a relatively low concentration, whereas low drainage rates allow chloride concentrations to build. The chloride mass-balance (CMB) method is based on this model.

The unsaturated zone CMB method for estimating drainage is derived from a water-budget equation for a column of soil extending to a depth greater than that of the zero-flux plane:

$$P + Q^{sw}_{on} = ET + Q^{sw}_{off} + D \qquad (7.7)$$

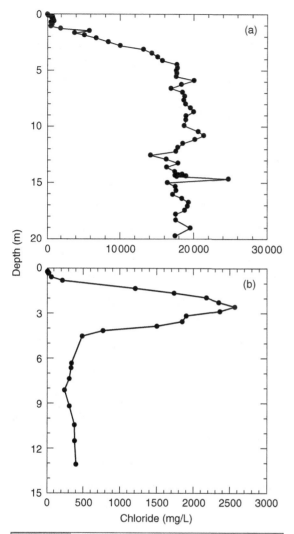

Figure 7.5 Subsurface distribution of chloride beneath native vegetation at sites in (a) the Murray Basin, Australia (reprinted from *Journal of Hydrology*, v. III, Cook et al. (1989), Figure 3, copyright (1989), with permission from Elsevier); and (b) the Chihuahuan Desert, West Texas (reprinted from *Journal of Hydrology*, v. 128, Scanlon (1991), Figure 2, copyright (1981), with permission from Elsevier).

where P is precipitation, Q^{sw}_{on} is surface flow onto the column (including irrigation), Q^{sw}_{off} is surface flow off of the column, D is drainage out of the bottom of the column (potential recharge), and it is assumed that there is no subsurface flow through the sides of the column and that steady-state conditions exist, so there is no change in water storage within the column. A chloride mass-balance equation can

be obtained by associating a chloride concentration with each component of the water-budget equation:

$$PC_p + Q^{sw}_{on}C^{sw}_{on} = Q^{sw}_{off}C^{sw}_{off} + DC_{uz} + M_{app} \quad (7.8)$$

where C_p, C^{sw}_{on}, C^{sw}_{off}, and C_{uz} are chloride concentrations in precipitation, surface-water flow onto the column, surface-water flow off of the column, and unsaturated zone pore water at the base of the column. M_{app} is the rate of chloride mass application to the column in dry form and accounts for additional sources of chloride that can be natural (e.g. atmospheric dry deposition) or related to human activity (e.g. agricultural chemicals or road salt). A number of assumptions are inherent in Equation (7.8) including no change in the amount of chloride stored within the column, no plant uptake of chloride, no additional sources of chloride in the column (such as dissolution of rocks), no subsurface sinks of Cl (such as by adsorption), and vertical downward movement of water and chloride within the column. Equations (7.7) and (7.8) are readily modified to account for other terms that contribute or remove water or chloride from the column.

The CMB estimate of drainage is obtained by rearranging Equation (7.8):

$$D = (PC_p + Q^{sw}_{on}C^{sw}_{on} - Q^{sw}_{off}C^{sw}_{off} - M_{app})/C_{uz} \quad (7.9)$$

Application of Equation (7.9) requires measurements of chloride concentrations in precipitation, surface water, and unsaturated-zone pore water, along with an estimate of dry-source application rates. The large number of terms in Equation (7.9) complicates application of the CMB method in areas where many of the terms are nonnegligible. The method is most often applied to estimate natural drainage in arid and semiarid regions where irrigation, runon, and runoff are negligible and the chloride mass-balance equation can be simplified to:

$$D = \frac{C_p P + M_{app}}{C_{uz}} = \frac{PC_p^*}{C_{uz}} \quad (7.10)$$

where C_p^* is effective chloride concentration (the sum of total wet and dry chloride deposition rate divided by precipitation).

The numerator on the right-hand side of Equation (7.10) represents total chloride deposition at land surface; under natural conditions this includes chloride in precipitation and dry fallout. Rates of both wet and dry deposition can be measured with modified precipitation collectors. Unfortunately, however, only limited historical data on these rates are available, an important source of uncertainty for applying the CMB method. The National Atmospheric Deposition Program (https://nadp.isws.illinois.edu; accessed December 13, 2009) provides isopleth maps of annual average chloride concentrations in precipitation across the US for 1994 through to the present (data for some sites in the network are available from as early as 1978). No such data network exists for chloride deposition in dry fallout, so these rates are largely unknown. Chloride concentrations in precipitation display a distinct spatial trend of decreasing concentrations with increasing distance from oceans. Dry deposition rates are more spatially variable because of local sources such as wind-blown dusts and salts. In addition, winds can remove dust and salts from an area, so net deposition may differ from measured total deposition. Dettinger (1989) found dry and wet deposition rates to be similar in Nevada. In southern California, dry Cl deposition can be up to four times that of wet Cl deposition (J. A. Izbicki, US Geological Survey, pers. comm., 2003). Among the many CMB studies that have been conducted, there appears to be little consistency in approaches for estimating dry chloride deposition rates. Simple assumptions are often made, such as dry deposition is negligible (Edmunds et al., 2002) or dry deposition is equal to wet deposition (Dettinger, 1989; Nolan et al., 2007; Gates et al., 2008; Healy et al., 2008).

As an alternative to direct measurement of chloride deposition, a one-time profile sampling of soils and measurement of natural pore water ^{36}Cl/Cl concentration ratios (i.e. ratios that are not affected by nuclear weapons testing) can be used to estimate long-term chloride deposition at a site. The ^{36}Cl approach for estimating chloride input involves dividing the natural ^{36}Cl fallout at a site, which varies according to latitude and can be obtained from Phillips (2000) or Moysey et al. (2003), by the measured natural ^{36}Cl/Cl ratio. This approach has been used at several sites in the United States (Phillips et al., 1988; Scanlon and Goldsmith, 1997). Scanlon (2000) found uniform natural ^{36}Cl/Cl concentration ratios in unsaturated-zone pore water over a wide range in depths in different regions in the southwestern United States. The implication of these results was that chloride deposition rates have remained fairly stable over time.

Sediment samples for chloride analyses are generally collected from boreholes. Water content is determined and chloride is extracted with deionized water as described in Section 7.2.1. Chloride analyses can be conducted with a specific ion electrode or ion chromatography. In areas where electrical conductivity (Ec) is highly correlated with chloride concentrations, Ec measurements can be conducted and calibrated against chloride measurements. Chloride concentration per unit of dry sediment mass is converted to chloride concentration in pore water by dividing by the gravimetric water content and multiplying by the pore-water density.

Numerous applications of the unsaturated zone CMB method are found in the literature (Dettinger, 1989; Phillips, 1994; Flint et al., 2002); these were mostly conducted in arid and semiarid regions. Gates et al. (2008) applied the method with data from 18 boreholes to examine spatial variability in potential recharge rates in desert terrain in northern China; average potential recharge was estimated to be about 1 mm/yr. Edmunds et al. (2002) applied the CMB method to sites in northern Nigeria; estimated potential recharge rates were between 14 and 49 mm/yr. Nolan et al. (2007) applied the method in humid and subhumid regions of the eastern United States, requiring consideration of additional chloride sources, as given in Equation (7.9).

Land-use change, such as replacement of native vegetation with agricultural crops, may result in a change in drainage rate through the unsaturated zone. Such an occurrence violates the steady-state assumption of the CMB method, although after a period of time a new steady-state system may develop. In southern Australia, Walker et al. (1991) found that the

CMB approach underestimated drainage when steady-state conditions were not quickly reestablished following land-use change; for such cases, they suggested drainage be estimated with Equation (7.4). In contrast, Scanlon *et al.* (2007) found the standard CMB method to be applicable at sites in the southern High Plains of Texas that were converted from native vegetation to crops. In this region, the chloride bulge that accumulated beneath native vegetation was displaced downward in the unsaturated zone in response to increased drainage under cropland, indicating piston-like flow.

The peak-displacement method also can be used to estimate the change in drainage rate resulting from a land-use change. Soil-core samples were obtained and analyzed from fields under native eucalyptus and from nearby agricultural fields in Victoria, Australia (Allison and Hughes, 1983). Equation (7.3) was used with Δz_r set equal to the difference in depth of the chloride front under native vegetation and that under cropland (1 to 1.5 m); Δt was set equal to the time between land-use change and profile sampling, 65 years. Multiplying the calculated velocities by an average water content of 0.2 produced drainage estimates of 3 to 5 mm/yr beneath the agricultural field.

Unsaturated zone chloride concentration-depth profiles are useful qualitative tools for identifying areas where drainage is occurring in arid and semiarid regions. In the Southern High Plains in Texas, Scanlon and Goldsmith (1997) found much lower chloride concentrations beneath playas than beneath interplaya settings, indicating that the playas are an area with much higher infiltration rates. The profiles can also serve as archives of past climates. Bulge-shaped chloride profiles (Figure 7.5b) in semiarid settings in the southwestern United States are attributed to high water fluxes during the Pleistocene when the climate was cooler and wetter, followed by greatly reduced fluxes during the Holocene period (Scanlon, 1991; Phillips, 1994; Tyler *et al.*, 1996). Results of numerical modeling of water and chloride movement within thick unsaturated zones over a period of 15 000 years support this hypothesis (Walvoord *et al.*, 2002a,b; Scanlon *et al.*, 2003).

The time required to accumulate chloride or the age of chloride and, by implication, that of the water at a particular depth, z_1, may be useful information in some studies. The chloride mass-balance age, t_1, can be calculated by dividing the integrated mass of chloride in the unsaturated zone from the surface to z_1 by the annual chloride input flux:

$$t_{1(z1)} = \frac{\int_0^{z_1} C_{uz}\theta dz}{PC_P^*} = \frac{\int_0^{z_1} C_{uz}\rho_b dz}{PC_P^*} \quad (7.11)$$

where C_{uz} is chloride concentration in terms of mass of chloride per unit mass of dry soil, and ρ_b is soil bulk density. The chloride mass-balance age is independent of soil-water content, and its calculation, therefore, is not affected if samples dry out prior to analysis.

In analyzing the CMB method, several points should be considered. Uncertainties in chloride deposition rates result in corresponding uncertainties in estimated water fluxes because of the linearity of Equation (7.10). Chloride profiles are generally insensitive to preferential flow if chloride deposition is constant over time. Unsaturated zone concentration-depth profiles of bomb-pulse tracers such as ^{36}Cl and ^3H are generally more affected by preferential flow (Scanlon, 2000). Drainage estimates can be obtained with a one-time collection of soil samples (although many depths need to be sampled), and chloride analysis is relatively inexpensive. There is no minimum drainage rate that can be estimated, so the method is widely used in arid regions. Because estimated drainage rates are inversely related to chloride concentration, the maximum rates that can be estimated are limited by the analytical detection limits for chloride. For this reason and because chloride sources other than atmospheric deposition may need to be considered, the unsaturated zone CMB method is not often applied in humid settings.

Example: Impacts of land use change on groundwater recharge in the southern High Plains and the Murray Basin

Environmental chloride has been used to estimate impacts of changing from natural ecosystems to agricultural ecosystems on drainage rates through the unsaturated zone in the Southern High Plains in the United States (Scanlon *et al.*, 2007) and in the Murray Basin in Australia (Jolly *et al.*, 1989). Both locations are characterized by high chloride concentrations under native vegetation, which indicate no drainage during the last 10 000 to 30 000 years in the Southern High Plains and drainage rates less than 0.1 mm/yr in the Murray Basin. Downward displacement of the high chloride concentrations under increased recharge beneath cropland and pastures is used to estimate the change in recharge rates. In the case of the Southern High Plains, piston displacement of the preexisting chloride allows recharge rates to be estimated with the CMB method, Equation (7.9). The chloride bulge was displaced by a distance of 7.7 m (Figure 7.6a). The average chloride concentration in the flushed portion of the profile is 5.3 mg/L (Scanlon *et al.*, 2007). An average water flux of 24.4 mm/yr was estimated by using mean annual precipitation of 461 mm/yr and an effective chloride concentration in precipitation of 0.28 mg/L. The peak displacement method can also be applied. Assuming that the change in land use occurred 75 years before samples were obtained, an apparent velocity of 103 mm/yr is calculated. With an average volumetric water content of 0.19, Equation (7.3) provides a drainage estimate of 19.5 mm/yr.

The chloride concentration peak was displaced downward by about 2.7 m at a site in the Murray Basin (Figure 7.6b) over a period of 9 years since native vegetation was removed and replaced with dryland agricultural crops (Jolly *et al.*, 1989). Chloride concentrations under the native vegetation were fairly uniform at about 10 000 mg/L at depths greater than 2 m. Applying the CMB method (Equation 7.9) with average rainfall of 370 mm and an effective chloride concentration in precipitation of 2.3 mg/L produces an average annual drainage estimate of about 0.08 mm. The peak displacement method was used to examine the drainage rate in the cleared field. The 2.7 m movement of the chloride peak over 9 years corresponds to an apparent velocity of 0.3 m/yr. At an average volumetric water

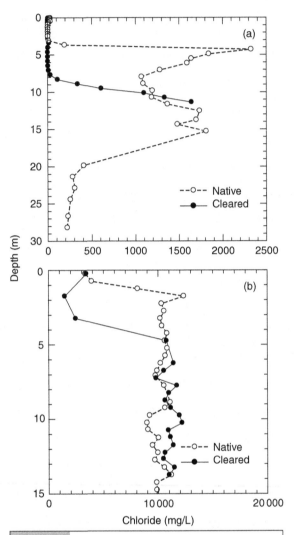

Figure 7.6 Comparison of chloride profiles beneath native vegetation and cropland at sites in (a) the southern High Plains, Texas (Scanlon *et al.*, 2007) and (b) the Murray Basin, Australia (reprinted from *Journal of Hydrology*, v. 111, Jolly *et al.* (1989), Figures 1 and 2, copyright (1989), with permission from Elsevier).

content of 0.15, this velocity corresponds to a drainage rate of 45 mm/yr.

Stable isotopes of oxygen and hydrogen

The stable isotopes of oxygen (^{18}O) and hydrogen (^{2}H or D, deuterium) are valuable tracers in many types of hydrologic studies (Kendall and McDonnell, 1998). They can be used in some regions to identify the source and timing of recharge, and they can provide valuable information on evaporation rates and flow processes in the unsaturated and saturated zones. However, attempts to use unsaturated-zone stable isotope concentration profiles to quantify drainage rates have met with little success. Concentrations of these isotopes in water, which are usually given as ratios of a standard and are referred to as del values, depend on temperature. Because of this temperature dependence, seasonal variations in del values of precipitation are common (Gat, 1996). If these variations persist in infiltrated water as it moves downward through the unsaturated zone, then the depths of the cyclical patterns can be used with the profile method (Equation (7.1)) to estimate drainage rates. The best results (i.e. the most discernible variations in del values through the soil profile) have been obtained in temperate climates where drainage through the unsaturated zone occurs during a large part of the year (Barnes and Allison, 1988). But even under apparently ideal conditions, patterns of del values with depth do not always occur (Sharma and Hughes, 1985), perhaps because of molecular diffusion. Regions where drainage occurs only in a single season would not be expected to show much variation in del values within the unsaturated zone. Detailed sampling and stable-isotope analysis of precipitation is required if one were to use stable-isotope concentration profiles to estimate drainage rates.

Del values are also affected by evaporation, due to fractionation, once precipitation arrives at land surface. In arid regions, evaporation of soil water can produce a unique isotope signature; subsequent infiltration events may transport that front downward, and the front could serve as an event marker for estimating drainage rates when using the profile method (Barnes and Allison, 1983).

7.2.3 Historical tracers
Tritium
The distribution of bomb-pulse tritium (^{3}H) in the unsaturated zone can be used to estimate drainage rates with any of the methods described in Section 7.2, the tracer-profile, peak-displacement, or mass-balance method. Tritium

is a good tracer because, being contained in a water molecule, it is conservative and because concentrations can be accurately measured. Tritium was widely used as an unsaturated-zone tracer in the 1970s and 1980s in all climatic regions. Tritium use in unsaturated zone studies since about 2000, however, has largely been limited to arid and semiarid regions. Because of relatively thin unsaturated zones and high drainage rates in more humid regions, tritium, chlorine-36, and other historical tracers produced by nuclear testing have been flushed from unsaturated zones in many areas.

Tritium has a half-life of 12.32 years (Lucas and Unterweger, 2000) and is naturally produced by cosmic-ray neutrons interacting with nitrogen in the upper atmosphere. Subsurface production of ^{3}H is low (Lehmann et al., 1993). Natural atmospheric ^{3}H concentrations prior to nuclear testing in the 1950s were estimated from vintage wines at 3 to 6 tritium units (TU) for Europe and North America (Kaufman and Libby, 1954) and at 1 to 3 TU for southern Australia (Allison and Hughes, 1977). (One tritium unit is equivalent to one tritium atom in 10^{18} hydrogen atoms or 0.118 becquerels per liter.) Atmospheric tritium concentrations at Ottawa, Canada, increased from these levels to more than 400 TU during above-ground nuclear testing (International Atomic Energy Agency, 1983) that began in 1952 and peaked in 1963 (Figure 7.1). Because most nuclear tests were conducted in the northern hemisphere, ^{3}H concentrations were much greater in the northern than in the southern hemisphere: 3278 TU at Ottawa, Canada in 1963 vs. 38 TU at Kaitoke, New Zealand in 1964 (Solomon and Cook, 2000). Information on ^{3}H input in precipitation is available from a worldwide network of measurement stations; online data are maintained by the International Atomic Energy Agency (http://www-naweb.iaea.org/napc/ih/IHS_resources_ISOHIS2.html; accessed April 1, 2009).

Pore water for tritium analysis can be obtained from soil cores or, in humid regions, from suction lysimeter or other pore water samplers. Azeotropic or vacuum distillation is required to collect water from soil cores (Hendry, 1983; Ingraham and Shadel, 1992). Few commercial, academic, or government laboratories are capable of doing the water extraction and tritium analysis. Tritium can be analyzed by radiometric methods (liquid scintillation or gas proportional counting) or by helium-3 (^{3}He) ingrowth (Clarke *et al.*, 1976; Schlosser *et al.*, 1989). Radiometric methods require about 20 mL of sample and have a detection limit of about 6 TU. Electrolytic enrichment of samples can reduce that detection limit to about 0.1 TU; however, the enrichment process requires much larger volumes of water, up to hundreds of milliliters (Ostlund and Dorsey, 1977). The ^{3}He ingrowth method of measuring ^{3}H concentrations involves degassing of the water sample under vacuum and sealing and storing the water in a special gas-tight container for time periods ranging from 1 month to 1 year, resulting in detection limits down to 0.1 TU or less (Clarke *et al.*, 1976). The ^{3}He ingrowth method for analyzing ^{3}H in unsaturated pore-water samples should not be confused with the ^{3}H/^{3}He groundwater dating method (Section 7.3.3). Dating groundwater with ^{3}H/^{3}He requires isolation of the ^{3}He from the atmosphere, which only occurs below the water table (Solomon *et al.*, 1992).

The presence of ^{3}H levels greater than about 0.5 TU in a water sample generally indicates that some of the sample is from post-1952 precipitation (assuming maximum prebomb ^{3}H of 7 TU and a residence time of 55 years or 4.5 half lives). Many studies have estimated unsaturated zone drainage rates on the basis of depth of bomb-pulse tritium (Dincer *et al.*, 1974; Aranyossy and Gaye, 1992; Heilweil *et al.*, 2006; Lin and Wei, 2006). The tracer-profile method (Equation (7.1)) and peak-displacement method (Equation (7.3)) can be applied by using the depth of the peak concentration or the center of mass (Equation (7.2)). In each case, Δt is set equal to the time between peak atmospheric concentration of ^{3}H (1963–4) and sampling time. Heilweil *et al.* (2006) applied the profile method on the basis of depths of peak concentrations in 11 boreholes in southwestern Utah. Gvirtzman and Magaritz (1986) used differences in ^{3}H concentrations in irrigation water (summer) and rainfall (winter) to identify 14 years of recharge

cycles in an unsaturated zone tritium-depth profile at a site in Israel. The peak-displacement method was applied for each of the 14 concentration peaks. Normally, this method requires sampling at two different times. However, a single sampling date sufficed for this study because of the seasonal variability of tritium input at land surface; in effect, 14 tracer experiments were analyzed with a single sampling date. The average vertical velocity for all peaks was 0.7 m/yr, and drainage was estimated as 8% of precipitation and irrigation.

The use of tritium in the mass-balance approach requires modification of Equation (7.6) to account for radioactive decay (Allison *et al.*, 1994; Cook *et al.*, 1994; Heilweil *et al.*, 2006):

$$D = \int_0^\infty C_{uz}(z)\theta(z)dz \Big/ \sum_{i=1}^\infty w_i C_{pi} \Delta t e^{-i\lambda} \qquad (7.12)$$

where $C_{uz}(z)$ is now tritium concentration at the time of sampling, i refers to number of years prior to the time of sampling, w is equal to precipitation during year i (P_i) divided by long-term mean annual precipitation, C_{pi} is tritium concentration in precipitation i years prior to the time of sampling, Δt is 1 year, and λ is the decay constant for tritium (0.0565 yr^{-1}). The decay correction allows concentrations to be expressed relative to the time of sampling. The mass-balance equation for tritium has been expressed in an alternative form by Heilweil *et al.* (2006):

$$D = \frac{M_{UZ}}{M_{PPT}} P \qquad (7.13)$$

where M_{UZ} is ^{3}H mass in the unsaturated zone at the sampling site, equal to the numerator on the right-hand side of Equation (7.12); M_{PPT} is ^{3}H mass input to the site from precipitation, equal to the summation in the denominator of Equation (7.12) with each term multiplied by P_i, and P is mean annual precipitation. The difference between ^{3}H mass in precipitation and the unsaturated zone represents ^{3}H and, therefore, water that was evapotranspired. Heilweil *et al.* (2006) found that the tritium mass-balance drainage estimates were less than estimates determined with the tritium profile method, perhaps because of seasonal variability in tritium concentration in precipitation.

Bomb-pulse ^{3}H can also provide some insight into flow processes in the unsaturated zone. Penetration of bomb-pulse ^{3}H to about 12 m in the Negev Desert was attributed to preferential flow in fractured chalk (Nativ *et al.*, 1995). Tritium concentrations beneath playas in the Texas High Plains provide examples of piston and preferential flow (Wood and Sanford, 1995; Scanlon and Goldsmith, 1997). The tritium concentration-depth profile beneath a playa in the southern part of the Texas High Plains is smooth and has a single peak (Figure 7.7), indicating predominantly piston-like flow. In contrast, a profile beneath another playa to the north contains multiple peaks, the deepest at 21 m, suggesting preferential flow.

Example: Inner Mongolia

Unsaturated-zone tritium concentration profiles were determined with soil cores collected in 1988 and 1997 in a loess-covered region of Wudan County, Inner Mongolia (Lin and Wei, 2006). The site was free of vegetation, and mean annual precipitation was 360 mm. The 0.1 to 0.4 m sample intervals to depths of 15 to 20 m provide an unusually detailed profile of ^{3}H concentration (Figure 7.8). The profiles are noteworthy for several reasons: the distinct peaks, the high concentrations, the fact that tritium mass (when corrected for decay) is about equal for the two sampling dates, and the fact that the 1997 profile shows more dispersion (a feature that suggests non-piston-type flow).

Peak concentrations were 338 TU at the 6.3 m depth for the 1988 profile and 230 TU at the 10.4 m depth for the 1997 profile (all values are decay corrected to 1997). These high ^{3}H concentrations indicate that this region received relatively high ^{3}H fallout from nuclear weapons testing. Lin and Wei (2006) estimate that maximum atmospheric tritium concentration in northern China may have been as high as 4000 TU.

Pore-water velocities were estimated from the penetration depths of the ^{3}H peaks at 0.25 m/yr (1988) and 0.3 m/yr (1997). Mean annual drainage rates of 38 and 45 mm/yr were calculated by the profile method (Equation (7.1)) with an average volumetric water content of

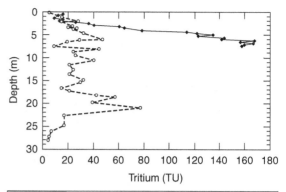

Figure 7.7 Representative tritium profiles beneath playas. The smooth, shallow profile, beneath a site in the southern Texas High Plains, suggests predominantly piston-like flow (Wood and Sanford, 1995); the deeper profile, from a site further north, has multiple peaks suggesting preferential flow (Scanlon and Goldsmith, 1997).

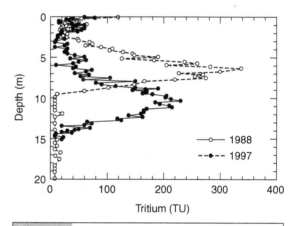

Figure 7.8 Subsurface distribution of tritium in profile CHN88 sampled in 1988 (peak depth 6.3 m) and profile CHN97 sampled in 1997 (peak depth 10.4 m) in the Loess Plateau, Inner Mongolia (reprinted from *Journal of Hydrology*, v. 328, Lin and Wei (2006), Figure 2, copyright (2006), with permission from Elsevier).

0.15. These fluxes represent 11 to 13% of mean annual precipitation, a substantial percentage for semiarid regions.

The peak-displacement method (Equation (7.3)) can also be used to estimate tracer and water velocity from the Lin and Wei (2006) data. The peak was displaced 4.1 m over the course of 9 years, equivalent to a velocity of about 0.46 m/yr. Multiplying this velocity by volumetric water content (0.15) produces a drainage estimate of 69 mm/yr. The difference in results of the two methods can perhaps be explained by the different time intervals over which the drainage estimates are made. The peak-displacement method is averaged over 9 years, whereas the tracer-profile method is averaged over 25 years (1988) and 34 years (1997). The fact that average annual precipitation for 1989 through 1997 (about 390 mm) was slightly higher than the long-term average could also contribute to the difference.

Chlorine-36

Chlorine-36 (half-life of 301 000 years) is produced naturally in the atmosphere by cosmic-ray spallation of ^{36}Ar and neutron activation of ^{35}Cl (Bentley et al., 1986). ^{36}Cl was also introduced into the atmosphere by neutron activation of ^{35}Cl as a result of nuclear weapons testing between 1952 and 1958 in the Pacific Ocean (Figure 7.1). As a tracer, ^{36}Cl is similar to ^3H in several respects; both have bomb-pulse sources, both are conservative in nature, and data interpretation is similar. Much of the discussion on tritium in the previous section is pertinent for ^{36}Cl as well. There are some important differences, though. ^{36}Cl is nonvolatile and its movement in the subsurface is restricted to liquid phase water movement, whereas tritium can move in the liquid and vapor phases (Scanlon, 1992). Tritium may be taken up by plant roots and returned to the atmosphere via evapotranspiration; plants roots tend to reject ^{36}Cl and chloride in general. The peak in atmospheric ^{36}Cl concentration occurred in 1957, about 6 years prior to that of tritium. Tritium has been much more widely used than ^{36}Cl in hydrologic studies, perhaps because historically there have been few laboratories capable of analyzing for ^{36}Cl. Sample preparation and analysis of ^{36}Cl are also expensive (Scanlon, 1992).

The ^{36}Cl concentration-depth profile must be measured to locate the depth of the peak or center of mass; the soil-water content depth profile also must be determined. Sampling procedures for ^{36}Cl are similar to those for chloride. Soil samples are usually collected throughout the interval of interest in the unsaturated zone, and chloride is extracted with deionized water, as described in Section 7.2.1. A total of 1 to 10 mg of chloride is required for the analysis. In areas of low chloride concentrations (which usually are areas of high drainage rates), hundreds of grams of soil may be required to provide sufficient amounts of chloride for ^{36}Cl analysis. The analysis is done by accelerator mass spectrometry; the cost for chemical preparation of the extract sample and ^{36}Cl analysis can be up to $1000 per sample. Currently (2010), there are three laboratories where ^{36}Cl can be analyzed: Purdue University (http://www.physics.purdue.edu/primelab/; accessed April 2, 2009), Lawrence Livermore National Laboratory (https://cams.llnl.gov/; accessed April 2, 2009), and the Australian National University (http://wwwrsphysse.anu.edu.au/nuclear/group.php?id=6; accessed April 2, 2009).

Drainage rates can be estimated from the depth distribution of ^{36}Cl by using the methods described above for tritium (Scanlon, 1992; Liu et al., 1995; Tyler et al., 1996). Prych (1998) compared drainage estimates based on ^{36}Cl measurements (using the profile method and penetration depth of the center of mass) with those based on the chloride mass balance; the estimates based on ^{36}Cl were higher by a factor of about 10 at sites in eastern Washington State. The reason for this discrepancy is not clear. Cook et al. (1994) compared drainage estimates from ^{36}Cl mass-balance and chloride mass-balance methods at sites in southern Australia; they found variability in results, but no consistent trend between estimates by the two methods. The surprising penetration of bomb-pulse ^{36}Cl to depths exceeding 400 m at Yucca Mountain, Nevada, has been attributed to preferential flow through faults or fractures in the volcanic rock (Campbell et al., 2003).

7.2.4 Applied tracers

Applied tracers are generally used in areas of relatively high rates of infiltration and drainage because tracer movement is evaluated over fairly short times (from days to one or more years). Tracers are usually applied on land surface as a single pulse, and they are transported downward through the unsaturated zone by infiltration of precipitation. The infiltration rate should be high enough to move the tracer through the root zone quickly so as to minimize the waiting time between tracer application and profile sampling. The infiltration rate required to transport the tracer through the root zone can be estimated from Equation (7.3) (Walker, 1998). For example, if a typical root zone is 1 m thick and the average water content is 0.1 m^3/m^3, a water flux of 100 mm/yr through the base of the root zone would transport the applied tracer through the root zone in 1 year. Water fluxes of this magnitude are common in humid regions and in semiarid regions where focused flow occurs (e.g. beneath streams, lakes, or irrigated fields).

Tracers can also be applied at depths below the zero-flux plane where downward flow is relatively steady (Athavale and Rangarajan, 1990). This approach may present some logistical difficulties (i.e. a borehole or trench may be required), but it avoids the complications of dealing with changing water contents within the root zone, and the profile can be sampled more quickly after application. Care should be taken that the injection process does not alter the system being evaluated (particularly for structured soils) and result in recharge rates that are an artifact of the tracer-injection process.

Anions, in general, are very useful hydrologic tracers (Herczeg and Edmunds, 2000). Bromide is perhaps the most popular applied tracer because it has low natural background concentrations relative to those of other anions such as chloride and nitrate (Davis et al., 1985). Bromide does not undergo chemical transformations, whereas nitrate can be lost by denitrification or produced by mineralization and nitrification. Nitrate is also extracted from soil by plants. Bromide can be taken up by plants (Kung, 1990a; Bowman et al., 1997), but to a much lesser degree. Plant uptake of chloride is very low. All anions, including bromide, may move slightly faster than water because of anion exclusion (repulsion of anions from negatively charged solid particles) (Gvirtzman et al., 1986). Bromide toxicity is low (Flury and Papritz, 1993). Bromide concentrations can be measured by a

specific ion electrode or by ion chromatography in the laboratory (Frankenberger et al., 1996). Bromide can be applied as Ca(Br)$_2$, KBr, or NaBr in solid or dissolved form; a typical application amount is 15 g Br/m^2 (Fisher and Healy, 2008). Use of NaBr should be avoided in fine-grained soils because of the potential swelling of clays.

Although introduction of ^{3}H to the environment is prohibited in many countries, it has been widely used as an applied tracer in India and China (Rangarajan and Athavale, 2000). Tritium is introduced below the root zone at the base of injection boreholes. Soil samples are obtained from boreholes drilled 1 to 2 years after injection. Water is extracted from the soil samples and analyzed for ^{3}H. Drainage rates determined with the peak-displacement method for 25 study areas in India ranged from 24 to 198 mm/yr, or 4 to 20% of precipitation (Rangarajan and Athavale, 2000). A study on the Hebei Plain of China involved injection of bromide (100–200 g NaBr) and ^{3}H in testholes at 39 sites to examine the influence of soil type, land use, and agricultural practices on recharge rates (Wang et al., 2008). The tracers were injected at depths of 0.6 to 1 m; 1 or 2 years after injection, soil samples were collected from the depth of injection down to the water table (or to a maximum depth of 5 m). Water was extracted from the soil samples and analyzed for tracers. Drainage rates were estimated with the peak-displacement method; they ranged from 0.0 to 1.0 mm/d or 0 to 43% of water input (precipitation and irrigation). Bromide was found to move deeper than tritium, on average; anion exclusion was cited as a possible explanation.

Tracer application is sometimes accompanied with water application by sprinkler or ponding (Bowman and Rice, 1986; Ghodrati and Jury, 1990). These types of experiments are designed to study infiltration rates associated with irrigation or extreme precipitation events or, alternatively, to examine subsurface flow processes. Dye tracers are particularly useful for identifying preferential flow paths in the unsaturated zone (Ghodrati and Jury, 1990; Kung, 1990b; Delin et al., 2000). Dye tracer experiments can be conducted over relatively short periods of time, days to weeks. Dyes, such as rhodamine

WT, are applied evenly over a test plot that is typically one to several square meters in surface area. The plot is usually irrigated, although not in all studies. After a set period of time or after a certain amount of water has been applied, the soil is carefully removed in layers of predetermined thickness (e.g. 200 mm) and detailed photographs are used to document tracer distribution in each exposed layer. The process is repeated layer by layer until a final depth is reached. The process produces a quasi-three-dimensional picture of tracer distribution in the subsurface. Dye tracer experiments can provide useful qualitative information on the importance of preferential flow paths, but they cannot be used by themselves to determine quantitative flow rates for individual flow paths.

Many other chemicals and isotopes have been used as applied tracers. These include stable isotopes (McConville *et al.*, 2001), lithium (Rosqvist and Destouni, 2000), ^{15}N (Nielsen *et al.*, 1997), and microspheres (Burkhardt *et al.*, 2008). Kung *et al.* (2000) used three fluorobenzoic acids as tracers in a study of preferential flow through agricultural soils; water samples for tracer analysis were obtained from tile drains in that study.

Studies described in this section all used soil cores for tracer analyses. Fixed-depth instruments (Section 7.2.1) can also be used to track the movement of applied tracers within the unsaturated zone. Fixed-depth instruments allow for high-frequency data collection. Analysis of fixed-depth instrument data is concerned with determining tracer arrival time at specific depths (as opposed to determining tracer penetration depth at specific times, which is the aim of soil core sampling). The profile method (Equation (7.1)) and peak-displacement method (Equation (7.3)) can both be applied with fixed-depth instrument data. For both methods, the depth of penetration, z_T, is known, and Δt, the time between tracer application and arrival, must be determined. Tracer arrival time, $t_T(z_1)$, can be set to the arrival time of the peak concentration or the arrival time of the center of tracer mass:

$$t_T(z_T) = \int_0^\infty tC(z_T)\theta(z_T)dt \Big/ \int_0^\infty C(z_T)\theta(z_T)dt \qquad (7.14)$$

For the peak-displacement method, velocity is determined as distance traveled divided by the travel time. For example, if tracer concentration data were collected at depths z_1 and z_2, velocity is calculated as:

$$v = (z_2 - z_1)/[t_T(z_2) - t_T(z_1)] \qquad (7.15)$$

and drainage can be calculated as velocity times volumetric water content.

7.3 | Groundwater tracers

Approaches for using groundwater tracer data to estimate recharge are theoretically similar to those based on unsaturated-zone data. There are some differences, however, in terms of underlying assumptions and implementation. Unsaturated-zone tracers provide a point estimate, in terms of space, of drainage or potential recharge; groundwater samples provide an estimate of actual recharge that is integrated over some larger but often poorly defined area. Unsaturated-zone tracer analyses rely on the assumption of vertical downward flow, an assumption whose validity is seldom questioned. Analyses of groundwater tracer data generally require no assumptions with regard to flow within the unsaturated zone (thus, estimates are usually not affected by the presence of preferential flow paths in the unsaturated zone). However, age-dating methods require assumptions with regard to the travel paths that groundwater takes from the water table to a sampling point. Many tracers used in unsaturated-zone studies are also used in groundwater studies, including chloride, ^3H, and ^{36}Cl. Other tracers are specific to saturated-zone applications (e.g. dissolved gas tracers, such as chlorofluorocarbons (CFCs), SF_6, and carbon-14, ^{14}C). Contaminants in groundwater can sometimes be used as tracers. Groundwater samples are usually more easily obtained than unsaturated-zone pore-water samples, and there is a wealth of groundwater chemistry data available in national databases. Tracer data collected in both the unsaturated and saturated zones can be used to calibrate water-flow and solute-transport models for estimating recharge.

Profile, peak-displacement, and mass-balance methods can be applied with groundwater tracer data as described in Section 7.2. Groundwater age-dating methods can also be used to estimate recharge. The profile and peak-displacement methods, described in Section 7.3.3, are based on assumptions of a pulse tracer input (such as bomb-pulse ^{3}H or ^{36}Cl) and vertical groundwater flow. Groundwater samples need to be collected at several depths to define the concentration-depth profile. These requirements preclude application of profile and peak-displacement methods at many sites. The mass-balance method is often applied with groundwater data, most often on the basis of chloride or tritium concentrations. The approach is similar to that described for unsaturated-zone application in Section 7.2.3 and is discussed in Section 7.3.2. Groundwater age-dating techniques are described in the following section. Applied tracers are used in many groundwater studies to determine groundwater velocities and travel paths. However, applied tracers are not commonly used to quantify recharge rates, and, therefore, they are not addressed herein.

7.3.1 Age-dating methods

Age refers to the elapsed time since a tracer moved across the water table and became isolated from the atmosphere, its presumed source. We commonly refer to the age of a groundwater sample; however, it is the age of the dissolved tracer, not the water itself, that is calculated from tracer concentrations. It is more appropriate to use the term *apparent age* when referring to water. Differences may exist between the actual and apparent ages of water because various physical and chemical processes can affect tracer concentrations. These processes include mixing of water, adsorption of tracers onto solids, biodegradation of tracers, entrapment of air, and the presence of alternate tracer sources (especially for CFCs). Plummer (2005) discussed these complications in some detail and described techniques for assessing their importance (e.g. measuring concentrations of a suite of noble gases in the water sample).

Age-dating methods rely on a simple equation to estimate recharge (Böhlke, 2002):

$$R = \varphi v_v \tag{7.16}$$

where φ is porosity and v_v is vertical water velocity at the water table. Apparent age dates of groundwater, determined at one or more sampling points below the water table, and an assumption on the travel path of water from the water table to the sampling points are used to estimate v_v. The following paragraphs provide an analysis of assumptions related to the use of apparent groundwater ages to estimate recharge. Subsequent sections provide specific information for some of the more widely used tracers. Additional details can be found in Kendall and McDonnell (1998), Cook and Herczeg (2000), and the US Geological Survey CFC Laboratory website (http://water.usgs.gov/lab; accessed April 10, 2009).

A number of compounds that have been introduced to the atmosphere by human activity since the 1930s (e.g. ^{3}H, CFCs, SF$_6$, and ^{85}Kr) are used to date groundwater with ages less than 50 to 60 years. Tritium was discussed in Section 7.2.3. The other compounds are gases whose atmospheric concentrations change over time (Figure 7.2). The apparent groundwater age is determined by analyzing the concentrations of these gases in groundwater and then finding (after possible corrections) the year, from Figure 7.2, whose concentration matches that of the sample. That year is referred to as the recharge date; apparent age is the difference between sampling date and recharge date.

This age-dating approach relies on the assumption that tracer gases become dissolved in recharging water right at the water table and that gas concentrations in the unsaturated zone above the water table at the time of recharge are equal to atmospheric levels. Thus, time lags for the movement of gases from land surface to the water table are ignored. This assumption may be appropriate for relatively thin unsaturated zones where depth to the water table is a few meters or less. But for thicker unsaturated zones, in which diffusion is the dominant gas-transport mechanism (Weeks et al., 1982), this assumption can introduce important errors into the calculation of groundwater ages. Cook and Solomon (1995) determined that the time lag associated with diffusion of CFCs through a coarse-textured unsaturated zone is 1 to 2 years for a 10 m water-table depth and 8 to 15 years for a 30 m water-table depth. In practice, recharge dates for systems with thick unsaturated zones may need to be adjusted to account for this lag time. The tritium/helium-3 (^{3}H/^{3}He) method, which does not require information on atmospheric concentrations, and the use of radionuclide tracers with long half lives, such as ^{14}C, to date very old waters are not affected by this lag time.

Any groundwater sample represents a distribution of different ages, particularly in heterogeneous media (Maloszewski and Zuber, 1982; Weissmann et al., 2002). Mixing of water of different ages occurs naturally in groundwater systems due to molecular diffusion and hydrodynamic dispersion. Mixing can also be an artifact of sampling procedure. For purposes of estimating recharge, minimal mixing is desired. Sampling locations and sampling procedures need to be selected with this concept in mind. In areas of recharge, deeper water within an aquifer is generally older, and therefore more mixed, than shallower waters. Sampling depths near the water table should minimize natural mixing. To avoid artificial mixing while sampling, small screened intervals are desirable so that only a small volume of the aquifer is sampled, approaching as near as possible a point sample. Delin et al. (2000) used screen lengths of 30 mm. Screen intervals of a meter or more provide an integrated (as opposed to a point) sample, with waters of multiple ages being possibly mixed together. Tracer concentrations determined on samples collected from large screened intervals may be useful for determining proportions of old and young waters and identifying groundwater flow paths, but water samples with mixed ages are not ideal for estimating recharge rates. Analysis of multiple gas tracers can be used to assess the degree of mixing that a sample may have experienced (Aeschbach-Hertig et al., 1999).

The collection of groundwater samples for tracer analysis often involves detailed procedures and special collection vessels to avoid exposing the sample to contamination from the atmosphere. Specific procedures may be required for each tracer. The CFC laboratory of the US Geological Survey provides sampling instructions for all of the tracers mentioned above and several other tracers (http://water.usgs.gov/lab; accessed April 10, 2009). It is best to contact the laboratory that will be doing the analysis for sampling instructions prior to initiating the sampling. Improper sampling is a major cause of sample rejection.

Estimation of recharge from age dating of groundwater relies on the simple assumption of piston flow (Cook and Solomon, 1997); young water moves downward displacing older water with little mixing or dispersion. Recharge rates must be relatively high to use dissolved gas concentrations for age dating; Cook and Solomon (1997) suggested a minimum of about 30 mm/yr, and Böhlke (2002) suggested a minimum of about 100 mm/yr. At lower rates, dispersion dominates advection and the validity of the piston-flow assumption is questionable.

Determining a value for v_v in Equation (7.16) requires an assumption on the way in which ages are distributed with depth in an aquifer. Cook and Böhlke (2000) provided equations for estimating velocity and recharge for piston flow in generic aquifer types. For a hypothetical unconfined aquifer of constant thickness, Z, underlain by an impermeable layer, groundwater age contours are horizontal when recharge is uniform (Figure 7.9). In this highly idealized flow system, age increases with logarithm of depth from zero at the water table to infinity at the base of the aquifer (where vertical velocity equals zero), and recharge can be calculated from an age measurement, t, at a given depth below the water table, z, as (Böhlke, 2002):

$$R = \varphi\, Z \ln(Z/(Z-z))/t \qquad (7.17)$$

where ln is natural logarithm. The dependence of age on logarithm of depth has been found to be a reasonable assumption in a number of studies (Figure 7.10), but other assumptions are possible as well. Solomon et al. (1993) determined that ages increased linearly with depth (i.e. flow was essentially vertical) within the top 15 m of

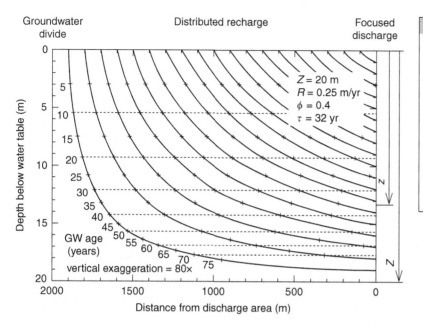

Groundwater divide

Distributed recharge

Focused discharge

$Z = 20$ m
$R = 0.25$ m/yr
$\phi = 0.4$
$\tau = 32$ yr

GW age (years)

vertical exaggeration = 80×

Depth below water table (m)

Distance from discharge area (m)

Figure 7.9 Groundwater flow paths and groundwater age distribution in a cross section of a hypothetical unconfined aquifer (Böhlke, 2002). Flow lines (solid) and age isochrons (dotted) were calculated by assuming homogeneous aquifer properties and a uniform recharge rate, R, with parameters listed on the figure (Z is aquifer thickness, φ is porosity, and τ is average groundwater age).

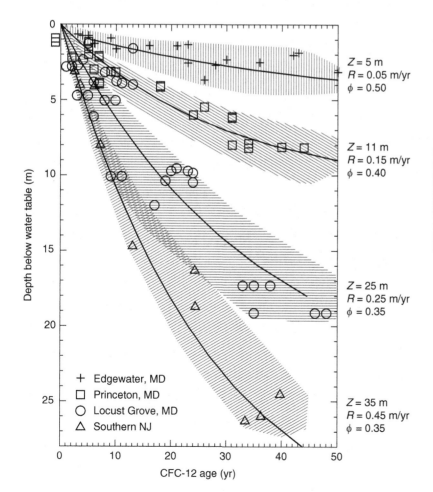

$Z = 5$ m
$R = 0.05$ m/yr
$\phi = 0.50$

$Z = 11$ m
$R = 0.15$ m/yr
$\phi = 0.40$

$Z = 25$ m
$R = 0.25$ m/yr
$\phi = 0.35$

$Z = 35$ m
$R = 0.45$ m/yr
$\phi = 0.35$

Depth below water table (m)

+ Edgewater, MD
□ Princeton, MD
○ Locust Grove, MD
△ Southern NJ

CFC-12 age (yr)

Figure 7.10 Apparent groundwater ages derived from CFC-12 concentrations in groundwater samples collected from unconfined aquifers in the eastern United States (Böhlke, 2002). Shading highlights trends in different aquifers. Recharge was calculated by using Equation (7.17) with parameters listed on right side of the figure (Z is aquifer thickness, φ is porosity, and R is average estimated recharge).

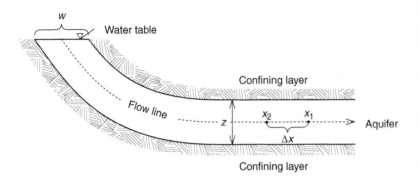

Figure 7.11 Schematic showing hypothetical aquifer cross section to illustrate calculation of groundwater velocity from Equations (7.19) and (7.20). w is width of recharge area where aquifer is unconfined; Z is thickness of confined part of aquifer; x_1 and x_2 are sampling points; and Δx is distance between x_1 and x_2.

the silty sand aquifer at the Sturgeon Falls study site, Ontario, Canada; Equation (7.3) was used to estimate recharge. Groundwater ages can be used to estimate recharge rates in less idealized systems as well, if the recharge location and flow pathway for the sampled water are known. Groundwater flow models are useful tools for identifying groundwater flow paths and assessing the validity of assumptions on age distributions and tracer transport in general (Robertson and Cherry, 1989; Engesgaard *et al.*, 1996; Zhu, 2000; McMahon *et al.*, 2004).

When mixing of ages is apparent in groundwater samples, whether due to natural processes or artifacts of sampling, Equations (7.16) and (7.17) are not directly applicable. However, age-date data can be used in an age mass-balance approach to estimate average recharge for the aquifer:

$$R = \varphi Z / \tau \qquad (7.18)$$

where τ is average age or residence time of water in the aquifer (Böhlke, 2002). Equation (7.18) is analogous to the chloride mass-balance Equation (7.10). Many of the considerations discussed for chloride mass-balance methods are relevant here as well.

Age dates can also be used to estimate horizontal velocity, v_h, in a confined aquifer, in which flow lines are horizontal (Figure 7.11):

$$v_h = (Age_1 - Age_2) / \Delta x \qquad (7.19)$$

where Age_i is apparent groundwater age at point x_i along a flow line at the mid distance between confining layers and Δx is the distance between points x_1 and x_2. A mass-balance approach,

similar to the flow-net analysis described in Section 6.3.3, can then be used to relate v_h to v_v in the unconfined regions of these aquifers, where recharge is occurring (Cook and Böhlke, 2000):

$$wv_v = v_h Z \qquad (7.20a)$$

$$v_v = v_h Z / w \qquad (7.20b)$$

where w is width of recharge area and Z is thickness of the confined part of the aquifer. Equation (7.20a) simply states that, within a vertical cross section aligned with flow lines of the aquifer, the vertical flow of water across the water table is equal to the horizontal flow through the aquifer. Inherent to the equation are assumptions of steady flow and no additional sources or sinks of water movement to or from the aquifer. Recharge can be calculated by inserting the value of v_v from Equation (7.20b) into Equation (7.16).

7.3.2 Natural environmental tracers
Chloride
The groundwater chloride mass-balance (CMB) method is similar to the unsaturated-zone CMB method described in Section 7.2.2; the column used for derivation of Equations (7.8) to (7.10) now extends to the water table, and in those equations drainage (D) is replaced by recharge (R) and chloride concentration in unsaturated-zone pore water (C_{uz}) is replaced by chloride concentration in groundwater (C_{gw}). Equation (7.10) takes the form:

$$R = \frac{C_p P + M_{app}}{C_{gw}} = \frac{P C_p{}^*}{C_{gw}} \qquad (7.21)$$

This approach was first used to estimate recharge in a study in Israel (Eriksson and Khunakasem, 1969). An inherent assumption in Equation (7.21) is that all of the chloride within the aquifer is derived from atmospheric deposition. However, additional chloride sources can be accommodated by use of Equation (7.9). Estimation of chloride deposition rates are subject to the same consideration discussed in Section 7.2.2.

There are some differences between application of the CMB method with groundwater data as opposed to unsaturated-zone data. The groundwater CMB method is not sensitive to mechanisms of flow through the unsaturated zone, whereas estimates of drainage provided by the unsaturated zone CMB method cannot account for flow through preferential flow paths (Wood, 1999). Collecting groundwater samples for chloride analysis requires much less effort than collecting samples from within the unsaturated zone. Analytical techniques for determining chloride concentrations are identical, ion-specific meter or ion chromatography, regardless of whether groundwater or unsaturated zone pore water is being analyzed. Large volumes of groundwater can be collected and concentrated, thereby reducing analytical detection limits and extending the maximum rate of recharge that can be determined with Equation (7.21). In addition, extensive data on chloride concentrations in groundwater are available through the US Geological Survey (http://waterdata.usgs.gov/nwis/qw; accessed April 9, 2009), the US Environmental Protection Agency (http://www.epa.gov/storet/; accessed April 9, 2009), and other international, national, state, and local agencies. These data greatly facilitate application of this method.

Wood and Sanford (1995) used Equation (7.21) to estimate recharge to the Ogallala aquifer in the southern High Plains of Texas on the basis of more than 3000 measurements of chloride concentrations in groundwater. The average concentration was 25.2 mg/L. Mean annual precipitation was 485 mm/yr, and the effective chloride concentration in precipitation was estimated at 0.58 mg/L. An average recharge rate of 11 mm/yr was calculated. In this same region,

Scanlon and Goldsmith (1997) used unsaturated zone tracer methods and determined that drainage was negligible in interplaya areas but could be as high as 120 mm/yr beneath playas.

Carbon-14

Carbon-14 (^{14}C) is a radioactive isotope with a half life of 5730 years. It is formed naturally from nitrogen-14 (^{14}N) by cosmic-ray bombardment. It was also produced as a result of nuclear bomb testing in the 1950s and 1960s. Because of its long half life, ^{14}C can be used to date a much wider range of water ages than can be dated with tritium, CFCs, or SF_6. The ^{14}C age of water is given by:

$$t = -(1/\lambda)\ln([^{14}C]/[^{14}C_0]) \qquad (7.22)$$

where λ is the decay constant for ^{14}C and $[^{14}C]$ and $[^{14}C_0]$ are the measured and initial concentrations of dissolved ^{14}C in groundwater, respectively. Equation (7.22) applies if radioactive decay is the only process occurring and there are no carbon exchanges.

The scale used to represent ^{14}C concentrations is percent modern carbon (pmc) with 100 pmc equal to atmospheric ^{14}C concentrations in 1950. Nuclear bomb testing in the 1950s and 1960s increased ^{14}C concentrations up to 200 pmc. Concentrations of ^{14}C that exceed 100 pmc represent post-1950s water. For the purposes of age dating, measurements of ^{14}C are made on dissolved inorganic carbon (DIC), which includes dissolved carbon dioxide (CO_2) and bicarbonate (HCO_3^-). Concentrations are generally measured by Accelerator Mass Spectrometer (Zhu, 2000), which requires about 10 mg of carbon. The amount of water required for ^{14}C analysis can be determined from knowledge of DIC concentrations.

As with other age-dating tracers, the ^{14}C age refers to the age of the ^{14}C rather than the age of the water itself. In a simple, open system, all carbon within an aquifer is derived from recharge water with atmospheric ^{14}C concentrations. At any point in such an aquifer, water age is assumed equal to the uncorrected ^{14}C age. However, there often are additional sources or sinks of ^{14}C that must be taken into account. These include dissolution and precipitation of carbonate minerals, carbon isotopic exchange between mineral and dissolved phases, and oxidation of organic matter (Wigley *et al.*, 1978; Plummer *et al.*, 1994). A variety of models have been developed to correct ^{14}C ages of groundwater on the basis of these reactions (Plummer *et al.*, 1994; Kalin, 2000). Stable carbon isotope ratios ($\delta^{13}C$) often are helpful in identifying different sources of carbon contributing to groundwater DIC.

If ages are estimated at two points along a groundwater flow path, groundwater velocity along that flow path can be calculated by dividing the distance between the points by the difference in calculated ages between the points. Recharge was estimated with ^{14}C concentrations by Vogel (1967) as described in Cook and Herczeg (1998). The reduction in ^{14}C concentration from 82.2 pmc at the 11 m depth to 74.4 pmc at the 67 m depth in an unconfined aquifer in Holland corresponded to a groundwater age of 821 years at the 67 m depth. Assuming an aquifer thickness of 70 m, a porosity of 0.4, and ^{14}C apparent ages, a recharge estimate of 0.1 m/yr (Equation (7.17)) provided the best fit to the data (Cook and Herczeg, 1998). McMahon *et al.* (2004) used ^{14}C data to calibrate a groundwater flow model for a section of the central High Plains aquifer in Kansas; simulated results based on steady recharge rates of 0.8 to 8 mm/yr throughout the Holocene agreed well with measured apparent groundwater ages.

Example: Middle Rio Grande Basin, New Mexico

The primary sources of recharge in the Middle Rio Grande Basin (7900 km^2 area) in New Mexico are losing ephemeral streams along the mountain front, underflow from mountain blocks, and losing reaches of the Rio Grande. Hydrochemistry data were used to develop a conceptual understanding of recharge sources and processes (Plummer *et al.*, 2004). ^{14}C ages were determined for 200 groundwater samples. The age data were used to calibrate a groundwater flow model for the purpose of estimating recharge rates (Sanford *et al.*, 2004). Little correction of ^{14}C ages was required for this siliciclastic system,

and groundwater flow was simulated with a MODFLOW model (McDonald and Harbaugh, 1988). Travel times to observation wells in the basin were simulated by using particle tracking in MODPATH (Pollock, 1994) and compared with ^{14}C ages. MODFLOW and MODPATH were calibrated by using nonlinear regression methods in the UCODE parameter-estimation program (Poeter and Hill, 1998) to minimize differences between measured and simulated ages. Total recharge to the basin was estimated to be 2.15 m^3/s, with the Rio Grande contributing 0.78 m^3/s. The total recharge rate is considerably lower than the previous recharge estimate of 4.86 m^3/s obtained from a basin-wide groundwater flow model calibrated with heads and discharge estimates alone. The dramatic revision of recharge rates underscores the value of tracers in hydrologic studies.

7.3.3 Historical tracers
Chlorofluorocarbons

Chlorofluorocarbons (CFCs) are organic compounds that are entirely anthropogenic in origin. They have been manufactured since the 1930s and used primarily as propellants for aerosol cans, as refrigerants, and as solvents. Although nontoxic and noncarcinogenic, CFCs have been shown to contribute to depletion of stratospheric ozone. Hence, worldwide production of CFCs was banned in industrial countries by the year 2000 and in developing countries by the year 2010 under the Montreal Protocol on Substances that Deplete the Ozone Layer signed in 1987 and amended in 1990 and 1992. Production of CFCs in the United States was halted on January 1, 1996 (Plummer and Busenberg, 2000).

High atmospheric concentrations, low analytical detection limits, and generally conservative natures are features that make CFCs useful as hydrologic tracers. Dating groundwater with CFCs was first proposed by Thompson *et al.* (1974), Schultz *et al.* (1976), and Thompson and Hayes (1979). A brief overview of the use of CFC concentrations in groundwater to estimate recharge is provided here. More detailed descriptions can be found in Cook and Solomon (1997) and Plummer and

Busenberg (2000) and at the US Geological Survey CFC lab website (http://water.usgs.gov/lab; accessed April 10, 2009).

CFCs that are used to date young (less than about 60 years in age) groundwater include CFC-11 (CCl_3F), CFC-12 (CCl_2F_2), and CFC-113 ($C_2Cl_3F_3$). Atmospheric concentrations of CFCs (Figure 7.2), estimated from production and release data before 1977 (McCarthy *et al.*, 1977) and from a global measurement network since then (Cunnold *et al.*, 1994), increased from the 1930s until the mid- to late-1990s, when concentrations began declining. Low CFC degradation rates result in long residence times in the atmosphere (approximately 44 years for CFC-11, 180 years for CFC-12, and 90 years for CFC-113) and relatively homogeneous concentrations within each hemisphere (Plummer and Busenberg, 2000).

CFC concentrations can be detected to 0.3 picogram per kilogram (pg/kg) of water by using a purge and trap method with a gas chromatograph equipped with an electron capture detector. Concentrations of CFCs in air during recharge are calculated from those in groundwater according to Henry's Law:

$$Pa_i = C_i / K_i \qquad (7.23)$$

where i refers to the ith CFC compound, Pa_i is the partial pressure of air in equilibrium with water, C_i is concentration in water, and K_i is Henry's Law constant. Values for K_i are dependent on recharge temperature (temperature at the water table at the time of recharge), salinity, and pressure (Warner and Weiss, 1985; Bu and Warner, 1995). The mixing ratio, x_i, is determined from the partial pressures:

$$x_i = Pa_i / (P - P_{vapor}) \qquad (7.24)$$

where P and P_{vapor} are total atmospheric and water-vapor pressure, respectively, and x_i is in units of volume of CFC per unit volume of air. The mixing ratio of the CFC is used with the atmospheric growth curve (Figure 7.2) to determine the time of recharge or recharge date. The age of the water is the difference between sampling and recharge dates. Independent ages can be calculated for each of the CFC compounds and also for ratios of different CFCs.

Several processes can affect the measurement of CFC concentrations and calculation of groundwater ages. These include the entrapment of excess air as percolating water moves past the water table, other sources of CFCs (such as from a sewage plume), and loss of CFCs by degradation or adsorption. The measured concentration of a CFC in groundwater, $[CFC_{meas}]$, is influenced by all of these sources and sinks:

$$[CFC_{meas}] = [CFC_{eq}] + [CFC_{exc\ air}] + [CFC_{cont}] - [CFC_{loss}] \qquad (7.25)$$

where the bracketed terms are concentrations and the subscripts eq, exc air, cont, and loss refer to CFC mass due to equilibrium with the atmosphere, entrapped excess air, contamination, and losses related to microbial degradation and adsorption, respectively (Plummer and Busenberg, 2000). $[CFC_{eq}]$ is the concentration needed for calculating groundwater age, and to determine that concentration, the magnitude of other terms on the right side of Equation (7.25) must be evaluated. Solubilities of CFCs decrease with increasing temperature. An uncertainty of $\pm2°C$ in recharge temperature would result in a ±3 year uncertainty in age for water recharged prior to 1990. For water recharged after 1990, CFC ages are much more sensitive to recharge temperature because atmospheric concentrations have changed little since then. Overestimation of recharge temperatures underestimates apparent ages. $[CFC_{eq}]$ is also a function of barometric pressure, which can be estimated from recharge elevation.

Most groundwater is supersaturated with air (Busenberg and Plummer, 2000). The excess air (i.e. dissolved air concentrations in excess of equilibrium solubility) is generally attributed to the entrapment of air during recharge. Excess air concentrations are typically less than $3\ cm^3$ STP/kg of water (Wilson and McNeill, 1997), but concentrations as high as $18\ cm^3$ STP/kg have been reported (Glynn and Busenberg, 1996). If excess air is ignored, dissolved gas concentrations will be overestimated and groundwater ages will be underestimated. For water recharged prior to 1990, the effect of excess air on CFC concentrations is negligible, but for younger water the effect can be important.

Age dates determined by other gas tracers tend to be more sensitive to excess air concentrations. Recharge temperature and the amount of excess air can be estimated by measuring concentrations of noble gases, such as neon, argon, and krypton, and solving a set of nonlinear equations (Aeschbach-Hertig *et al.*, 1999). Other approaches to determining these unknowns involve measurement of dissolved nitrogen and argon (Heaton and Vogel, 1981).

Contamination is a critical issue that limits the use of CFCs in some areas. Sources of CFC contamination in groundwater include effluent from septic tanks and from leaking sewer lines, industrial waste water, and recharge from rivers that have effluent from sewage treatment plants (Plummer and Busenberg, 2000). Local contamination of air is also associated with industrial and residential regions. Contamination is obvious when CFC concentrations exceed concentrations in equilibrium with normal hemispherically averaged air. The presence of contaminant sources of CFCs precludes their use for age dating, although CFCs can be useful for tracking the movement of groundwater contamination plumes (Schultz *et al.*, 1976).

CFCs are stable in groundwater under aerobic conditions. The stability of CFCs has been confirmed by concurrence of groundwater ages calculated from CFC-11, CFC-12, CFC-113, and other dating methods (Dunkle *et al.*, 1993; Szabo *et al.*, 1996; Ekwurzel *et al.*, 1994). However, CFC degradation occurs under anaerobic conditions. The different CFC compounds are not equally susceptible to degradation. CFC-11 and CFC-113 are generally degraded under sulfate reducing and methanogenic conditions, whereas CFC-12 is quasi-stable (Plummer and Busenberg, 2000; Cook *et al.*, 1995). Adsorption can also result in reduction of CFC concentrations and overestimation of groundwater ages; however, Plummer and Busenberg (2000) suggested that adsorption does not seem to be important for CFC-11 and CFC-12 in most groundwater systems. Few data are available on adsorption of CFC-113.

Samples of groundwater for CFC analyses must be isolated from contact with the atmosphere. Simple 125 mL glass bottles are used for sampling. After purging the well, the sample bottle is placed in a 2 to 6 L beaker. The outlet line of the pump is placed in the bottom of the bottle, the bottle is allowed to overflow into the beaker, and the beaker is allowed to overflow as well. After at least 2 L has flowed out of the beaker, a cap (rinsed in the beaker) is screwed onto the bottle while still submerged. Once capped, the bottle is removed from the beaker and dried, and the cap is secured with electrical tape. Five samples are required from each well (US Geological Survey CFC lab, http://water.usgs.gov/lab; accessed April 10, 2009).

Recharge rates of 0.05 to 0.45 m/yr were estimated on the basis of CFC-12 concentrations in samples obtained from several unconfined aquifers in the eastern United States (Böhlke, 2002). Recharge rates were calculated assuming a linear increase in age with depth (Equation (7.1)) near the water table and an exponential increase with depth (Equation (7.17)) for profiles at greater depth (Figure 7.10). At the Sturgeon Falls site in Ontario, Canada, Cook *et al.* (1995) used CFC-12 data to estimate average recharge at 130 mm/yr, a value in good agreement with that predicted with a groundwater flow model. High organic matter content and anaerobic conditions at that site resulted in degradation of CFC-11. Retardation of CFC-113 was also noted.

CFC data have also been used to calibrate groundwater flow models (Reilly *et al.*, 1994; Portniaguine and Solomon, 1998). Hydraulic heads are sensitive to the ratio of recharge flux to hydraulic conductivity, whereas ages are sensitive to the ratio of recharge flux to porosity. Combining head and age data in an inverse model (Section 3.5) allows estimation of recharge flux and other hydraulic parameters (Portniaguine and Solomon, 1998).

Sulfur hexafluoride

The use of sulfur hexafluoride (SF_6) to date groundwater and estimate recharge is similar to the use of CFCs described in the previous section. Although primarily of anthropogenic origin (electrical insulator), SF_6 also exists naturally in some igneous, metamorphic, and sedimentary rocks and in hydrothermal fluids. The natural background atmospheric concentration of SF_6 is estimated to be 0.054 ± 0.009 pptv (Busenberg and Plummer, 2000). Atmospheric concentrations of SF_6 have increased about 7% per year from background levels in about 1960 to more than 4 pptv in the late 1990s (Figure 7.2). The long residence time of SF_6 in the atmosphere (estimated to range from 1935 to 3200 years) results in uniform atmospheric concentrations. Concentrations of SF_6 in groundwater can be used to date water from 1970 to the present in a manner similar to that described for CFCs. Prior to 1970, SF_6 concentrations were extremely low making it very difficult to date the water. Ratios of SF_6 to CFCs have also been used to date groundwater.

As with CFCs, the measured SF_6 concentration, $[SF_{6meas}]$, includes the mass of SF_6 due to equilibrium with the atmosphere at the time of recharge, entrapped excess air, contaminant sources, and losses to biodegradation and adsorption:

$$[SF_{6meas}] = [SF_{6eq}] + [SF_{6excair}] + [SF_{6cont}] + [SF_{6loss}] + [SF_{6terr}]$$
$$(7.26)$$

where subscripts are as defined for Equation (7.25) and the term $[SF_{6terr}]$ accounts for natural subsurface (terragenic) sources of SF_6. $[SF_{6eq}]$ is the concentration needed for age dating, so the other interfering terms on the right side of Equation (7.26) must be evaluated. Although SF_6 solubility varies markedly with temperature (3.5% per °C), the rapid increase in atmospheric SF_6 concentration with time results in low sensitivity of ages to uncertainties in recharge temperature. For example, a 1 to 2°C uncertainty in recharge temperature would result in an error of less than 0.5 years in calculated age. Similarly, calculated ages are only slightly sensitive to recharge elevation; an uncertainty of ±300 m would result in an age uncertainty of ±5 years.

If entrapped excess air is not accounted for, SF_6 concentrations will be overestimated, and groundwater ages will be underestimated. Errors in estimated ages from excess air are potentially greater for SF_6 than for CFCs because of the lower solubility of SF_6 relative to CFCs. The Henry's Law constant for SF_6 is 55- and 13-times lower than those for CFC-11 and CFC-12, respectively.

Estimated errors in age from underestimation of excess air by 1 cm³ STP/kg water range from 1 to 2.5 years. Local anthropogenic sources of SF_6 in groundwater have not been reported, probably because SF_6 is not generally contained in household products. Ho and Schlosser (2000), however, found high atmospheric SF_6 concentrations in the vicinity of New York City. SF_6 seems resistant to both aerobic and anaerobic biodegradation and does not adsorb significantly to organic matter (Wilson and Mackay, 1996; Busenberg and Plummer, 2000).

Analysis of SF_6 concentrations in groundwater is by gas chromatograph with electron capture detector; accuracy is 1 to 3% (Busenberg and Plummer, 2000). Samples are collected in 1 L glass bottles. After the well to be sampled is purged, the outflow line from the pump is inserted into the bottom of the sample bottle, and the bottle is allowed to overfill. It is recommended that at least 2 L of water be allowed to overflow before screwing the cap onto the bottle. There should be no head space, and after drying, the bottle cap should be secured with electrical tape.

SF_6 concentrations were measured in groundwater samples from multiple wells at a field site in Locust Grove, Maryland, where concentrations of other tracers were also measured (CFCs, ³H/³He, ⁸⁵Kr) (Busenberg and Plummer, 2000). SF_6 ages agreed to within 3 years with those based on CFC-11 and CFC-113 in 80 to 85% of analyses. Discrepancies resulted largely from the inability to date groundwater older than 1970 with SF_6 and problems with dating post-1993 water with CFCs because of flattening of the CFC growth curve.

Tritium and tritium/helium-3

Many studies have used ³H concentrations in groundwater to estimate recharge rates. The tracer-profile, peak-displacement, and mass-balance methods, as described for unsaturated-zone application in Section 7.2, can also be applied with tritium concentrations in groundwater. These methods invoke the assumption of vertical groundwater flow for some distance below the water table and require measurement of the vertical profile of tritium concentrations in groundwater. Vertical flow in the saturated zone is a more tenuous assumption than vertical flow in the unsaturated zone and, therefore, requires closer scrutiny. The assumption is likely to be more valid for shallower sampling depths (Figure 7.9) or in highly permeable aquifers. Knott and Olimpio (1986) identified the peak tritium concentration at a depth of about 36 m below the water table in an unconfined sand and gravel aquifer on Nantucket Island, MA. An apparent vertical velocity of 1.89 m/yr was calculated, and Equation (7.1) was applied with a porosity of 0.36 to derive an average annual recharge rate of 0.68 m/yr (68% of annual precipitation). Water levels measured in piezometers confirmed a downward head gradient. Robertson and Cherry (1989) also applied the profile method with tritium concentrations in groundwater at the Sturgeon Falls study site in Ontario. Peak tritium concentrations were found at depths of 8 to 12 m; assuming a porosity of 0.35, an average recharge rate of 150 mm/yr was calculated with Equation (7.1). The validity of the vertical flow assumption at this site was confirmed by results from a groundwater-flow model (Robertson and Cherry, 1989).

As the elapsed time since peak atmospheric tritium concentration increases, tritium concentrations in groundwater become less useful for estimating recharge because dispersion and radioactive decay make it difficult to determine the location of the bomb-pulse concentration peak, particularly in the southern hemisphere where bomb tritium levels were about an order of magnitude lower than those in the northern hemisphere (Allison and Hughes, 1977).

The tritium/helium-3 (^{3}H/^{3}He) method for determining apparent groundwater ages does not require information on atmospheric tritium concentrations, nor does it require detailed profiling to identify the location of the bomb-pulse tritium concentration peak. The use of ^{3}H/^{3}He to date water is briefly summarized here; detailed descriptions can be found in Solomon and Cook (2000). Tritium (^{3}H) decays by beta emission to the noble gas isotope, ^{3}He. ^{3}He that is generated by decay of tritium in a water sample is referred to as *tritiogenic* helium-3 (^{3}He$_{trit}$). Knowledge of the rate of ^{3}H decay allows the age of a water

sample, t, to be determined from the concentration ratio of ^{3}He$_{trit}$ to ^{3}H (Schlosser *et al.*, 1989):

$$t = \frac{1}{\lambda} \ln \left[\frac{[^3He_{trit}]}{[^3H]} + 1 \right] \qquad (7.27)$$

where λ is the ^{3}H decay constant; $[^3He_{trit}]$ and $[^3H]$ are ^{3}He$_{trit}$ and ^{3}H concentrations, respectively; and $[^3He_{trit}]$ is expressed in equivalent tritium units (1 cm^3 ^{3}He STP per kg of water = 4.0177 × 10^{11} TU). Equation (7.27) relies on the assumptions that groundwater is isolated from the atmosphere, that there is no loss of ^{3}He$_{trit}$ through diffusion into the unsaturated zone, and that there is no hydrodynamic dispersion or mixing. Estimated losses of ^{3}He by diffusion to the unsaturated zone range from 1% for a recharge rate of 150 mm/yr to 20% for a recharge rate of 30 mm/yr (Solomon and Cook, 2000). Therefore, the reliability of the recharge rates decreases with decreasing recharge rates.

Determining the concentration of tritiogenic helium-3, $[^3He_{trit}]$, is not always straightforward. As with other dissolved gases, there can be multiple sources of ^{3}He in groundwater. $[^3He_{trit}]$ can be calculated in a manner similar to that used for CFCs and SF$_6$ (Solomon and Cook, 2000):

$$[^3He_{meas}] = [^3He_{trit}] + [^3He_{eq}] + [^3He_{exc\ air}] + [^3He_{terr}] \qquad (7.28)$$

where $[^3He_{meas}]$ is ^{3}He concentration measured in the water sample and subscripts for the other terms are analogous to those in Equations (7.25) and (7.26). Concentrations of ^{3}He in equilibrium solubility with the atmosphere depend on the temperature and to a lesser degree on salinity of the water and the ambient pressure during recharge. Recharge temperature, $[^3He_{exc\ air}]$, and $[^3He_{terr}]$ can be estimated by measuring concentrations of noble gases, as with CFC dating (Aeschbach-Hertig *et al.*, 1999). Additional details on segregating the different sources of ^{3}He in groundwater are provided in Schlosser *et al.* (1988, 1989). The solubility of ^{3}He is much less than that of CFCs; therefore, ^{3}H/^{3}He ages are more sensitive to excess air than CFC ages. Although ^{3}H decay begins in the unsaturated zone, ^{3}He is only preserved in the saturated zone; therefore, ^{3}H/^{3}He age is zero at the water table (Solomon and Cook, 2000). If the water table fluctuates, the seasonal low water table should be used as the point of zero age (Solomon et al., 1993).

Analysis of water samples for ^3He is by mass spectrometry, with special procedures for sample preparation. Tritium analysis is as described in Section 7.2.3. Laboratories that currently (2010) perform these analyses include the Lamont-Doherty Earth Observatory (http://www.ldeo.columbia.edu/environmental-tracer-group; accessed April 17, 2009) and the University of Miami (http://www.rsmas.miami.edu/groups/tritium/order-tritium.html; accessed April 17, 2009). The cost of analysis for ^3He, ^3H, and noble gases is about $1000. Detailed sampling procedures are required to ensure that samples remain isolated from the environment. These procedures can be obtained from the Lamont-Doherty or University of Miami websites or from the US Geological Survey CFC laboratory website (http://water.usgs.gov/lab; accessed April 17, 2009).

The $^3H/^3He$ method can be used to date waters up to about 30 years in age. Recharge rates of 100 to 1000 mm/yr have been estimated with the method (Schlosser et al., 1989; Solomon et al., 1995). The upper limit on recharge rates depends on the vertical resolution of sampling and the accuracy of estimating $^3H/^3He$ ages. The reliability of $^3H/^3He$ ages decreases in young water because the levels of $^3He_{trit}$ are low, making it difficult to distinguish from other atmospheric sources of ^3He. In an example by Solomon and Cook (2000), 2-year-old groundwater with an initial value of 20 TU would contain low levels of $^3He_{trit}$ (2.39 TU), equivalent to about 10% of the atmospheric solubility value and similar to the value of ^3He from excess air with excess air of 1.0 cm^3 (STP)/kg. $^3H/^3He$ can be used to determine the fraction of old (pre-1960s) water in bimodal mixtures of young water (single age) and old water by comparing the estimated tritium input function ($^3H + ^3He_{trit}$) with historical records of ^3H in precipitation.

Example: Sturgeon Falls, Ontario

Tritium, $^3H/^3He$, and CFCs were each used to estimate recharge at the Sturgeon Falls study site in Ontario, Canada (Solomon et al., 1993; Cook et al., 1995). The site is vegetated with boreal forests; mean annual precipitation is 990 mm. All three dating approaches were applied on the basis of concentrations in groundwater samples collected from piezometers with 0.15 m long screens at depths ranging from 0.6 to 21 m.

Tritium concentrations were determined for samples obtained in 1986, 1990, and 1991. Samples obtained in 1986 were analyzed for $^3H/^3He$; samples obtained in 1993 were analyzed for CFCs. The 1986 ^3H peak occurred at a depth of 8.7 m (Figure 7.12a). The time between the peak in atmospheric ^3H concentration and sampling (23 years) and the penetration depth of 8.7 m produces an apparent vertical velocity of 0.38 m/yr. A recharge rate of 132 mm/yr is calculated by Equation (7.1), assuming a porosity of 0.35. The 1991 peak occurred at a depth of 10.6 m (Figure 7.12a); application of either Equation (7.1) or (7.3) produces an identical recharge estimate.

Concentrations of neon and nitrogen in the samples were used to estimate the average recharge temperature at 2°C and the average amount of excess air at 3 cm^3/kg. Concentrations of Ne and ^4He were used to estimate $[^3He_{terr}]$. $[^3He_{trit}]$ was determined with Equation (7.27); both $[^3He_{exc\ air}]$ and $[^3He_{terr}]$ were important components that could not be ignored. An average vertical velocity of 0.465 m/yr was calculated with $^3H/^3He$ ages from the upper three sampling points in the profile. Applying Equation (7.16), this velocity corresponds to a recharge rate of 163 mm/yr.

CFC-12 behaved more conservatively than the other CFC tracers at this site (Figure 7.12b). CFC-11 appeared to be degraded in the highly organic unsaturated zone and at depth, and CFC-113 seemed to be retarded relative to CFC-12 and ^3H due to adsorption. The effect of 3 cm^3/kg of excess air was less than 2% on calculated ages and thus was ignored. CFC-12 concentrations at 9.8 m (59 pg/kg) and 11.9 m (23 pg/kg) indicated recharge years of 1966 and 1959 corresponding to ages of 27 to 34 years. Rather than using the CFC-12 ages to calculate a vertical velocity, the ages were used to calibrate a two-dimensional

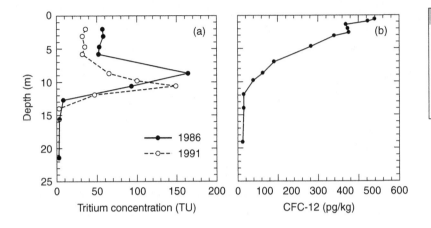

Figure 7.12 Concentration-depth profiles of (a) tritium for 1986 and 1991 (Solomon et al., 1993) and (b) CFC-12 for 1993 (Cook et al., 1995) in groundwater at the Sturgeon Falls site, Canada.

cross-sectional groundwater flow and solute transport model for the site. A recharge rate of 130 mm/yr produced model results that closely matched the measured CFC-12 ages.

7.4 | Discussion

Tracer methods can be used in many regions to estimate recharge. They provide estimates that are integrated over the time interval from when the tracer was introduced to the hydrologic system to when samples are obtained. Natural environmental and historical tracers provide information on cumulative drainage or recharge over periods of years to millennia. These tracers are perhaps the most useful tools available for assessing impacts of long-term climate variability and land use change on recharge rates. Chloride is particularly useful for quantifying low recharge rates in arid and semiarid regions. Applied tracers generally provide estimates over much shorter time scales, days to one or two years. Drainage estimates within the unsaturated zone, determined by tracer or other methods, are for a specific point in space, i.e. they are point estimates. Recharge estimates based on groundwater data are representative of a somewhat larger area, although the size of that area is largely unknown and likely varies with location within an aquifer. Regardless of whether estimates are generated from unsaturated-zone data or groundwater data, multiple data-collection points are usually required to determine average drainage or recharge rates and to assess the variability of those rates.

Tracer methods can serve as useful complements to other methods for estimating recharge. Recharge estimates using environmental or historical tracers often require only a single field visit to collect soil or water samples; therefore, tracer methods can be cost effective relative to water-budget methods (Chapter 2) and unsaturated-zone physical methods (Chapter 5) that require instrument installation and maintenance. Tracer techniques generally fail to provide information on recharge mechanisms and on the timing of recharge. Because sample collection is not time consuming, reconnaissance tracer sampling of soil or groundwater can be used to identify representative sites for application of more detailed studies of flow processes and the temporal variability of fluxes through the unsaturated zone. In addition, groundwater ages generated by tracer methods can be used to improve calibration of groundwater flow models (Chapter 3) that previously relied solely on hydraulic head data for calibration.

In most tracer studies, the distribution of a tracer in space or time within the subsurface is directly measured; however, information on tracer input and the mechanisms of tracer movement within the subsurface must often be inferred on the basis of little or no data. The accuracy of tracer-based flux estimates depends on the extent to which the assumptions are valid for a particular system. Uncertainties in measurement of tracer concentrations, estimation of tracer input, and assumptions about water and tracer transport processes are the most important limitations on the use of tracers to estimate recharge rates (Scanlon, 2000). Where possible, multiple tracers should be used to better constrain recharge estimates and reduce uncertainties.

Heat tracer methods

8.1 | Introduction

The flow of heat in the subsurface is closely linked to the movement of water (Ingebritsen et al., 2006). As such, heat has been used as a tracer in groundwater studies for more than 100 years (Anderson, 2005). As with chemical and isotopic tracers (Chapter 7), spatial or temporal trends in surface and subsurface temperatures can be used to infer rates of water movement. Temperature can be measured accurately, economically, at high frequencies, and without the need to obtain water samples, facts that make heat an attractive tracer. Temperature measurements made over space and time can be used to infer rates of recharge from a stream or other surface water body (Lapham, 1989; Stonestrom and Constantz, 2003); measurements can also be used to estimate rates of steady drainage through depth intervals within thick unsaturated zones (Constantz et al., 2003; Shan and Bodvarsson, 2004). Several thorough reviews of heat as a tracer in hydrologic studies have recently been published (Constantz et al., 2003; Stonestrom and Constantz, 2003; Anderson, 2005; Blasch et al., 2007; Constantz et al., 2008). This chapter summarizes heat-tracer approaches that have been used to estimate recharge.

Some clarification in terminology is presented here to avoid confusion in descriptions of the various approaches that follow. *Diffuse recharge* is that which occurs more or less uniformly across large areas in response to precipitation, infiltration, and drainage through the unsaturated zone. Estimates of diffuse recharge determined using measured temperatures in the unsaturated zone are referred to as *potential recharge* because it is possible that not all of the water moving through the unsaturated zone will recharge the aquifer; some may be lost to the atmosphere by evaporation or plant transpiration. Estimated fluxes across confining units in the saturated zone are referred to as *interaquifer flow* (Chapter 1). *Focused recharge* is that which occurs directly from a point or line source, such as a stream, on land surface. Focused recharge may vary widely in space and time. If the water table intersects a stream channel, estimates of stream loss are called actual recharge, or just recharge. If the water table lies below the stream channel, estimates are referred to as potential recharge. For simplicity, all vertical water fluxes are referred to as *drainage* throughout this chapter. Whether the estimated quantity represents actual or potential recharge or drainage depends on the circumstances of each individual study.

8.2 | Subsurface heat flow

Heat flow within the subsurface is driven by two phenomena. High temperatures in the Earth's interior produce a flow of heat outward from the core toward land surface. Near land surface, diurnal, seasonal, and climatic trends in solar radiation produce temporal trends in

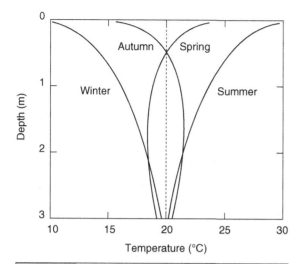

Figure 8.1 Hypothetical soil-temperature profiles at four times during a year (reprinted from *Fundamentals of Soil Physics*, Hillel (1980), Figure 12.4, copyright (1980), with permission from Elsevier).

energy exchange between the atmosphere and the soil surface. Parsons (1970) distinguished between two zones in the subsurface: the *surficial zone*, where temperatures fluctuate over time in response to energy exchange at the surface, and the deeper *geothermal zone* that is largely isolated from the influence of climate. Within the geothermal zone, heat flows at a relatively steady rate toward the surface as indicated by a vertical profile (called the *geothermal profile* or *geothermal gradient*) of increasing temperatures with depth, typically at rates of 0.01 to 0.02°C/m. The depth to which the surficial zone extends varies. Short-term energy exchanges at land surface result in measurable temperature changes to depths of 0.1 to 0.3 m on a daily basis and to about 10 m on an annual basis (Anderson, 2005). The magnitude of diurnal or seasonal fluctuations decreases with depth from land surface (Figure 8.1). Water movement affects temperature profiles within both the surficial and geothermal zones. Within the geothermal zone, a temperature–depth profile at a single point in time can be used to generate an estimate of steady, vertical water flux. In contrast, estimates of water flux within the surficial zone usually rely on tracking changes in surface and subsurface temperatures over time;

these estimates can be for steady or nonsteady water flux.

The theory of heat flow in geologic media and its relation to groundwater systems is presented in a number of texts (Bear, 1972; de Marsily, 1986; Domenico and Schwartz, 1998; Ingebritsen *et al.*, 2006) and is briefly described here. Water vapor movement, evaporation, and condensation are important heat transport processes in geothermal reservoir engineering (Ingebritsen *et al.*, 2006) and in diffuse water movement through unsaturated zones of arid regions (Philip and de Vries, 1957; Scanlon and Milly, 1994; Walvoord *et al.*, 2002a, b; Constantz *et al.*, 2003). However, for the purposes of this text and for using heat as a tracer, most recharge studies are concerned only with the movement of liquid water. So water-vapor transport will not be addressed herein.

The heat-flow equation for single-phase water movement (within the temperature range for liquid water) is given by Healy and Ronan (1996):

$$\nabla \cdot [K_T(\theta) + \theta C_w D_H]\nabla T - \nabla C_w q T + q^* C_w T^*$$
$$= \partial[\theta C_w + (1-\phi)C_s]T / \partial t \quad (8.1)$$

where K_T is thermal conductivity, θ is volumetric moisture content, C_w is heat capacity of water, D_H is the hydrodynamic dispersion tensor, T is temperature, q is specific flux (drainage rate, if vertical flow is assumed), q^* is flux from a source or to a sink at temperature T^*, φ is porosity, and C_s is heat capacity of the dry rock matrix. The two primary mechanisms for heat flow within the subsurface are conduction and advection; these are represented by the first and second terms, respectively, on the left-hand side of Equation (8.1). Conductive heat flow is the movement of heat (or energy) from areas of higher temperature to areas of lower temperature, a mechanism analogous to molecular diffusion. Advective heat flow is the transport of heat due to water movement. For example, water, with a temperature of T_1, infiltrating and percolating through an unsaturated zone where initial temperatures were less than T_1, is a source of heat to the subsurface. Distinguishing the advective component of heat transport from the conductive

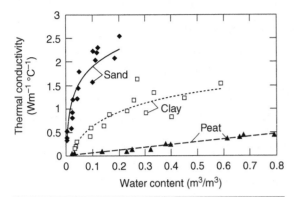

Figure 8.2 Thermal conductivity of soils as a function of volumetric water content; points were experimentally determined (de Vries, 1966), and lines are empirical fits to the data for the three soil types (Stonestrom and Blasch, 2003).

component allows determination of specific flux.

Solution of Equation (8.1) requires information on thermal conductivity, heat capacities of water and rock matrix, water flux, and boundary conditions, in addition to measured temperatures. Thermal conductivity, the ability of a material to conduct heat, varies with soil texture and moisture content (Figure 8.2) and ranges from approximately 0.2 W/m°C for a dry soil to 2.5 W/m°C for a saturated soil (Stonestrom and Blasch, 2003). A number of methods are available for measuring K_T in the laboratory or the field (Bristow, 2002); values for K_T also can be estimated on the basis of moisture content and other physical properties such as texture and mineral content (Campbell, 1985).

Heat capacity of water, C_w, is the product of water density, ρ_w, and specific heat, c_w. A typical value for C_w is 4.2×10^6 J/m³°C (Carslaw and Jaeger, 1959); this value varies little over the temperature range from 0 to 100°C. C_s is the product of solid density and specific heat. On the basis of known heat capacities of constituent materials, de Vries (1966) presented a simple equation to calculate C_s:

$$C_s = x_w C_w + x_o C_o + x_m C_m + x_a C_a \qquad (8.2)$$

where x is volume fraction, C is heat capacity of the specific constituents, and subscripts w, o, m, and a indicate water, organic, mineral, and air, respectively. Typical values for C_s fall in the range of 1.1×10^6 to 3.2×10^6 J/m³°C (Stonestrom and Blasch, 2003).

Estimates of specific flux in the vertical direction (i.e. drainage, D) can be obtained by simultaneously solving Equation (8.1) and the Richards equation (Section 3.3.2) for water movement. Ranges in values of thermal properties are much less on a relative basis than the range in magnitudes of hydraulic conductivity, a parameter in the Richards equation. Hence, analytical and numerical solutions to Equation (8.1) and the Richards equation are typically much more sensitive to uncertainty in values of hydraulic conductivity than to uncertainty in values of thermal properties (Constantz *et al.*, 2003), a feature that underscores the usefulness of subsurface temperature data for estimating water fluxes.

Analytical solutions to the one-dimensional vertical form of Equation (8.1) for vertical flow are available. Stallman (1965) developed a solution to predict temperatures at any depth and time under the assumption of a sinusoidal variation in surface temperature. Silliman *et al.* (1995) adapted that solution to predict steady exchange between a stream and an aquifer. Keery *et al.* (2007) and Schmidt *et al.* (2007) presented alternate solutions for the same problem. Bredehoeft and Papadopulos (1965) developed an analytical solution for the case of steady vertical flow through a semiconfining layer.

Numerical models offer more flexibility than analytical models. They typically are not bound by assumptions of one-dimensional steady flow, constant boundary conditions, or uniform thermal and hydraulic properties. Lapham (1989) developed a numerical model to solve the one-dimensional heat-flow equation for the case of a stream connected to an aquifer. A multitude of numerical models are available that simultaneously solve Equation (8.1) and the groundwater flow equation (Section 3.5) or the Richards equation for variably saturated flow; these include VS2DH (Healy and Ronan, 1996), HYDRUS (Simunek *et al.*, 1999), FEHM (Zyvoloski *et al.*, 1997), HST3D (Kipp, 1997), SUTRA (Voss and Provost, 2002), and TOUGH2 (Pruess *et al.*,

1999). Anderson (2005) provided a more extensive list and a summary of selected models.

Whether analytical or numerical models are used, the general approach for determining water flux is similar and follows the general model calibration guidelines described in Section 3.2 and those presented by Niswonger and Prudic (2003). Thermal properties are usually held constant, and flux (or hydraulic conductivity and/or other hydraulic properties in the case of a numerical model that solves a flow equation in addition to Equation (8.1)) is adjusted between model applications until simulated temperatures agree with measured temperatures. Model calibration can be accomplished manually (Lapham, 1989; Ronan *et al.*, 1998) or with a parameter estimation program such as PEST (Bartolino and Niswonger, 1999; Prudic *et al.*, 2003).

8.2.1 Temperature measurements

A variety of sensors are available for measuring temperature within the subsurface. These include thermistor, thermocouple, resistance type, and integrated circuit devices. Stonestrom and Blasch (2003) reviewed the advantages and limitations of the different types of sensors. Sensor accuracy of 0.1 to 0.2°C is sufficient for many studies within the surficial zone; an accuracy of about 0.01°C is desirable for measuring temperature profiles within the geothermal zone. External data loggers allow temperatures to be electronically recorded at fixed time intervals; some newer temperature sensors have internal recording capabilities.

Temperature sensors can be emplaced more or less permanently at specific depths within the subsurface with wire leads extending to land surface. *Fixed-depth* sensors are usually set up to record temperature at fixed time intervals and thus are capable of providing detailed time histories of temperature for specific locations. Expense, however, may limit the number of locations that can be monitored, and the sensors cannot be recalibrated after emplacement. Temperatures can also be logged within observation wells or piezometers at discrete points in time. Logging, usually a manual operation, consists of taking temperature measurements at specific depths with a single sensor. Because many depths are easily measured, logging can provide detailed temperature–depth profiles at specific times. Logging with a single sensor offers economy and allows for recalibration of the sensor, but the procedure is time consuming. The detail provided on the time history is dictated by the frequency with which the logging is performed. Although there are many exceptions, heat transport studies within the surficial zone usually are based on fixed-depth temperature measurements, whereas investigations within the geothermal zone generally rely on temperature logging.

Recently developed distributed temperature sensors (DTS) hold great promise for future application in hydrologic studies (Selker *et al.*, 2006a, b; Lowry *et al.*, 2007). The sensors determine temperature by measuring the scattering of light along a fiber-optic cable. By laying out the cable in a stream channel or hanging it down an observation well, temperatures can be measured over length increments of less than 1 m and at frequencies greater than once per minute. DTS provide the spatial detail of temperature logging and temporal detail of fixed-depth sensors. Lowry *et al.* (2007) used the sensors to obtain streambed temperatures at 1 m and 1 min intervals along a 650 m stream reach in northern Wisconsin; temperatures were averaged over 15 min periods, and a measurement accuracy of ±0.03°C was determined. DTS systems are expensive, and the cable is fragile. It is anticipated that the systems will come into widespread use as these problems are resolved.

Remote sensing of stream temperatures with forward-looking infrared technology from helicopter platforms can provide detailed information on spatial patterns of stream temperatures (Mertes, 2002; Loheide and Gorelick, 2006). However, helicopter flights are expensive, and multiple flights are required to obtain temporal patterns in temperatures.

8.3 | Diffuse recharge

8.3.1 Diffuse drainage in the geothermal zone

Temperatures within the geothermal zone increase linearly with depth if there is no water

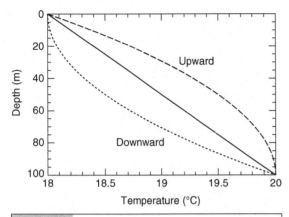

Figure 8.3 Hypothetical vertical temperature gradients within interval of geothermal zone of uniform thermal properties for cases of no vertical water movement (solid line), steady downward water movement (dotted line), and steady upward water movement (dashed line). Depth is measured from top of interval.

movement and if thermal conductivity is uniform (Figure 8.3). In reality, thermal conductivity is rarely uniform, but under steady heat-flow conditions, within each layer of uniform properties, a constant vertical temperature gradient exists. The occurrence of upward or downward water movement adds an advective component to heat flow, thus altering the temperature gradient. Groundwater moving upward carries with it heat from deeper in the profile, enhancing heat flow toward land surface and producing a convex pattern of higher temperatures relative to the conduction-only condition (Figure 8.3). Water at land surface has a cooler temperature, on average, than that in the subsurface, so water movement downward from land surface tends to inhibit the upward flow of energy, resulting in a concavity in the temperature profile (Figure 8.3).

Analysis of temperature–depth profiles to estimate rates of vertical groundwater movement was first proposed by Suzuki (1960) and Stallman (1965). Perturbations from the profile expected under pure conductance are assumed to be related to water percolation, and models are used to determine the flux rate that produces the best agreement between simulated and measured temperatures. Bredehoeft and Papadopulos (1965) developed an analytical

solution to Equation (8.1) for the special case of steady vertical water flow through a semiconfining bed of uniform hydraulic and thermal properties. The approach requires measurement of the temperature–depth profile across the semiconfining bed and an estimate of average thermal conductivity. The measured profile is plotted in dimensionless form and matched to a type curve. Once the appropriate type curve is selected, vertical velocity can be calculated. Flux is then determined as velocity times porosity. Numerous studies have used the Bredehoeft and Papadopulos (1965) approach for estimating vertical rates of water flow (Sorey, 1971; Silliman and Booth, 1993; Constantz *et al.*, 2003).

The Bredehoeft and Papadopulos (1965) solution was developed for estimating flow through a semiconfining bed under saturated conditions. However, the approach is applicable over any interval of the saturated zone that is dominated by steady, vertical water movement. It is also applicable in certain intervals of thick unsaturated zones where water percolates downward at a steady rate (see the discussion on the unit-hydraulic gradient method in Chapter 5; many of the assumptions for that method are also applicable here) and hydraulic and thermal properties (including moisture content) are uniform (Constantz *et al.*, 2003). Shan and Bodvarsson (2004) developed an analytical solution for the case of steady, vertical flow through multilayered geologic media; this solution is applicable for both the saturated and unsaturated zones. Using temperature logs that spanned a 400 m thick segment of unsaturated zone at borehole SD-12 at Yucca Mountain, Nevada, Shan and Bodvarsson (2004) estimated that water was percolating through the unsaturated zone at the rate of about 15 mm/yr. Numerical models of water and heat transport also have been used to determine the drainage rates that are required to reproduce temperature profiles in the unsaturated zone (Kwicklis, 1999; Flint *et al.*, 2002; Constantz *et al.*, 2003). Numerical models may require more effort to apply, but they are not limited by assumptions inherent in analytical solutions to the heat-flow equation (as described in Section 8.2).

Several factors can complicate the use of temperature–depth profiles to estimate drainage rates. Horizontal water movement within the saturated zone can affect measured profiles; Lu and Ge (1996) extended the Bredehoeft and Papadopulos (1965) analytical solution to account for horizontal groundwater flow. Heat-transport processes other than conduction and advection may be important within the unsaturated zone. Heat flow due to evaporation, condensation, vapor-phase transport, and advective soil-gas movement are commonly assumed to be negligible in unsaturated zones of humid and semihumid settings. However, these processes can account for a substantial portion of total heat flow within unsaturated zones of arid regions (Scanlon and Milly, 1994; Thorstenson et al., 1998; Walvoord et al., 2002a, b; Constantz et al., 2003). Analyses of temperature–depth profiles in arid regions generally employ sophisticated numerical models that can account for these additional processes.

Example: Frenchman Flat, Nevada

Frenchman Flat is a sediment-filled basin within the Basin and Range Province in southern Nevada. Temperature sensors were emplaced at depths of 30.5, 61, 91.5, 152.5, and 213.5 m in borehole PW-1 that extended from land surface down to the water table at a depth of about 240 m (Constantz et al., 2003). The borehole was then sealed with grout, and temperatures were recorded over time. Measured temperatures increased in a fairly linear fashion with depth (Figure 8.4a) at a rate of about 0.012°C/m.

The Bredehoeft and Papadopulos (1965) solution was used to calculate temperature at any depth, $T(z)$:

$$T(z) = T_0 + (T_L - T_0)f(B, z/L) \qquad (8.3)$$

$$f(B, z/L) = (1 - e^{Bz/L})/(1 - e^B) \qquad (8.4)$$

$$B = v_z C_w L / K_T \qquad (8.5)$$

where z, depth, is equal to 0 at the top of the measurement interval and increases downward, T_0 is temperature at the top of the interval, L is the thickness of the interval (183 m for this example), T_L is temperature at the bottom of the interval, and v_z is vertical velocity. To

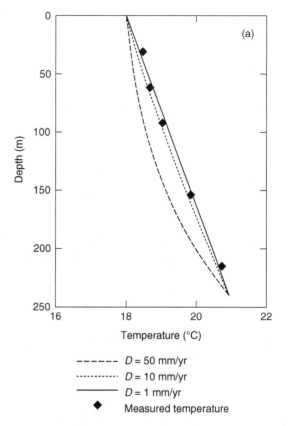

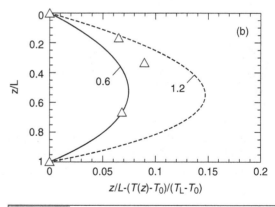

Figure 8.4 (a) Measured temperatures and those predicted from Equation (8.3) for a steady vertical drainage rate (D) of 1, 10, and 50 mm/yr for the unsaturated zone underlying Frenchman Flat, NV (Constantz et al., 2003). (b) The same measured temperatures (triangles) and those predicted from Equation (8.3) for values of B of 0.6 and 1.2 plotted in the dimensionless form suggested by Stallman (1967).

apply the method, type-curve matching is used. Data are plotted in dimensionless form with z/L on the y axis and $(T(z) - T_0)/(T_L - T_0)$ on the x axis. Type curves given in Figure 2 of Bredehoeft and

Papadopulos (1965) are then compared with the data and B is selected from the curve that best matches the data. The value of v_z is then determined from Equation (8.5). The drainage rate is equal to the product of v_z and volumetric moisture content. As an alternative to the type-curve matching, a parameter estimation routine, such as PEST (Doherty, 2004), can be applied to Equations (8.3) and (8.4) to determine the B value that produces the best match with measured temperatures.

Constantz et al. (2003) determined temperature profiles for drainage rates of 1, 10, and 50 mm/yr (Figure 8.4a). Results for 1 and 10 mm/yr are nearly linear with depth and appear to provide a reasonable match to measured temperatures. Results for 50 mm/yr are more nonlinear with depth; the higher flux rate results in a concave profile. The authors could conclude only that the magnitude of the vertical flux appeared to be less than 10 mm/yr.

Stallman (1967) proposed a slight modification to the Bredehoeft and Papadopulos (1965) approach that allows for finer resolution of low vertical velocities. Instead of plotting dimensionless temperature on the x axis, Stallman (1967) plotted dimensionless depth minus dimensionless temperature $[z/L - (T(z) - T_0)/(T_L - T_0)]$. The data from PW-1 are replotted in this format (Figure 8.4b), and although the data points do not match theoretical curves exactly, the curves for B = 0.6 and B = 1.2 completely bracket the data. Velocities calculated from Equation (8.5) are 18 and 36 mm/yr downward for B equal to 0.6 and 1.2, respectively, assuming $K_T = 0.8$ W/m°C (Constantz et al., 2003). So the modified approach produces a more refined range of drainage estimates, between 1.8 and 3.6 mm/yr, assuming an average moisture content of 0.10 (Reynolds Electrical and Engineering Company, 1994).

Analysis of temperature data can provide a quick and easy estimate of diffuse drainage rates within the geothermal zone, but the method is limited by uncertainty in values of thermal conductivity and by assumptions of uniform thermal conductivity and moisture content. Stallman (1967) stated that sensitivity for the method is about 0.1 mm/d when applied over

5 m thick semiconfining beds with temperatures measured to an accuracy of 0.001°C. As demonstrated by this example, however, greater sensitivity can be obtained when measurements are made over much wider depth intervals.

8.3.2 Diffuse drainage in the surficial zone

The majority of applications of heat to trace diffuse drainage rates have centered on the geothermal zone, where drainage rates are assumed constant in time. A few studies have examined the use of heat to trace diffuse drainage rates within the surficial zone. Cartwright (1974) identified groundwater discharge and recharge zones on the basis of differences in soil temperatures at a depth of 1 m between winter and summer. Groundwater discharge tends to moderate temperature fluctuations; areas with small temperature differences were assumed to be dominated by discharge. Recharge zones were those areas that displayed large differences. This qualitative approach was extended to provide quantitative estimates of infiltration based on differences (over periods of weeks to years) in soil temperatures at single or multiple depths down to 1 m (Tabbagh et al., 1999; Bendjoudi et al., 2005; and Cheviron et al., 2005). The abundance of available soil-temperature data and the ease of application make these methods attractive, but there are complications. Water percolating downward within the surficial zone may never reach the water table; it may be withdrawn by plants and returned to the atmosphere by evapotranspiration. Shallow water-table depths may impact the soil temperature profile and affect the calculations. Cheviron et al. (2005) examined six 3-year periods for a site in central France. The predicted infiltration rates were used with a water-budget equation to estimate recharge. For four of the periods, estimated recharge rates were deemed acceptable; for the other two periods, estimates were unacceptable because total recharge exceeded precipitation. With further research, these methods may prove to be useful in recharge studies.

8.4 | Focused recharge

Heat-tracer techniques, when applied in the surficial zone, are most often used to estimate rates of exchange between surface and groundwaters. The discussion here will be on streams, with the understanding that identical principles hold for other surface-water bodies. Temperatures in natural streams fluctuate on daily and annual patterns in response to fluctuations in air temperature and net radiation. Stream temperatures can also be affected by human activities, such as dam releases and cooling operations at industrial facilities. Variability in heat exchange between a stream and the subsurface is displayed by fluctuations in time and space in subsurface temperatures.

Temperature envelopes in sediments underlying the center of a stream for an annual cycle for losing and gaining streams are shown in Figure 8.5. (The temperature envelope is defined by annual minimum and maximum temperatures at each depth.) Water percolating downward beneath a losing stream carries heat with it, thus contributing an advective component to the heat flow. On an annual basis, surface-water temperatures fluctuate over a much wider range than groundwater temperatures; hence, a deeper, wider temperature envelope is expected beneath a losing stream relative to that beneath a gaining stream. The temperature envelope beneath a gaining stream is condensed in terms of depth and width because the discharging groundwater has a relatively constant temperature. Temperature envelopes for diurnal cycles display patterns similar to those in Figure 8.5, only with less temperature difference at any depth. The depths to which temperature envelopes extend beneath streams vary with exchange rates, climate, and sediment properties. For diurnal patterns, depths can range from a few tenths of a meter to several meters. To monitor diurnal temperature patterns, sensors are usually installed at fixed depths, and measurements are typically recorded every 15 to 60 min. The envelope for annual temperatures can extend past a depth of 15 m (Bartolino and Niswonger, 1999); annual temperature trends are usually measured by logging observation wells on a weekly or monthly schedule.

Numerical or analytical solutions to the Equation (8.1) can be used to simulate trends in subsurface temperatures. Whereas methods applied for estimating diffuse drainage attempt to match spatial patterns in temperature that are steady in time, methods applied for focused drainage attempt to match temperature fluctuations over time at specific locations. Diurnal temperature cycles are usually analyzed (Ronan et al., 1998), but the approach can also be used to analyze annual or seasonal temperature trends (Lapham, 1989; Bartolino and Niswonger, 1999) and to study infiltration events that may occur over periods of hours to days on ephemeral streams (Blasch et al., 2006; Hoffmann et al., 2007). Surface-water temperature and temperature at one or more depths within the diurnal or annual temperature envelope must be measured.

Numerical simulations usually consider one-dimensional vertical flow, but two-dimensional simulations also have been used (Prudic et al., 2003, 2007; Essaid et al., 2008), and three-dimensional simulations are possible. The upper boundary condition consists of the streambed and is usually treated as a specified pressure head and temperature boundary. Boundary

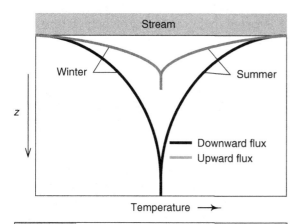

Figure 8.5 Hypothetical annual temperature envelopes beneath gaining and losing streams (after Stonestrom and Constantz, 2003). The envelope is defined by the annual minimum and maximum temperatures at all depths (z). The envelope is wider and deeper for a losing stream because surface-water temperatures fluctuate more over a year than do groundwater temperatures.

heads and temperatures are set equal to measured values and are allowed to vary over time, usually over periods of about 15 min for matching diurnal patterns and days or weeks for matching annual patterns. The bottom boundary can be represented by a specified pressure head (corresponding to a measured water level) or a free-drainage condition and a specified or variable temperature. Simulated temperatures are compared with measured temperatures at various depths. Because simulation results are more sensitive to hydraulic properties of sediments than to thermal properties, thermal properties are held constant. Model calibration consists mainly of varying hydraulic properties, either manually or automatically, to obtain agreement between simulated and measured subsurface temperatures. Fluxes to or from the stream are calculated by the calibrated model.

Analytical solutions to the heat-flow equation also can be used to estimate the rate of exchange of ground and surface waters through a streambed (Silliman *et al.*, 1995; Keery *et al.*, 2007; Schmidt *et al.*, 2007). These methods require temperature histories in the streambed and at least one depth beneath the stream. Hatch *et al.* (2006) developed an approach that does not require streambed temperatures. The method requires temperature histories at two depths and analyzes the attenuation of the phase and amplitude of the diurnal temperature pattern between the two depths. Assuming a sinusoidal upper temperature boundary, the Stallman (1963) solution to the heat-flow equation is applied; the solution is optimized to find the drainage rate that best reproduces the attenuation of the temperatures between the two measurement depths.

Distributed temperature sensors described in Section 8.2.1 have been employed in a few hydrologic studies. Lowry *et al.* (2007) used a DTS system to obtain temperature measurements along a 650 m reach of Allequash Creek in northern Wisconsin. Quantitative estimates of exchange between surface and groundwater could not be made, but on the basis of the daily range in measured temperatures, zones of groundwater discharge to the stream could be identified. Westhoff *et al.* (2007) used a DTS system to measure stream

temperatures along a 580 m reach of stream in central Luxemburg. These data were used to calibrate a stream energy-budget model that generated estimates of groundwater discharge rates along the reach. Model results were similar to those determined with a stream water-budget method, but processes that are not included in the model, such as shading of the stream and sudden changes in weather conditions, can affect measured stream temperatures.

Example: Rillito Creek, Tucson, Arizona

Rillito Creek in southeastern Arizona displays a flow pattern that is typical for many streams in the arid and semiarid southwestern United States. For most days of the year there is no flow in the stream. Following periods of prolonged rainfall or snowmelt, water flows in the stream for periods of several hours to several days. Ephemeral streams such as Rillito Creek are often an important source of recharge for underlying groundwater systems. An extended effort to quantify drainage rates from the creek was undertaken by Hoffmann *et al.* (2003, 2007) and Blasch *et al.* (2006). Several different methods were used to estimate the rate of movement of water from the stream to the underlying aquifer, including a heat-transport method, a stream water budget, chemical and isotopic tracers, and a temporal gravity method.

A two-dimensional array of temperature and water-content sensors was installed on a transect beneath and perpendicular to the stream channel (Hoffmann *et al.*, 2007). Sensors were installed at seven depths (ranging from 0.5 to 2.5 m) at four points on the transect, separated by a distance of 3 m. Data indicated that horizontal water movement was insignificant, so a one-dimensional water- and heat-transport simulation using the VS2DH model (Healy and Ronan, 1996) was constructed for each of the four locations for each period of streamflow. Simulated temperatures were in good agreement with temperatures measured in April 2001 (Figure 8.6); the model-calculated infiltration rates were similar for the four locations, averaging 0.32 m/day for this period. The authors calculated an average volumetric flow rate per unit length of stream channel by multiplying the average infiltration rate by channel width. The calculated number, 0.09 m³/s/km, is about half of that estimated by the stream water-budget method (Hoffmann *et al.*, 2007), a discrepancy attributed to the fact that channel width at the transect location was less than the average width for the reach over which the water-budget method was applied.

Example: Trout Creek, north-central Nevada

Trout Creek in north-central Nevada is typical of streams in the Basin and Range Province of the western United States (Prudic *et al.*, 2003; 2007). Throughout much of its reach, streamflow is ephemeral; most days of the year there is no flow. Following periods of extended snowmelt or rainfall, streamflow can occur for periods of several days. Because of limited groundwater supply and increased demands for water from

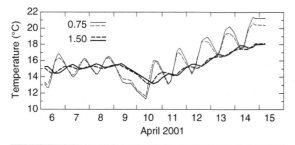

Figure 8.6 Measured (solid line) and simulated (dashed line) temperatures at depths of 0.75 and 1.50 m beneath Rillito Creek in April 2001 (after Hoffmann *et al.*, 2007).

mining, agricultural, and municipal interests in the region, a study was conducted to estimate loss rates from Trout Creek (Prudic *et al.*, 2007). Stream water-budget and heat-transport methods were applied. Streambed temperatures were monitored at seven locations. At three of these locations, temperatures were monitored at depths of 0.1, 0.2, 0.5, 1.0, and 1.5 m beneath the stream channel.

At one of the measurement sites, infiltration was estimated for a 7 day period in April 2000 by using the VS2DH model for simulating water and heat transport in the subsurface. The model was set up to simulate water and heat flow in a two-dimensional section perpendicular to the stream channel. Lateral boundaries extended 15 m on either side of the stream to allow horizontal flow within the unsaturated zone. The water table was set at a fixed depth of 30 m, consistent with water levels in a nearby observation well. Stream stage was assumed constant throughout the simulation. Stream temperatures were set equal to measured temperatures, which were allowed to change every 30 min.

The PEST parameter estimation program (Doherty, 2004) was used to determine the value of saturated hydraulic conductivity that produced the best agreement between simulated and measured temperatures at depths of 0.2, 0.5, and 1 m. All other hydraulic properties, as well as the thermal properties, were held constant and are found in Prudic *et al.* (2007).

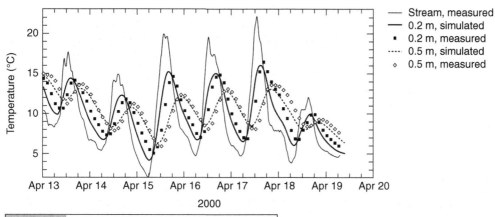

Figure 8.7 Measured and simulated temperatures at depths of 0.2 and 0.5 m beneath Trout Creek in April 2000 (after Prudic *et al.*, 2003).

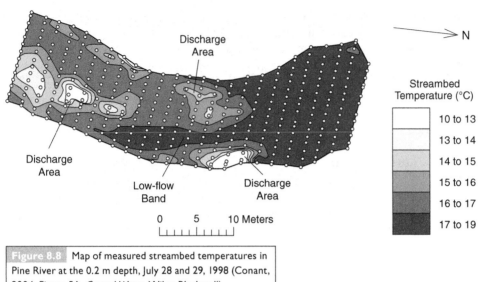

Figure 8.8 Map of measured streambed temperatures in Pine River at the 0.2 m depth, July 28 and 29, 1998 (Conant, 2004, Figure 5A, *Ground Water*, Wiley-Blackwell).

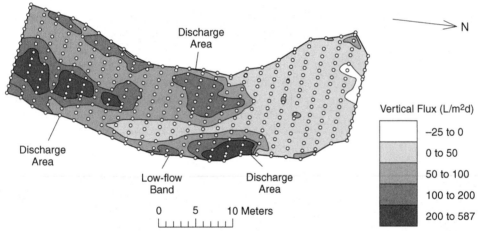

Figure 8.9 Map of calculated vertical fluxes across Pine River streambed, July 28 and 29, 1998; positive values indicate groundwater discharge to the river (Conant, 2004; Figure 8A, *Ground Water*, Wiley-Blackwell).

A vertical hydraulic conductivity value of 0.56 m/day produced simulated results shown in Figure 8.7. The calculated infiltration rate was 0.46 m/day, which is close to the value calculated with a stream water-budget method (Prudic *et al.*, 2007).

Example: Pine River, Angus, Ontario, Canada

Conant (2004) measured streambed temperature at a depth of 0.2 m on a grid of about 400 points over a 60 m reach of the Pine River in Angus, Ontario, Canada, to investigate variability of streambed temperatures and fluxes. Temperatures varied over a range of almost 10°C for measurements made in July 1998 (Figure 8.8). The river is primarily a gaining stream over this reach. Rates of vertical exchange between groundwater and the stream were calculated with the Darcy method for the 34 locations on the grid where piezometers were installed. Hydraulic conductivity was determined from slug tests. A regression equation was developed to relate estimated fluxes to measured streambed temperatures. The regression equation was then used to estimate vertical fluxes at all points on the temperature grid (Figure 8.9). Fluxes varied from slightly less than 0 (negative flux is movement from the river to the subsurface) to more than 500 L/m²d. Figure 8.9 clearly illustrates the variability in exchange rates across a streambed and highlights the limitation of using single point estimates, such as obtained from a seepage meter measurement or from a heat-transport approach, to estimate those exchange rates. Point estimates of exchange rates are very useful for describing variability in those rates over space and time, but, as demonstrated in this study, single point estimates may not be adequate for describing average flux across a stream transect or reach.

Example: Middle Rio Grande, Albuquerque, New Mexico

Vertical temperature profiles and groundwater-level altitudes were measured in piezometers installed along the banks of the Rio Grande at four locations in central New Mexico (Bartolino and Niswonger, 1999). The data were used to calibrate VS2DH simulations of water and heat flow between the river and the underlying aquifer. The objective of the study was to generate estimates of recharge to the aquifer. The Rio Grande is an important regional source of water for agricultural and municipal uses. Temperatures were measured in 0.15 m depth increments on seven dates between September 1996 and August 1998 (Figure 8.10). The wide seasonal temperature envelope extends unusually deep to depths greater than the 15 m maximum depth of measurement.

One-dimensional models of water and heat transport were constructed for each of the four locations to simulate the seasonal trends in groundwater levels and temperatures. Thermal and hydraulic properties, with the exception of hydraulic conductivity, were obtained from the literature on the basis of lithology and bulk density of soil samples. The bottom boundary of the simulated column consisted of fixed heads and temperatures that were determined from measured values (interpolated between measurement times). The PEST parameter estimation program was used to determine values of vertical hydraulic conductivity and top boundary

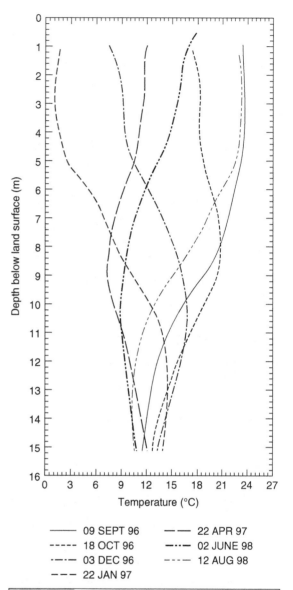

Figure 8.10 Groundwater temperature profiles adjacent to the Rio Grande at a site near Albuquerque, New Mexico (after Bartolino and Niswonger, 1999).

temperatures that produced the best agreement between measured and simulated water levels and temperatures. Final results showed that simulated temperatures differed from measured values by 0 to 2.1°C (most differences were less than 0.5°C). Calculated drainage rates, averaged over the study period, were between 2×10^{-7} and 4×10^{-7} m/s.

8.5 | Discussion

Analytical and numerical models that simulate heat flow are useful tools for estimating drainage rates within the subsurface. Models are calibrated so that results mimic patterns of measured temperatures with the objective of identifying the advective component of heat flow. A key reason for the success of this approach is that thermal properties of geologic material vary within a range that is much smaller than the range over which hydraulic conductivity varies (Constantz et al., 2008). So models can be used to determine the values of hydraulic conductivity (or flux) that produce the best agreement with measured temperatures. For estimation of focused drainage, model results are compared with temporal fluctuations in temperatures, usually diurnal or seasonal. For estimation of diffuse drainage, steady water and heat flow are assumed, and model results are compared with a temperature–depth

profile. Analytical models are often more easily applied than numerical models, but numerical models have a greater ability to simulate complex hydrologic systems because they are less limited by assumptions.

Temperatures within the geothermal zone represent long-term average conditions, reflective of a large block of the subsurface. These temperatures respond only slowly to any changes in surface conditions. Temperature–depth profiles within the geothermal zone have been used to predict past ground-surface temperatures and to identify periods of climate change (Clow, 1992; Huang et al., 2000; Beltrami et al., 2005). Estimated drainage rates within the geothermal zone reflect diffuse, steady drainage.

Temperatures within the surficial zone fluctuate on diurnal and seasonal patterns. Temperature-based estimates of focused recharge from streams are generally for a single point in space. Estimates based on diurnal temperature patterns are considered point estimates in time; those based on seasonal patterns

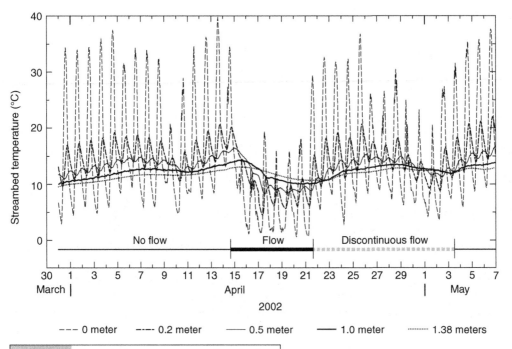

Figure 8.11 Streambed temperatures at various depths beneath Trout Creek over periods of no channel flow, continuous flow, and discontinuous flow (after Prudic et al., 2007).

represent a seasonal or annual average. Stream losses are best reported in units of volume of water loss per unit length of stream per time (L^2/T). In this manner, the total amount of loss from a stream reach can be determined from knowledge of reach length. Many studies report streambed infiltration or exfiltration rates as one-dimensional vertical flux rates (L/T); these rates are useful for determining hydraulic conductivity for input to a groundwater flow model, for example, or for examining the variability of hydraulic conductivity or flux across a streambed (Conant, 2004; Genereux *et al.*, 2008; Kennedy *et al.*, 2008). However, one-dimensional infiltration rates by themselves do not allow for calculation of total stream loss rates.

Temperature measurements provide other useful hydrologic information in addition to allowing direct quantification of recharge rates. Streambed temperature measurements offer an economical means for monitoring flow duration in intermittent streams (Constantz *et al.*, 2001; Prudic *et al.*, 2003, 2007; Blasch *et al.*, 2004). The presence of water in a stream channel substantially reduces the magnitude of diurnal temperature fluctuations in channel sediments (Figure 8.11). Streamflow duration is an important factor in recharge from intermittent streams (Besbes, 2006; Dahan *et al.*, 2007; Niswonger *et al.*, 2008). Becker *et al.* (2004) formulated a streamflow hydrograph separation routine based on measured surface and groundwater temperatures (Section 4.6.1). Cartwright (1970) used differences between measured temperatures at depths of about 160 m with those predicted solely on the basis of conductive geothermal heat flow to map recharge and discharge areas within the Illinois Basin. Measured temperatures were less than predicted in recharge areas and greater than predicted in discharge areas. Reiter (2003) used temperature logs in observation wells to determine the influence of faults on groundwater flow in the Albuquerque Basin.

Linking estimation methods to conceptual models of groundwater recharge

9.1 | Introduction

The selection of appropriate methods for estimating groundwater recharge should be tied to a conceptual model of recharge processes; assumptions inherent in any method must be consistent with that conceptual model. The emphasis of Chapters 2 through 8 was on estimation methods. Various categories of methods were described and systematically analyzed with particular attention to underlying assumptions. The objectives of this final chapter are to illustrate how methods for estimating recharge are tied to conceptual models and to provide some broad guidelines for selecting methods.

Section 9.2 provides a reexamination of the conceptual model development discussed in Chapter 1 in light of the information provided in the intervening chapters. A comparison of the various families of methods is provided in Section 9.3; tables summarize recharge processes, space and time scales of applicability, and the relative expense and complexity of methods. Section 9.4 contains discussions of conceptual models of recharge processes that have been developed and used within different groundwater regions of the United States. Also included in Section 9.4 are discussions of methods that have been applied in support of those conceptual models and a necessarily brief sampling of recharge studies that have been conducted within each region. The chapter concludes with

final thoughts related to future developments in estimating groundwater recharge.

9.2 | Considerations in selecting methods for estimating recharge

The first consideration in selecting methods for estimating recharge is the goal of the study (e.g. to assess water resources of a basin, to provide model input, to evaluate potential for groundwater contamination). Forming an initial conceptual model of the hydrologic system is the next step. The conceptual model identifies the prominent recharge mechanisms, provides initial estimates of recharge rates, and serves as a guide for the selection of methods and of locations and times for data collection. Recharge processes are largely controlled by climate, geology and soils, topography, hydrology, and vegetation and land use; these factors need to be considered in formulating the conceptual model. Analysis of existing data and the results of recharge studies in similar areas can help to shape the initial conceptual model. The conceptual model usually evolves over time as data are collected and analyzed; new information and interpretations may support revision of the conceptual model or suggest the application of alternative approaches. It is important to identify recharge mechanisms because some methods are specifically designed to estimate focused recharge, whereas others are designed for estimating diffuse recharge exclusively.

Spatial and temporal variability are important both in constructing a conceptual model and in selecting and applying methods. Diffuse recharge can occur somewhat uniformly across the area of interest at specific times during the year, but patterns of land use or soil permeability can result in a nonuniform distribution of recharge. Focused recharge occurs at specific points in a watershed; a stream can be a source of focused recharge during part of the year but a sink for groundwater discharge at other times of the year.

Space and time considerations must also be tied to study objectives. What types of recharge estimates are desired? Long-term averages? Annual or monthly values? Historical rates of recharge for periods prior to human development? Is an average estimate required for an entire aquifer or watershed? Or is information on spatial variability needed for an aquifer vulnerability study? Answers to these questions will help guide selection of appropriate methods.

Each estimation method is associated with specific space and time scales. The spatial and temporal scales of interest in a study need to be matched with those of the selected estimation techniques. If one is interested in recharge estimates for a large watershed (say in the order of hundreds of square kilometers), then methods that integrate over large areas (e.g. water-budget methods, streamflow hydrograph analysis, groundwater-flow modeling) might be preferable to methods that provide an estimate at a point in space (e.g. the zero-flux plane method). This is not to say that point estimates are not useful in large basins. Integration of multiple point estimates within a watershed can produce a meaningful watershed-wide estimate if sufficient data points are available. New modeling tools, such as combined watershed/ groundwater flow models (Section 3.6), can provide estimates of recharge at discrete locations and times in a watershed, while also providing an average estimate for the entire watershed.

In term of time scales, some methods provide recharge estimates for individual precipitation events. Other methods produce estimates that might be averaged over a single year, multiple

years, several decades, or even centuries and millennia. If a study site is in an area of shallow water table where recharge occurs consistently on a seasonal basis, then methods based on tracers in the unsaturated zone may be of only limited use. Bomb pulses of tritium and chlorine-36 most likely would have been flushed out of the unsaturated zone in these areas. If variability in interannual recharge was also a concern at this site, then natural environmental and historical groundwater tracers would likewise be of little use, because these tracer methods provide an average recharge rate that is integrated over several years.

Time constraints of the study are also important. If recharge estimates need to be developed in a short time (months), then techniques based on long-term monitoring (several years) are clearly inappropriate if the requisite data are not available. Tracer techniques may be suitable for a short-term study because they usually require only a one-time sampling and provide an estimate of recharge averaged over multiple years.

As with time constraints, financial constraints often are a controlling factor in selecting methods. Analyses of unsaturated-zone cores for tritium and chlorine-36 can be very useful in some arid regions. However, the cost of obtaining the samples (especially from thick unsaturated zones) and analyzing the samples in the laboratory may be prohibitively expensive for some studies. One point should be clear, though. The cost of applying a method does not necessarily correlate with the accuracy or appropriateness of the method. The fact that one method costs considerably more than a second method should not imply that the results of the first method are any better than those of the second. Similarly, just because a method is very inexpensive to apply does not mean that there is no value in using the method. Any method that is affordable and easy to apply merits consideration.

The anticipated accuracy of estimates obtained with any method should be evaluated as thoroughly as possible before applying a method. This is no simple task. It is difficult to apply formal error analysis because there are several sources of error, and they come into play in all segments of the estimation process. The most serious errors arise due to formulation of an incorrect conceptual model of recharge processes. Development of a conceptual model and estimation of recharge rates are iterative processes. Therefore, it is natural that revisions be made to the conceptual model as more information becomes available. Continual reevaluation of the conceptual model is critical.

Inappropriate application of a method and violations of one or more assumptions of a method are other common sources of error. For example, estimates of recharge by the water-table fluctuation method are based on rises in groundwater levels. As explained in Chapter 6, water levels can rise because of factors other than the occurrence of recharge. Failure to account for rises due to changes in barometric pressure or entrapped air or changes in pumpage will lead to inaccuracies in predicted recharge rates. Measurement errors also contribute to overall error. The Darcy method requires measurement or estimation of hydraulic conductivity; however, hydraulic conductivity is highly variable in space and is also very difficult to measure accurately. Errors also can accrue if spatial and temporal variability are not properly accounted for.

Water-budget methods and modeling approaches are amenable to classical error analysis, as described in Chapters 2 and 3. Such analyses are useful, but they have to be viewed with some caution because they are based on the assumption that the conceptual model is correct. If the conceptual model is incorrect, the error analysis is meaningless. Application of multiple methods will not necessarily improve estimation accuracy in a quantitative sense. However, having multiple estimates of recharge is beneficial in several respects. Consistency among estimates lends support to validity of the estimates. Inconsistency among estimates can provide insight on measurement errors and invalid assumptions and thus indicate in which manner the conceptual model could be revised. Many methods can be applied by drawing on the vast amount of existing data without the need to collect additional data.

9.3 | Comparison of methods

Previous chapters have presented details of individual methods. Here the main features of all methods are discussed and summarized in Tables 9.1 and 9.2. Information provided in these tables represents the views of the author and should not be viewed as strict guidelines. The tables can assist readers in identifying potentially useful methods and pointing them toward the chapters that contain details on the methods. The information contained in the tables is not all self explanatory. A full understanding of the content of the tables requires a careful reading of this section.

Table 9.1 indicates whether a method can be applied for focused and/or diffuse recharge estimates. Table 9.1 also indicates whether a method estimates an actual recharge rate (R), a rate of drainage (D) through the unsaturated zone, or base flow (B). According to the definitions given in Chapter 1, the term *recharge* applies to water that actually arrives at the water table. The term *drainage* applies to a vertical water flux within the unsaturated zone beneath the zero-flux plane (bottom of the root zone). Draining water will eventually become recharge when it arrives at the water table. Methods that use data collected in the unsaturated zone usually generate estimates of drainage. For methods that assume the recharge is steady in time, such as the Darcy unsaturated-zone unit-gradient method, drainage and recharge are equivalent terms. Water-budget and modeling methods can generate estimates of either recharge or drainage, depending on assumptions made for specific applications.

Table 9.1 contains information on time scales, i.e. the intervals over which a recharge estimate can be calculated. For example, a water budget of a soil column can be calculated on an event (or daily) basis or on the basis of weekly data. But it is generally impractical to calculate a recharge estimate based on average annual values for precipitation and evapotranspiration. Annual, multiannual, and decadal estimates of recharge can still be obtained with the water-budget method by performing the necessary

Table 9.1. Time scales for application of individual methods for estimating groundwater recharge. "Event" refers to a time period that may be less than but not more than 1 day. "Steady" indicates that a method can provide a time-invariant estimate of recharge. Under "Data collection frequency", 0 means that existing data are used and no new data need to be collected, 1 means that data need to be collected at only one time, and m implies that data must be collected multiple times. Additional details are discussed in the text. UZ is unsaturated zone. WB is water budget. GW is groundwater. WS is watershed. SW is surface water. HS is hydrograph separation. R is recharge. D is drainage through the unsaturated zone. B is base flow. CFC is chlorofluorocarbons. SF_6 is sulfur hexafluoride.

Method	Type		Recharge	Time scales								Data collection frequency
	Focused	Diffuse	Drainage or Base flow	Event/daily	Weekly	Seasonal	Annual	Multi-annual	Decadal	Millennial	Steady	
Water budget												
aquifer	X	X	R,B	X	X	X	X				X	m
soil column		X	R,D	X	X							m
watershed	X	X	R,D,B	X	X	X	X					m
stream	X	X	R,D,B	X	X							1,m
Models												
UZ soil WB		X	R,D	X	X							0,1,m
UZ Richards equation		X	R,D	X	X							0,1,m
watershed (WS)	X	X	R,D,B	X	X							0,1,m
GW flow	X	X	R,B	X	X	X	X				X	0,1,m
combined WS/GW	X	X	R,B	X	X							0,1,m
empirical	X	X	R			X	X				X	0
Darcy methods												
UZ		X	D	X	X							m
UZ unit gradient		X	D								X	1
GW	X	X	R	X	X	X						m
SW/GW	X		R	X	X	X						m

Table 9.1. (*Cont.*)

Method	Type		Recharge	Time scales								Data collection frequency
	Focused	Diffuse	Drainage or Base flow	Event/daily	Weekly	Seasonal	Annual	Multi-annual	Decadal	Millennial	Steady	
UZ/GW methods												
zero-flux plane	X	X	D	X	X							m
lysimeter		X	D	X	X							m
water-table fluctuation		X	R	X	X	X						m
Surface Water Based												
seepage meter	X		R,D,B	X								l
step-response function	X		R	X	X							m
flow duration		X	B					X	X			o
hydrograph separation (HS)		X	B					X	X			o
recession-curve displacement	X		R,B					X	X			o
chemical HS		X	B	X	X	X	X	X	X			m
tracer injection		X	B	X		X		X	X			m

Tracer methods

Method	Type		Recharge	Time scales								Data collection frequency
	Focused	Diffuse	Drainage or Base flow	Event/daily	Weekly	Seasonal	Annual	Multi-annual	Decadal	Millennial	Steady	
UZ chloride		×	D					×	×	×	×	—
UZ tritium		×	D					×	×			—
UZ chlorine-36		×	D					×	×			—
UZ applied	×	×	D	×	×	×	×	×				m
UZ heat		×	D		×	×						m
GW chloride	×	×	R						×	×	×	—
GW carbon-14	×	×	R						×	×	×	—
GW tritium	×	×	R					×	×			—
GW chlorine-36	×	×	R					×	×			—
GW CFC	×	×	R					×	×			—
GW SF$_6$	×	×	R					×	×			—
GW tritium/ helium-3	×	×	R					×	×			—
GW applied	×	×	R	×	×	×	×					m
SW/GW heat	×		R	×	×	×						m

Table 9.2. Space scales, relative expense, and complexity of application of individual methods for estimating groundwater recharge. Space scales are given as areas, with the exception of the stream water-budget method for which the scale refers to the length of stream channel over which measurements are integrated. Expense and complexity are ranked on a scale of 1 (lowest) to 5 (highest). Additional details are discussed in the text. UZ is unsaturated zone. WB is water budget. GW is groundwater. WS is watershed. SW is surface water. HS is hydrograph separation. CFC is chlorofluorocarbons. SF_6 is sulfur hexafluoride.

	Chapter	Space scales 1 m^2	10 m^2	100 m^2	1 ha	1 km^2	10^3 km^2	10^6 km^2	Relative expense	Relative complexity
Water budget										
aquifer	2				×	×	×	×	2 to 4	2 to 4
soil column	2	×							3	3
watershed	2				×	×	×	×	2 to 4	2 to 4
stream	4			×	×	×			5	4
Models										
UZ soil WB	3	×							2	2
UZ Richards equation	3	×	×	×	×				4	4
watershed	3				×	×	×	×	2 to 5	5
GW flow	3				×	×	×	×	2 to 5	5
combined WS/GW	3				×	×			4 to 5	5
empirical	3	×	×	×	×	×	×	×	1	1
Darcy methods										
UZ	5	×	×						5	5
UZ unit gradient	5	×	×						3	3
GW	6	×	×	×					2	2
SW/GW	4	×	×	×					3	3
UZ/GW methods										
zero-flux plane	5	×	×						5	4
lysimeter	5	×	×	×					1 to 5	1 to 5
water-table fluctuation	6		×	×					2	2
Surface water based										
seepage meter	4	×							2	2
step-response function	4			×	×	×	×		3	3
flow duration	4					×	×	×	1	1
hydrograph separation (HS)	4					×	×		2	3

Table 9.2. (*Cont.*)

		Space scales							Relative expense	Relative complexity
	Chapter	1 m^2	10 m^2	100 m^2	1 ha	1 km^2	10^3 km^2	10^6 km^2		
recession-curve displace-ment	4					×	×		2	3
chemical HS	4				×	×	×		5	4
tracer injection	4				×	×	×		4	3
Tracer methods										
UZ chloride	7	×	×						3	2
UZ tritium	7	×	×						3	3
UZ chlorine-36	7	×	×						4	3
UZ applied	7	×	×	×					4	3
UZ heat	8	×	×						3	3
GW chloride	7			×	×				2	2
GW carbon-14	7			×	×				3	3
GW tritium	7		×	×					3	3
GW chlorine-36	7		×	×					4	3
GW CFC	7		×	×					3	3
GW SF$_6$	7		×	×					3	3
GW tritium/ helium-3	7		×	×	×				5	4
GW applied	7		×	×					4	3
SWGW heat	8	×							3	3

calculations with smaller time intervals (e.g. daily or weekly) and summing over the period of interest. As discussed in Chapter 2, the interval over which calculations are made affects estimated recharge rates. For water-budget methods, shorter time intervals generally produce larger estimates of recharge (Section 2.2).

While it is possible to upscale estimates by summing estimates generated with shorter time intervals, it is generally not possible to take estimates derived from annual or multiannual data to develop estimates for shorter time intervals. Methods based on tracers typically provide a recharge estimate that is an average value for the time interval between tracer application and sampling. An estimate based on chlorofluorocarbon (CFC) concentrations in groundwater (Section 7.3.3) could provide an average recharge rate for perhaps the previous 20 years, but no information would be provided on the year-to-year variability in recharge over that period.

Individual environmental and historical tracers are listed in Tables 9.1 and 9.2 as opposed to the specific tracer methods (profile, mass balance, peak displacement, and age dating) discussed in Chapter 7. This was done for purposes of clarity; multiple methods can be applied with data for a single tracer, and the time scales listed for each tracer generally are valid for any of the methods. An exception to this rule occurs when using the unsaturated zone peak-displacement method for a tracer, such as chloride, under conditions of changing land use. In this case, times scales given for applied tracers are more appropriate. Individual applied tracers are not listed in Table 9.1 or 9.2 because, in general, time and space scales are similar for all applied tracers. The tritium/helium-3 method generally is applicable only when the recharge rate is at least 30 mm/yr because of diffusion of helium in groundwater (Section 7.3.3). Historical groundwater and unsaturated-zone tracers, such as tritium, CFCs, and sulfur hexafluoride (SF_6), are applicable only for specific ranges in groundwater ages and sampling dates (Chapter 7) depending on the history of atmospheric concentrations (e.g. SF_6 is only good for dating water recharged after about 1970).

A few additional points require clarification with regard to Table 9.1. The data collection frequency column indicates which methods require no data collection (these methods rely solely on the use of existing data), which methods (mostly tracer methods) require only a single field trip to collect data, and which methods require data collection at multiple times (this could mean continuous data recording or multiple field trips). Methods based on streamflow data, such as streamflow duration and hydrograph analysis, are usually based on mean daily streamflows. However, the estimates generated by these methods are averages over multiyear or decadal periods. Hydrograph-separation and recession-curve displacement methods are applicable over identical space and time scales. The Darcy unsaturated-zone unit-gradient method is based on the assumption that recharge is a steady process; the method cannot be used to estimate nonsteady recharge. Other methods (e.g. water-table fluctuation, zero-flux plane, step-response function, and streamflow recession-curve displacement method) are based on the assumption that all recharge occurs as discrete episodes in time; therefore, these methods do not account for recharge that occurs as steady flow.

Models deserve particular attention in any discussion of time scales. They can be applied over a multitude of time intervals, including periods for which no data may exist (e.g. when making predictions for future periods). Model inputs can be based on actual data or synthetic data (such as weather or climate trends) generated independently outside of the model. In addition to providing recharge estimates, models have many other uses (Chapter 3). They can be used to integrate diverse types of data and to explore various hypotheses on recharge processes; in so doing, models can help to shape conceptual models. Models can provide error or uncertainty bounds on recharge estimates that they generate. Sensitivity analysis can be used to identify model parameters that most influence calculated recharge rates, and, therefore, help guide data-collection efforts. The effects of temporal and spatial variability of model parameters on timing and locations of recharge can also be investigated.

Spatial scales over which methods are applicable are listed in Table 9.2. As with time scales, estimates made at small scales can be used to calculate estimates at larger scales, but unlike the case for temporal scales, upscaling is not simply a matter of summing up the smaller scale estimates. Upscaling, as described in Chapter 3, usually takes into account the variability in factors described in Section 9.2, such as climate and land use, that affect recharge rates. Downscaling is problematic; if a recharge estimate represents an integrated value over some area (e.g. from application of a method such as streamflow hydrograph separation), it is generally not possible to infer information on recharge processes at specific points within that area.

Whether an estimate can be called a point estimate in space is a matter of perspective. From the viewpoint of a watershed that drains 1000 ha, estimates of recharge generated over areas of 1 ha may be considered a point estimate. If the watershed area is only 1 ha, the area pertaining to a point estimate decreases equivalently. For discussion purposes, results from methods applied over the smallest spatial scale, 1 m², are considered to be point estimates. Whether a point estimate is appropriate for a study should be determined in light of study objectives. If aquifer vulnerability to contamination is of concern, point estimates are useful for describing spatial variability of recharge and perhaps for identifying areas that are influenced by preferential flow paths. If an average recharge rate within a watershed is required, methods that provide point estimates of recharge may need to be applied at a number of points in the watershed.

There are some additional items of note with regard to Table 9.2. The spatial scale for the stream water-budget method is reported in units of length of stream channel rather than in terms of area. This is a focused form of recharge, and the recharge rate of interest is the rate of volumetric loss of water to the subsurface per unit length of stream channel in units of L^2/T; the concept of contributing area does not apply. The presence of preferential flow paths within the unsaturated zone may adversely affect some methods. Methods that provide point estimates of recharge are particularly affected; results produced by them can vary substantially, depending on whether the measurement area coincided with a preferential flow path.

The expense and complexity of applying a method are presented in a relative sense in Table 9.2. Again, these are subjective ratings based on the author's experiences and should not be taken as strict rules. The rating system runs from 1 (lowest) to 5 (highest). This information is included to assist the reader in screening out inappropriate methods. Some methods can be applied at different levels of complexity and expense. Water-budget methods can be simple and inexpensive if using only existing

data, or they can be quite complex if individual water-budget components are measured in the field. Similarly, models that have been used for estimating recharge range from very simple to quite complex. Models can be expensive to apply if training is required, but experienced modelers can often generate results efficiently.

9.4 | Recharge characteristics of groundwater regions of the United States

Discussion of recharge processes in every conceivable hydrogeologic environment, although desirable, is beyond the scope of this chapter. Natural and human-affected hydrologic systems, and associated conceptual models of recharge processes, are too diverse and complicated to describe without in-depth analysis. The aim of this section is to describe one or more conceptual models of recharge processes that have been adopted within different groundwater regions of the United States. The intent is to illustrate concepts of where, when, and why recharge occurs in typical hydrologic, geologic, and climatic settings. Examples point out the types of methods that have been applied and provide insight into the range of recharge rates that have been obtained. Limited space permits only brief discussion of examples, but many references are provided to allow interested readers to pursue additional information.

Attempts at categorizing hydrogeological and climatic provinces for purposes of identifying regions of similar hydrologic response were discussed in Section 1.4.9. As concluded in that section, no system is ideal. For discussion in this section, the classification system of Thomas (1952), which consists of 10 groundwater regions (Figure 9.1), has been adopted, largely on the basis of its simplicity. This was a somewhat arbitrary decision; other classification systems could equally serve this purpose. Features on which the Thomas (1952) classification system is based, and which are most pertinent to the topic of groundwater recharge, include geology, the water-bearing properties of prominent aquifers, and the nature and location of recharge and discharge. Descriptions of hydrologic processes within the regions are drawn from Meinzer (1923), Thomas (1952), and Heath (1984), as well as from references cited for each region. Although the regions are uniquely defined, similar conceptual models of recharge processes are often adopted in multiple regions, and areas over which a method is applied can overlap regional boundaries. Recharge characteristics in urban settings, topics that pertain to all 10 regions, are discussed in Section 9.4.11.

Discussion of groundwater regions is limited to those within the United States only because of the author's familiarity with groundwater conditions in this country. Recharge processes in similar regions in other parts of the world are no doubt similar in many respects to those described below. So readers involved with recharge studies outside of the United States can still benefit from the material contained in the following sections.

No attempt is made to balance the discussion among the different regions. Some discussions are extended in length; others are brief. This could be because of the number of studies conducted in a region or the relative importance of groundwater resources in a region or simply the existence of well-documented studies. Examples were selected to illustrate a point or a method and represent only a small fraction of studies conducted. Details on all the individual methods cited in the following sections are provided in Chapters 2 through 8 and are summarized in Tables 9.1 and 9.2.

Seldom do publications of recharge studies chronicle the development of a conceptual model of a hydrologic system. Usually, only the final conceptual model and estimated recharge rates are reported, and the process by which the final conceptual model was arrived at remains largely unknown to the reader. An exception is the study of water movement through the unsaturated zone at Yucca Mountain, Nevada.

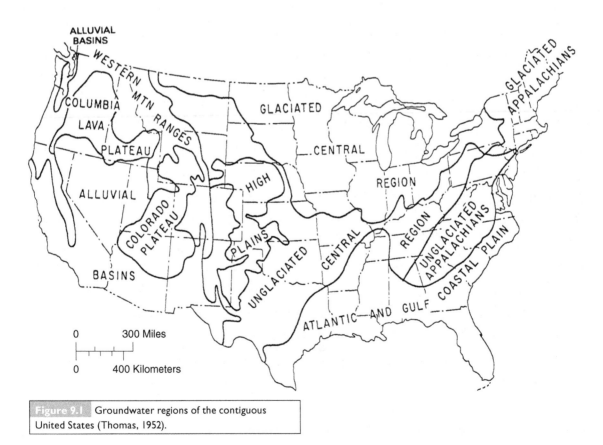

Figure 9.1 Groundwater regions of the contiguous United States (Thomas, 1952).

Flint *et al.* (2001a) provided an instructive description of how the conceptual model of water movement through the unsaturated zone evolved over time as new data were collected and analyzed. A brief overview of this work is presented in Section 9.4.2.

9.4.1 Western Mountain Ranges

The Western Mountain Ranges region consists of tall mountains and narrow intermontane valleys. The mountains generally consist of low-permeability rock with a thin covering of soil. Groundwater exists in the rock matrix, but most available water occurs in fractures. The aperture, density, and connectivity of the fractures control the storativity and transmissivity of the rock. Low yielding wells, screened in bedrock, are common for domestic water supplies. The most productive areas in this region are the intermontane drainages where alluvial deposits are present. The Salt Lake Valley in north central Utah, for example, contains important highly productive aquifers (Manning and Solomon, 2004).

Snowfall at high elevations is the primary source of water in this region. Snow accumulates during the winter, and water is released from the snowpack as temperatures rise in the spring. Recharge to the mountains occurs both in diffuse and focused forms. Water can infiltrate the mountain block or runoff. Mountain streams can be both gaining and losing streams over different reaches. Figure 9.2 shows a schematic of the hydrology of a mountain and adjacent basin aquifer. Mountain-front recharge refers to the accretion of water in a basin aquifer adjacent to the mountain (Wilson, 1980). That accretion occurs in two forms: focused recharge from streams that flow from the mountain and subsurface flow from the mountain. The latter form is sometimes referred to as mountain-block recharge (Wilson and Guan, 2004; Manning and Solomon, 2004). According to definitions given in Chapter 1, this phenomenon is interaquifer flow rather than recharge. The phenomenon is also referred to as underflow; this flow is referred to as subsurface mountain-block discharge in this chapter. Regardless of differences in terminology,

excellent discussions of hydrologic processes in mountain and basin settings can be found in Feth (1964), Wilson (1980), Wilson and Guan (2004), and Stonestrom and Harrill (2007).

Methods for estimating mountain-front recharge are discussed in some detail in Wilson (1980) and Wilson and Guan (2004). Because focused recharge from streams is an important process, methods such as water budget;

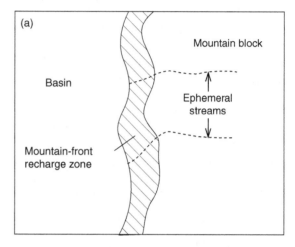

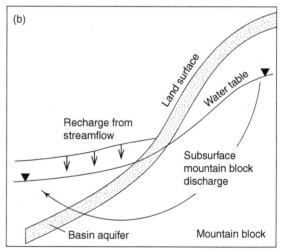

Figure 9.2 Schematic diagrams showing map view (a) and cross section (b) of mountain block and basin aquifer. Water flows from the mountain block to the aquifer as focused recharge from ephemeral streams that drain the mountain block (mountain front recharge) and as subsurface discharge, or interaquifer flow (after Wilson (1980) and Wilson and Guan (2004)).

chemical, isotopic, and heat tracer; Darcy; and modeling approaches are commonly applied. Water-budget methods (Feth *et al.*, 1966) and environmental groundwater tracers (Manning and Solomon, 2004) have been used to estimate subsurface mountain-block discharge to different parts of the Salt Lake Valley.

Increasing populations in the western United States underscore the need for sustainable groundwater supplies for the increasing number of households in mountainous areas. Groundwater in valley aquifers has historically been of high interest; recharge studies in mountain blocks, however, have garnered little attention (Wilson and Guan, 2004). Recent studies have begun to address mountain recharge directly, in spite of difficulties imposed by complex geologies and large differences in elevations. Bossong *et al.* (2003) used a watershed model and estimated base flow to be equivalent to 19 mm/yr (3% of precipitation) across a 120 km^2 area of the Turkey Creek watershed in Colorado. Elevations across the Turkey Creek watershed vary by more than 1000 m.

9.4.2 Alluvial Basins

The Alluvial Basins region consists of alternating basins and mountain ranges and covers much of the southwestern United States, as well as parts of Washington and Oregon. The region is dominated by broad valleys that are bordered by mostly low and narrow mountain ranges. This is in contrast to the Western Mountain Ranges in which the valleys are narrow and the high mountains are the dominant topographic feature. The coastal ranges in southern California are an exception to this description; although contained within the Alluvial Basins region, they actually resemble the Western Mountain Ranges in a topographic sense. Climate in this region is mostly arid to semiarid. Precipitation rates in the basins of Nevada and Arizona are typically in the range of 100 to 400 mm/yr (Heath, 1984).

The basins consist primarily of erosional debris and can store substantial amounts of water. Basin aquifers are the primary source of water supply in the region, serving even large metropolitan areas such as Albuquerque and Tucson. Recharge to these aquifers occurs as mountain-front recharge as described in Section 9.4.1. Ephemeral streams that drain adjoining mountains tend to lose their water to infiltration as they flow across valley floors. Susbsurface flow from mountain blocks is also an important source of water in parts of the region where mountains consist of permeable carbonate rock. Diffuse recharge from precipitation can occur in the valleys, as shown by Stephens and Knowlton (1986), but usually this is a small percentage of total recharge. In undisturbed watersheds, natural groundwater discharge is mostly by evapotranspiration, often through playas in the basin floors, or by groundwater flow to neighboring basins. Historically, groundwater discharged to streams in some areas, such as the Santa Cruz River in southern Arizona, but groundwater pumping has resulted in declining groundwater levels and a substantial reduction in groundwater discharge to streams in many areas (Webb and Leake, 2006).

Recharge processes for watersheds in the coastal mountains of California are generally represented by a slightly different conceptual model. Most recharge originates as precipitation in the mountains and either infiltrates or flows over the surface to streams in the valley floors. The streams are usually losing streams, providing water to underlying aquifers. However, unlike systems in more arid settings, these valley streams do not typically lose all of their flow to infiltration. A substantial amount of streamflow can flow directly into the ocean rather than becoming recharge, especially for streamflow generated by intense rainfall events.

Most methods for estimating recharge discussed in Section 9.4.1 are also applicable in the Alluvial Basins region. In particular, Wilson (1980) and Wilson and Guan (2004) discussed a number of approaches. Recharge to the Central Valley aquifer system in California (the largest groundwater system in the Alluvial Basins region) was estimated through application of a highly detailed groundwater-flow model (Faunt, 2009). Flint and Flint (2007) applied a soil water-budget model (Basin Characterization Model) to generate an average estimate of drainage beneath the root zone of 20 mm/yr

(7% of precipitation) over the Basin and Range carbonate-rock aquifer system in Nevada and parts of Utah. An empirical model based on annual precipitation and elevation developed by Maxey and Eakin (1949) is still extensively used in Nevada (Section 3.7). Unsaturated-zone tracer profiles of bomb-pulse tritium and chlorine-36 have been used to estimate rates of drainage through the unsaturated zone that are averaged over the time period between the 1950s and 1960s, when peak atmospheric concentrations occurred, and the time at which samples were collected (Phillips *et al.*, 1988; Scanlon, 1992). In particularly arid areas, the chloride mass-balance method has been used to estimate drainage rates through the unsaturated zone that are averaged over periods of hundreds to thousands of years (Phillips, 1994; Tyler *et al.*, 1996). Numerous additional studies of recharge to mountains and basin aquifers are described in Wilson (1980), Hogan *et al.* (2004), and Stonestrom *et al.* (2007).

In open basins that are traversed by perennial streams, focused recharge to underlying aquifers may support evapotranspiration by riparian phreatophytes, and thus not be available for human consumption. Culler *et al.* (1982) used a stream water-budget approach to determine evapotranspiration rates before and after phreatophytes were cleared from the Gila River floodplain in Arizona. In closed basins with no surface water outflow, groundwater is discharged to the atmosphere through evapotranspiration or to adjoining basins as groundwater flow. Nichols (2000) used evapotranspiration measurements and estimates of interbasin flow to generate water budgets and recharge estimates for basins in the Great Basin area of Nevada.

Alluvial valleys actually exist in many regions. In his description of groundwater regions of the United States, Heath (1984) listed Alluvial Valleys as a separate noncontiguous region. The hydrology of these valleys is similar, in some ways, to the hydrology described above; water flows to the valleys from adjacent uplands. Climate, however, plays an important role in the hydrology of these systems. In humid areas, recharge is predominantly diffuse, and

streams in alluvial valleys are usually sinks for discharging groundwater, as opposed to being sources of focused recharge in arid and semi-arid regions. A particular estimation method may be appropriate in one alluvial valley but inappropriate in another.

Example: Yucca Mountain, Nevada

Throughout the 1980s and 1990s, studies were conducted to assess the suitability of Yucca Mountain as a repository for high-level radioactive waste. Of major concern and interest was the amount of water that waste would be exposed to, i.e. the rate of water drainage through the unsaturated zone. Flint *et al.* (2001a) provided an excellent description of the conceptual model of water flow through the unsaturated zone and discussion of how the conceptual model evolved over time as more data and results of computer simulations became available. The Yucca Mountain study may be the most expensive and most detailed study ever conducted on water movement through an unsaturated zone.

The unsaturated zone at Yucca Mountain is approximately 500 m thick at the planned repository location and consists of layers of volcanic tuff, some of which are highly fractured, others of which have few fractures but are penetrated by faults. Average annual precipitation is about 170 mm. Initial estimates of drainage rates of less than 1 mm/yr were based on few data and on what turned out to be flawed hypotheses on the manner in which water moves through variably saturated fractured rock. Data collected during the study included concentrations of natural chemical and isotopic tracers in the unsaturated and saturated zones (in rock matrix and on fracture walls), soil-water content and chemistry, soil-gas chemistry, soil matric potential, subsurface temperatures, groundwater levels and groundwater chemistry, precipitation, evapotranspiration, and runoff. Results from simulation models were crucial for improving the understanding of water movement through the mountain. Applied models included groundwater-flow models, variably saturated flow models, water-budget and energy-budget models, watershed models, and various geochemical models. Many different

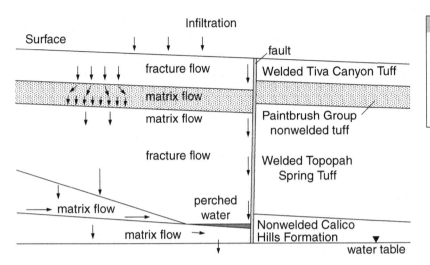

Figure 9.3 Conceptual model of water movement through the unsaturated zone at Yucca Mountain, Nevada (after Flint *et al.*, 2001b). Arrows indicate direction of water flow.

methods for estimating drainage rates through the unsaturated zone were applied, including water-budget, Darcy, zero-flux plane, modeling, and tracer methods (Flint *et al.*, 2002).

The conceptual model, as of 2000, suggests an average drainage rate of between 5 and 10 mm/yr, an order of magnitude greater than original estimates (Flint *et al.*, 2001a). The conceptual model also indicates that water moves in different ways through the different layers in the unsaturated zone (Figure 9.3). Matrix flow predominates in some layers, whereas in other layers water moves primarily through fractures. The approach taken in this study is instructive, and can be replicated at different scales for other studies – developing a conceptual model, using that model to guide data collection and analysis (including simulation models), revising the conceptual model, and repeating the loop.

9.4.3 Columbia Lava Plateau

The Columbia Lava Plateau is formed by extrusive volcanic rocks interbedded with or overlain by alluvium or lake sediments. The rocks, which are very permeable due to tubes and shrinkage cracks, make highly productive aquifers. The climate over most of the plateau is arid to semiarid, but mountains that border the region to the east and west can receive in excess of 2000 mm/yr of precipitation. Much of the natural recharge in the region occurs

as focused recharge from streams draining these mountainous areas. As one example, in most years, the entire flow from the Big Lost River in Idaho is lost to infiltration in the Snake River Plain in the vicinity of Lost River Sinks (Berenbrock *et al.*, 2007). Irrigated agriculture is the main industry; irrigation water is obtained from both surface and groundwater sources. Due to the permeable nature of the rocks, irrigation return flow, by drainage beneath the root zone and leakage from water delivery systems, is an important component of total recharge (Vaccaro and Olsen, 2007).

Because of the highly permeable sediments, water can move quickly from land surface to the water table, and there are few applicable methods for estimating recharge. Bauer and Vaccaro (1990) and Vaccaro (1992) used a water-budget/watershed model (the Deep Percolation Model) to estimate recharge for the entire Columbia Plateau regional aquifer system. Vaccaro and Olsen (2007) applied the Deep Percolation Model and another watershed model to the Yakima River watershed. Annual precipitation in the watershed ranges from about 175 mm in the lowland agricultural areas to about 3000 mm in upland forested areas. Recharge rates prior to the development of irrigation were estimated to range from 2 mm/yr to about 850 mm/yr; 97% of all recharge occurred in the uplands. With the onset of irrigation, total recharge for the area is estimated to have increased by more than 30%

over predevelopment totals as a result of return flow (Vaccaro and Olsen, 2007).

Nimmo *et al.* (2002b) described a large-scale tracer test above the Snake River aquifer in Idaho. Focused recharge from floodwaters of the Big Lost River were followed by means of a chemical tracer. Movement of the water in the subsurface was tracked by analyzing water samples obtained from numerous wells and piezometers. Estimates of recharge were not obtained, but the experiment provided insight into the rate and direction of water movement through the alternating layers of basalts and sediment that make up the 200 m thick unsaturated zone.

Estimates of natural diffuse recharge in more arid, nonirrigated areas of the plateau have been made for purposes of assessing potential mobility of buried waste material. A variety of approaches have been used to estimate drainage rates through the unsaturated zone near Hanford, Washington: Gee *et al.* (1992) used lysimeters; Prych (1998) applied the unsaturated-zone chloride mass-balance and chlorine-36 tracer profile methods; and Fayer *et al.* (1996) used lysimeters, the zero-flux plane method, and a numerical model. Estimated drainage rates were generally less than 2 mm/yr. Nimmo and Perkins (2008) also estimated rates of less than 2 mm/yr for a site in eastern Idaho by applying the Darcy unsaturated-zone unit-gradient method.

9.4.4 Colorado Plateau

This region consists of thick sedimentary strata (mostly interbedded shales, sandstones, and coal). The beds are roughly horizontal but may be folded, tilted, or broken by faults. The plateau has deeply incised streams. The climate in this region is semiarid, but mountainous areas may receive substantial precipitation. There are five regionally extensive bedrock aquifers of Mesozoic age, the most important of which may be the Navajo Sandstone. Recharge occurs mostly at higher elevations along margins of uplifts where aquifer rocks outcrop or are covered by a thin layer of permeable material and precipitation is greater than about 300 mm/yr (Freethey and Cordy,

1991). Focused recharge from streams is also important in some areas.

A variety of methods have been used to estimate recharge in this region. Watson *et al.* (1976) and Freethey and Cordy (1991) used modified versions of the Maxey-Eakin method (Maxey and Eakin, 1949; Section 3.7). Freethey and Cordy (1991) estimated average recharge for the upper Colorado River Basin to be 15 mm/yr. Heilweil and Freethey (1992) used an aquifer water-budget method for a part of the Navajo Sandstone, with measured or estimated groundwater discharges, to estimate recharge at 3 to 4 mm/yr. Recharge rates in the Black Mesa area of northeastern Arizona were estimated to be in the range of 13 to 19 mm/yr (4 to 6% of annual precipitation) on the basis of carbon-14 age dating and groundwater-flow modeling (Zhu, 2000) and at 10 mm/yr by using the groundwater chloride mass-balance method (Zhu *et al.*, 2003). Danielson and Hood (1984) applied the stream water-budget method and estimated focused recharge to the Navajo Sandstone from the Fremont River and other smaller streams to be about 2 500 000 m³/yr in the lower Dirty Devil River Basin in south-central Utah. Heilweil and Freethey (1992) reported similar rates of focused recharge from the North and East Forks of the Virgin River.

Heilweil *et al.* (2007) applied a variety of tracer techniques to estimate recharge to the Navajo Sandstone at Sand Hollow in southwestern Utah. The groundwater chloride mass-balance method was used to estimate a basin-wide average recharge of 8 mm/yr (4% of precipitation). Point estimates of unsaturated-zone drainage rates, determined with the unsaturated-zone tritium profile and tritium mass-balance methods at 18 borehole sites, ranged from 1 to 57 mm/yr; the wide range in estimates supports the hypothesis that recharge occurs primarily as focused recharge of runoff from bedrock outcrops and as direct infiltration in areas where surficial soils are coarse.

9.4.5 High Plains

The High Plains region, comprising an area of about 450 000 km² in the central Great Plains, is a remnant of a vast plain formed by sediments that were deposited by streams flowing eastward from the ancestral Rocky Mountains. The land surface is characterized by gently sloping plains that range in elevation from 2300 m in the west to 335 m in the east. Agriculture is the main industry, including dry-land and irrigated farming and ranching. Climate varies from arid to semiarid to subhumid with average annual precipitation ranging from 350 mm in the southwest to 800 mm in the northeast (McGuire *et al.*, 2003).

The High Plains aquifer is one of the most productive and most intensely developed aquifers in the United States. The unconfined aquifer consists of poorly sorted sand, silt, clay, and gravel. The large areal extent and thickness (up to 300 m) of the aquifer provide storage for as much as 3.7 trillion cubic meters of water (McGuire *et al.*, 2003). Water in the aquifer has accumulated over thousands of years. Recharge rates prior to the onset of development for irrigation in the 1940s were estimated to range from 0.8 to 26 mm/yr in the south and central parts and 2 to 39 mm/yr in the northern part of the aquifer (Luckey *et al.*, 1986; McMahon *et al.*, 2003). Depth to the water table is variable, ranging from 0 to about 150 m. Decades of pumping for irrigation have resulted in substantial water-table declines (in excess of 50 m) in parts of the south and central High Plains aquifer.

It is difficult to generalize about recharge mechanisms in an area as large as the High Plains. Undoubtedly, there are multiple sources of recharge throughout the region. Under rangeland vegetation, most recharge in the arid to semiarid southern High Plains occurs as focused recharge from surface-water bodies. The southern High Plains is extremely flat, and surface water tends to drain to ephemeral lakes and playas, which effectively serve as focused, natural recharge basins (Scanlon and Goldsmith, 1997; Gurdak and Roe, 2009). Diffuse recharge can occur under cropped land in the southern High Plains. Scanlon *et al.* (2007) applied the unsaturated-zone chloride mass-balance method at 19 locations in dry-land croplands and determined a median rate of drainage through the unsaturated zone of 24 mm/yr. McMahon *et al.* (2006)

used a similar approach at two sites under irrigation and estimated drainage rates of 17 and 32 mm/yr. Wood and Sanford (1995) applied the chloride mass-balance method by using groundwater concentrations and estimated an average recharge rate of 11 mm/yr for a large part of the area. Historically, groundwater discharge in the southern High Plains was to springs along the eastern escarpment. Falling groundwater levels have virtually eliminated discharge from most of these springs over the past several decades. Extraction of groundwater for irrigation is now the dominant discharge mechanism. Gurdak and Roe (2009) presented a review and synthesis of 175 publications on recharge in playa and interplaya settings, mostly within the southern High Plains.

Diffuse and focused recharge are both important in the central High Plains. Diffuse recharge rates are highest in areas with sandy soils, and rates tend to increase from west to east in tandem with the trend in annual precipitation (Luckey and Becker, 1999). Reported estimates of diffuse recharge under native rangeland vary from 5 mm/yr, as estimated with the unsaturated-zone chloride mass-balance method (McMahon *et al.*, 2006), up to 177 mm/yr in sandy areas with a shallow water table, as estimated with the water-table fluctuation method (Sophocleous, 1992). As with the southern High Plains, enhanced recharge is expected under agricultural crops. McMahon *et al.* (2003) used the unsaturated-zone tritium-profile method and estimated unsaturated-zone drainage rates of 21 and 116 mm/yr beneath two irrigated fields. Natural discharge before development was to gaining streams throughout the central High Plains. Many of these streams have dried up because of declining groundwater levels (Sophocleous, 2000) and are now ephemeral, losing streams that focus recharge when runoff from precipitation flows through stream channels. Results from a groundwater-flow model of the southern part of the central High Plains aquifer showed that recharge rates in 1997 were greater than those under predevelopment conditions (Figure 9.4) but that water storage within the aquifer had been reduced by about 18% (Luckey and Becker, 1999).

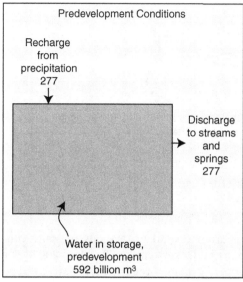

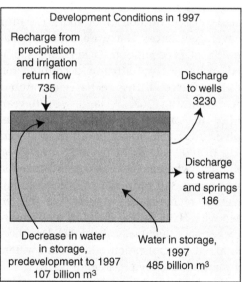

Groundwater budget in the southern part of the central High Plains aquifer during predevelopment and in 1997. Values without units are in million m³/yr (after Luckey and Becker, 1999).

In the northern High Plains, groundwater levels in many areas rose between 1950 and 2000 (McGuire *et al.*, 2003). Surface water is the primary source of irrigation water in these areas; irrigation return flow supplements natural recharge and has led to a rise in groundwater levels. Diffuse recharge is the dominant mechanism, but focused recharge from the

many perennial streams in the northern High Plains is important (Weeks *et al.*, 1988). Drainage rates within the unsaturated zone were estimated by using the tritium profile method (Figure 9.5) to be 70 mm/yr beneath undisturbed rangeland and 102 to 110 mm/yr beneath irrigated fields (McMahon *et al.*, 2006). At one of the irrigated sites, bomb-pulse tritium was not detected in the unsaturated zone at depths of 30 to 40 m but was detected in groundwater, at a depth of about 48 m. McMahon *et al.* (2006) concluded that recharge to the aquifer was occurring locally as focused recharge through nearby playas that collect surface runoff. Under natural conditions, drainage past the bottom of the root zone typically occurs from late fall into spring; however, excess irrigation can cause drainage to occur during the growing season.

9.4.6 Unglaciated Central Region

This expansive area extends from the Rocky Mountains eastward to the Appalachian Plateau and Valley and Ridge physiographic province and consists of rolling plains and plateaus underlain by sedimentary rocks (Heath, 1984). Dolomitic limestone and sandstone aquifers of moderate yield are common, and alluvial aquifers exist along major streams, but the region also contains some very productive aquifers, including the Edwards Limestone and the Trinity Group in Texas and the Ozark Plateaus aquifer system that lies mostly in Arkansas and Missouri. Large cave systems are a prominent feature of this region.

Recharge to the bedrock aquifers commonly occurs in upland areas where bedrock outcrops and can be diffuse or focused from streams. It is difficult to make broad statements as to the applicability of different methods for estimating recharge in this region because of the variability in climate (precipitation ranges from 400 to over 1200 mm/yr) and aquifer characteristics. Many approaches have been taken, including water-budget studies (Puente, 1978), tracer methods (Busenberg and Plummer, 1992), watershed and groundwater flow models (Czarnecki *et al.*, 2009), and streamflow hydrograph analysis (Hoos, 1990).

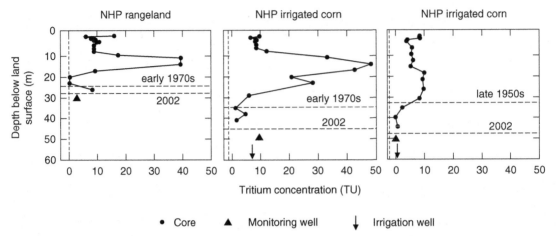

NHP rangeland   NHP irrigated corn   NHP irrigated corn

Figure 9.5 Profiles of tritium extracted from unsaturated zone sediment cores and tritium concentrations in groundwater from a rangeland and two irrigated sites in the northern High Plains (NHP). Vertical dashed lines represent the estimated maximum concentration of prebomb tritium in soil water in 2000. Horizontal dashed lines represent water-table depths at indicated times (McMahon et al., 2006).

Karst systems may occur in limestone, dolomite, and other carbonate rocks where solution cavities can act as preferential flow paths that focus recharge to and discharge from aquifers. Estimation of recharge rates in karst systems is particularly challenging; the flow paths are difficult to trace, and direct measurement of flow through a solution cavity is seldom possible (Bakalowicz, 2005). Recharge processes in karst systems have been studied worldwide. Natural and applied tracers are often used to obtain qualitative information on flow paths in karst systems (Greene, 1997; Katz et al., 1997). Streamflow and spring hydrograph analysis (Padilla et al., 1994; Pérez, 1997; Rimmer and Salingar, 2006) and chemical hydrograph analysis (Nativ et al., 1999; Lee and Krothe, 2001) have been used to determine base flow. Various modeling approaches have been used to estimate recharge in karst systems, including soil water-budget models (Hughes et al., 2008), groundwater flow models (Weiss and Gvirtzman, 2007), and watershed models (Ockerman, 2002, 2005, 2007).

The HSPF watershed model was used to estimate recharge to the Edwards and Trinity aquifer systems from several watersheds in south-central Texas by Ockerman (2002, 2005, 2007). For the Upper Cibolo Creek watershed, average annual recharge between 1992 and 2004 was estimated to be 98 million m³ or about 139 mm when averaged across the watershed, equivalent to 15% of precipitation (Ockerman, 2007). Focused recharge from streams accounted for 74% of all recharge; diffuse recharge accounted for the remaining 26%. A large year-to-year variability in streamflow and recharge was attributed to variable amounts of precipitation. Recharge in 1999 was estimated at 25 mm (6% of precipitation), whereas the recharge estimate for 1992 was about 340 mm (25% of precipitation). The average recharge estimate of 128 mm/yr obtained with HSPF for the Hondo Creek watershed (Ockerman, 2005) was about 22% less than that obtained by application of a stream water-budget method proposed by Puente (1978).

A soil-water budget model was used to estimate recharge over a large part of the central United States by Dugan and Peckenpaugh (1985); estimates of recharge in the Ozark area ranged between 178 and 380 mm/yr. Imes and Emmett (1994) and later Czarnecki et al. (2009) derived much lower estimates of recharge to the Ozark Plateaus aquifer system by using groundwater-flow models that were calibrated with measured water levels and base flows estimated from streamflow duration curves. Estimates of recharge determined by Imes and Emmett (1994) were between 34 and 41 mm/yr; those determined by Czarnecki et al. (2009) ranged from 2 to 89 mm/yr. In both modeling studies, attempts to use the recharge estimates of Dugan and Peckenpaugh (1985) resulted in excessively high computed water levels.

9.4.7 Glaciated Central Region

The Glaciated Central region, which overlies the north-central United States, is covered by drift deposited by continental glaciers. The drift can be as much as 300 m thick, and although it consists mostly of fine-grained rock debris, the area has many sand and gravel aquifers (principally, outwash plains and buried bedrock channels) that can yield substantial quantities of water (Heath, 1984). Groundwater also exists in the underlying limestone, dolomite, and sedimentary bedrock aquifers in many areas. Climate is subhumid to humid, except for the westward extent, which is categorized as semiarid.

Recharge to the glacial deposits is mostly diffuse, occurring over large areas in response to precipitation. Recharge can occur throughout the year, but most occurs in the spring, as soils thaw, and in the fall, after plant senescence and before soils freeze. Natural groundwater discharge from the glacial deposits is to streams, to plants, and to underlying bedrock aquifers.

Most methods described in this text have been applied at one time or another in some part of this region, although the use of unsaturated-zone tracer methods is generally limited to the more arid western areas. Rehm *et al.* (1982) applied the water-table fluctuation, Darcy, and flow-net methods (Section 6.3.3) to estimate recharge in a 150 km^2 area of central North Dakota on the basis of groundwater levels monitored in 175 wells and piezometers; recharge estimates varied in space (higher rates occurred where soils were more permeable and where runoff accumulated in surface depressions) and in time (most recharge occurred in spring). Average recharge for the entire area was estimated at between 10 and 40 mm/yr (2% to 9% of precipitation). Healy (1989) and Healy *et al.* (1989) applied multiple methods to estimate recharge at an 8 ha site in northwestern Illinois (Section 2.3); the unsaturated zone Darcy method produced an average estimate of 208 mm/yr (22% of precipitation), whereas

groundwater-flow modeling produced an estimate of 48 mm/yr. The discrepancy was attributed to uncertainty in values of unsaturated and saturated hydraulic conductivity.

Streamflow hydrograph analysis methods are commonly applied in this region because most streams are gaining streams. Holtschlag (1997) used streamflow hydrograph separation in conjunction with regression analysis to estimate average base flow across the lower peninsula of Michigan for the period 1951 to 1980 at about 211 mm, or about 28% of precipitation. Similar approaches were used to estimate base flow across the states of Ohio (Dumouchelle and Schiefer, 2002), where estimates ranged from 100 to about 300 mm/yr, and Minnesota (Lorenz and Delin, 2007), where estimates ranged from less than 50 to 290 mm/yr (Figure 3.17).

Szilagy *et al.* (2005) combined base-flow estimates derived from hydrograph-separation analysis with water-budget estimates of evapotranspiration of groundwater to derive estimates of total recharge for the state of Nebraska; recharge estimates averaged 48 mm/yr and increased across Nebraska from west, where rates were less than 30 mm/yr, to east, where rates exceeded 140 mm/yr. Recharge accounted for 3 to 26% of annual precipitation. The methods used by Szilagy *et al.* (2005) were applied statewide across multiple groundwater regions as defined in Figure 9.1. Methods are not limited to specific regions, and divisions between regions should not interfere with application of methods.

Models are commonly used to estimate recharge in this region, as is true in all regions discussed in this chapter. Groundwater-flow models (Mandle and Kontis, 1992; Yager, 1996; Juckem and Hunt, 2007) are used more than other models for this purpose, but recharge estimates have also been obtained from application of watershed models. Steuer and Hunt (2001) used a watershed model to examine the possible impacts of urbanization on the Pheasant Branch watershed near Madison, Wisconsin. Natural recharge rates, prior to development, were estimated to be about 220 mm/yr (25% of precipitation). Model simulations indicated that with urbanization of the watershed, the average recharge rate could be reduced by about 30%, conceivably impacting springs in the watershed that support diverse ecosystems (Hunt and Steuer, 2001).

9.4.8 Unglaciated Appalachians Region

The last three regions to be described are in the eastern part of the United States and all have humid climates. The Unglaciated Appalachians region encompasses the Piedmont and Blue Ridge physiographic provinces (Heath, 1984). The Piedmont consists of low, rounded hills and long ridges that trend northeast to southwest. The mountainous Blue Ridge area lies to the west; peaks there rise to altitudes of more than 2000 m. The mountains are rounded and are bordered by steep valleys containing well-graded streams. The region is underlain by crystalline and metamorphosed sedimentary rock of Precambrian and Paleozoic age. Typically, 10 to 20 m of unconsolidated clay-rich material lies above the bedrock. These deposits, referred to as saprolite, are derived from in-place weathering of the bedrock. Thin alluvial deposits are present in some valleys.

Groundwater exists in all of these deposits, yet, in general, none of these units can be called productive because they all have low permeabilities. Because of the very low porosity of the bedrock, groundwater is stored and transmitted through fractures. A well must be screened across fractures to produce water. Recharge occurs primarily as diffuse recharge on mountain ridges and in valleys above streams. Groundwater discharges to streams and springs and is consumed by evapotranspiration. Most streams in the area are gaining. Rutledge and Mesko (1996) and Nelms et al. (1997) used streamflow hydrograph separation to estimate base flow from over 200 streamflow gauging sites; estimates fell in the range of 250 to 1000 mm/yr. Plummer et al. (2001) and Busenburg and Plummer (2000) used multiple gas tracers to determine groundwater ages in this region. Rugh and Burbey (2008) applied chloride and bromide tracers to obtain qualitative information on flow paths and groundwater travel times through soil, saprolite, and fractured crystalline rock in western Virginia.

9.4.9 Glaciated Appalachians Region

The Glaciated Appalachians region coincides with the northeasternmost area of the US, covering most of New England, upstate New York, and parts of New Jersey and Pennsylvania. The climate is humid, with up to 1200 mm of annual precipitation. Bedrock in this region shares many of the hydraulic characteristics of the bedrock within the Unglaciated Appalachians region described in the previous section. The distinguishing feature of this region is the stratified drift aquifers that are found in principal valleys. These deposits of sand and gravel are disconnected but similar in terms of geology and hydrology. Underlying approximately 12% of the region, these aquifers are important sources of water in the region (Kontis et al., 2004). Water tables are generally found at shallow depths, often within a couple of meters of land surface. The aquifers are often hydraulically connected to perennial streams.

Several sources of recharge can be important (Figure 9.6). Water flowing from adjacent uplands, either within streambanks or as unchanneled flow, is usually an important source. Diffuse recharge occurs as precipitation that falls on the land surface overlying an aquifer, infiltrates, and percolates through the unsaturated zone to the water table. Interaquifer flow from and to adjoining groundwater systems also occurs. Under typical natural conditions, groundwater discharges to perennial streams; however, extraction of groundwater by humans can reverse this process. If the water table declines to an elevation lower than that of the stream, the stream becomes a source of recharge. Groundwater discharge by evapotranspiration is also common. High precipitation rates, permeable sediments, and shallow water-table depths result in relatively high rates of recharge. Kontis et al. (2004) provided a more thorough description of recharge processes in this region.

Streamflow hydrograph analysis methods are widely used for estimating recharge in this region, but many other methods have also been applied. Flynn and Tasker (2004) used the streamflow recession curve displacement method to estimate annual and seasonal rates of recharge

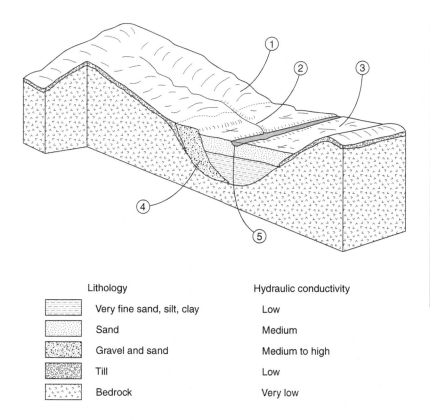

Figure 9.6 Physical setting and conceptual model of recharge processes for a valley-fill aquifer, northeastern United States (after Kontis *et al.*, 2004). Recharge processes include: (1) Recharge from unchanneled runoff of adjacent upland areas; (2) Recharge from seepage from tributary streams draining the uplands; (3) Recharge directly from precipitation; (4) Subsurface interaquifer flow to the aquifer from bedrock; and (5) Perennial streams can be sources of recharge under pumping conditions when groundwater levels fall below the stream.

Lithology		Hydraulic conductivity
	Very fine sand, silt, clay	Low
	Sand	Medium
	Gravel and sand	Medium to high
	Till	Low
	Bedrock	Very low

on the basis of data collected at 55 streamflow gauging sites on unregulated streams in New Hampshire. Average annual recharge was estimated at 530 mm; about 43% of that recharge occurred in spring, with the remainder occurring somewhat uniformly throughout the other seasons. Kontis *et al.* (2004) linked a groundwater-flow model of the drift aquifer system in Dover, New Jersey, with a water-budget analysis of adjacent uplands and found that approximately 32% of recharge to the aquifer came from upland sources. Knott and Olimpio (1986) applied the tritium-profile method, using tritium concentrations in groundwater, and estimated a recharge rate of 680 mm/yr (68% of precipitation) for a sand and gravel aquifer on Nantucket Island, Massachusetts. Mau and Winter (1997) compared results obtained from multiple streamflow hydrograph-analysis methods with data from the Mirror Lake watershed in New Hampshire. Tiedeman *et al.* (1997) used inverse modeling with a groundwater-flow model (Section 3.5) to estimate recharge in the Mirror Lake watershed in the range of 260 to 280 mm/yr.

9.4.10 Atlantic and Gulf Coastal Plain

The Atlantic and Gulf Coastal Plain covers an area in excess of 1 000 000 km^2 that stretches from the southern tip of New York to the southern tip of Texas. The area is relatively flat, reaching elevations of only 100 to 250 m above sea level along its inner margin. Groundwater occurs in unconsolidated sediments (sand, silt, and clay) that were originally transported by streams from adjacent uplands. The sediments can obtain substantial thickness along the seaward edge of the region and along the axis of the Mississippi embayment. In southern Louisiana, the sediments exceed 12 000 m in thickness (Heath, 1984). The geologic formations dip toward the sea or the axis of the Mississippi embayment. The climate is humid except for the southern tip of Texas, where it is semiarid to subhumid. Annual precipitation is 1400 mm/yr in Miami, Florida; 1600 mm/yr in New Orleans, Louisiana; and 1030 mm/yr in Baltimore, Maryland.

Recharge is predominantly diffuse, occurring in response to local precipitation, mainly in interstream areas where soils are permeable. Spring is typically the time of highest recharge rates, but recharge can occur throughout the year. Groundwater discharge is to streams and the sea, to extraction by humans, and, in areas with shallow water tables, to evapotranspiration.

Recharge rates vary depending on weather, soil properties, and land use. Many methods for estimating recharge have been applied; streamflow hydrograph analyses are widely used because most streams are gaining streams. Stricker (1983) applied streamflow hydrograph analysis to data from 35 streamflow gauging sites in South Carolina, Georgia, Alabama, and Mississippi; estimates of base flow ranged from 25 to 500 mm/yr. Base flow estimates for seven sites in North Carolina ranged from 300 to 320 mm/yr, as determined by the recession-curve displacement method (Coes *et al.*, 2007). Inverse groundwater-flow modeling techniques have been used to estimate recharge in studies of large aquifer systems (Faye and Mayer, 1997; Payne *et al.*, 2005). Fisher and Healy (2008) used a water-budget approach to estimate recharge at 315 mm/yr (32% of precipitation) in an agricultural field on the Delmarva Peninsula in Maryland. Using a soil-water budget model, O'Reilly (2004) determined that recharge at a site in central Florida was highly episodic, with essentially no recharge on most days and estimated rates of up to 80 mm/day in response to heavy spring rainfalls when the soil was initially saturated (Figure 3.4). Stewart *et al.* (2007) applied chemical streamflow hydrograph separation on the basis of specific conductance at 10 sites (Section 4.6.1). Historical groundwater tracers, such as CFCs and SF_6, have been used in a number of studies (Dunkle *et al.*, 1993; Szabo *et al.*, 1996; Plummer *et al.*, 1998).

Voronin (2004) describes a groundwater-flow model of the New Jersey coastal plain aquifer system in which recharge occurs in areas of aquifer outcrop. Recharge was treated as being constant in time but allowed to vary in space. Recharge estimates were obtained from a series of water-budget studies conducted in six subareas of the model domain. Because steady-state conditions were assumed, net recharge, or recharge minus groundwater evapotranspiration, was used for model input. Values of net recharge ranged from 150 to 500 mm/yr for the aquifers (approximately 13 to 42% of annual precipitation).

9.4.11 Recharge in urban settings

Urbanization of an area can greatly affect recharge rates, locations, processes, and even sources of water. Installation of storm and sanitary sewers on Long Island, New York, produced a decrease in base flows to streams (Prince, 1981; Spinello and Simmons, 1992) and presumably a decrease in recharge. But the long-held belief that increased impervious areas produced by buildings, roads, and parking lots lead to increased surface runoff and decreased recharge relative to natural conditions has been dismissed (Lerner, 2002). A number of studies have found that urbanization results in increased amounts of recharge (Appleyard, 1995; Fernando and Gerardo, 1999; Yang *et al.*, 1999), particularly for cities that import water from outside of the local groundwater system. Leaking water mains and sewers, storm-water drains, and surface runoff diversion and storage structures are all features that contribute to enhanced recharge. Increased runoff from built-up and paved areas may be funneled through a retention basin or infiltration gallery directly to the subsurface, resulting in a relocation of recharge areas and a transition from a slow, diffuse recharge process to a rapid, focused process. The large number of these features and the difficulty of measuring flows associated with them illustrate the complexity involved in developing a conceptual model for recharge in urban settings and in selecting appropriate estimation methods.

The large number of water conveyance pipes associated with urbanization represents a significant potential source of recharge. In Göteborg, Sweden, there are 16 km of water mains per square kilometer of urban area (Norin *et al.*, 1999). Lerner (2002) estimated that there are 50 km of water mains, sewers, and drains per square kilometer of urban area in the United Kingdom. Leakage rates of 20 to 25% are common for water mains (Lerner, 2002). Norin *et al.* (1999) estimated a leakage rate of 26% for Göteborg; leakage rates of up to 50% have been observed elsewhere (Lerner, 1986). In urban areas without sewers, domestic and industrial discharge may be the dominant source of recharge. The importance of leaks in water conveyance systems is not limited to urban areas; these leaks can be sources of recharge wherever water is transported long distances, particularly in areas of irrigation.

Many of the methods described in this text can be applied for estimating recharge in urban areas, but careful consideration is required. Urban hydrologic systems are generally more complex than natural systems. As with any study of natural systems, a recharge study of an urban area should be framed within the context of a conceptual model of the system. Developing a conceptual model is difficult because of the great number of potential sources of recharge. Point estimates of recharge will generally be of limited value because of the large spatial variability in recharge. Point methods can be useful, though, for determining representative flux rates at specific points of focused recharge.

Water-budget methods may be the most widely used techniques for studying recharge in urban settings. Water budgets can be determined for the groundwater system of interest, the watershed, or even for the water-delivery system. Lerner (2002) described how to conduct a water-budget study of a water-supply network. Basically, portions of the network are isolated during times of low water demand, and flow into and out of each portion is directly measured. Fernando and Gerardo (1999) analyzed the aquifer water budget for the city of Aguascalientes, Mexico; they determined that water-main leakage and infiltration of wastewater accounted for 30% of the total estimated recharge of 1040 mm/yr.

Groundwater tracers have also been used to study the contribution of human-induced urban recharge to the total recharge amount. Barth (1998) used boron isotopes and Seiler and Alvarado Rivas (1999) used stable isotopes to provide qualitative information on the contribution of human-induced urban recharge to total recharge. Other studies have been able to generate quantitative estimates of recharge by using tracers. Appleyard (1995), Davidson (1995), and Appleyard *et al.* (1999) used tritium and CFC concentrations in groundwater in conjunction with a water budget and determined that recharge in urban areas of Perth, Australia, is equivalent to up to 37% of precipitation, approximately double the amount of recharge in nearby nonurban settings.

Methods based on surface-water data have also been used for assessing the impacts of urbanization on base flow. The effects of urbanization on Long Island, New York, were studied with analysis of streamflow duration curves (Prince, 1981) and streamflow hydrograph-separation techniques (Spinello and Simmons, 1992). Brandes *et al.* (2005) and Meyer (2005) used streamflow hydrograph-separation techniques to investigate the effects of urbanization in the Delaware River watershed and in northeastern Illinois, respectively; both studies reported that urbanization had little impact on base flow.

Groundwater-flow models are useful for assessing past and current impacts of urbanization on recharge (Lerner, 2002). Models can also be used to predict future impacts and to evaluate the effects of various alternative management strategies that might be employed to minimize adverse effects. Yang *et al.* (1999) used numerical groundwater-flow, solute-transport modeling, and an end-member mixing analysis based on chloride, sulfate, and nitrogen concentrations to determine that precipitation, water-main leakage, and sewer leakage accounted for 30%, 65%, and 5%, respectively, of total recharge for Nottingham, United Kingdom.

9.5 | Final thoughts

Estimating recharge is one of the most difficult tasks in groundwater studies. Information contained in this book can provide direction in pursuit of that task. It is the author's intent to provide a framework within which methods for estimating recharge can be evaluated and their suitability for a particular study assessed. Many factors require consideration when determining the suitability of a method. The most important of these is the need for consistency between the conceptual model of recharge processes and assumptions underlying a method. The contents of this book can assist in this determination, but formulation of a sound conceptual model must be based on hydrologic insight on the part of the practitioner.

Water budgets of aquifers, watersheds, streams, soil columns, or other control volumes can form a basis for a conceptual model. Conceptual models are used to select methods, but estimates of recharge generated by different methods can also help refine the conceptual model. If estimates from various methods differ substantially, this difference may provide insight as to the validity of one or more assumptions and, in turn, may lead to a revision of the conceptual model. Good practice calls for application of multiple methods for estimating recharge. Many methods can be applied using existing data, and vast amounts of accessible data are available. Care is required, however. Although easy-to-access data facilitate application of many methods, inappropriate applications can be easily overlooked. Due diligence is required for the application of any method.

Models are useful tools for exploring hypotheses related to conceptual models and integrating and evaluating different kinds of data. Many readers may be reading this book because they are actively involved in a groundwater-flow modeling study and are looking for ways to evaluate recharge values used in their models. However, groundwater-flow models themselves are powerful tools for estimating recharge (Chapter 3). Other types of models are also useful in recharge studies, including watershed models (Bauer and Mastin, 1997; Bossong et al., 2003) and unsaturated-zone flow models

(Keese et al., 2005; Walvoord and Phillips, 2004). Models can provide predictions of how future changes in climate, land use, and other factors might affect recharge patterns. Models are also useful, in a qualitative sense, for identifying locations and times that are favorable for recharge.

Exact determination of recharge will never be possible because of the complex nature of recharge processes, but advances in estimation methods will occur in terms of improved data-collection techniques and tools that will facilitate access to and analysis of collected data. Recent years have seen remarkable improvements in the accuracy, durability, and recording capabilities of data sensors (e.g. water-level sensors, water-quality monitors, temperature loggers, and distributed temperature sensing systems). Geophysical techniques for sensing gravity and electromagnetic waves in the subsurface, from land surface and from remote platforms, will continue to develop and may eventually evolve into powerful tools for quantifying recharge rates. The potential of remote-sensing techniques to provide direct estimates of recharge is unknown, but information on land use, status of vegetation, and weather can be used in water-budget models for estimating recharge; the information may also be useful for refining conceptual models of recharge processes. Remote-sensing techniques are particularly valuable in developing countries that have minimal ground-based hydrologic data-collection networks. New analytical techniques for detecting dissolved gas concentrations in the part-per-trillion range may extend the period of use of existing tracers, such as SF_6, and may allow other compounds to be used as tracers. With more and more data becoming available, learning to properly interpret new information and to incorporate it into recharge estimation schemes continues to be a challenge. Improved modeling techniques and computer programs for analyzing time series of streamflow and stream and groundwater levels can help address this challenge.

References

Aeschbach-Hertig, W., Peeters, F., Beyerle, U. and Kipfer, R. (1999). Interpretation of dissolved atmospheric noble gases in natural waters. *Water Resour. Res.*, **35**, 2779–2792.

Ahuja, L. R. and El-Swaify, S. A. (1979). Determining soil hydrologic characteristics on a remote forest watershed by continuous monitoring of soil-water pressures, rainfall and runoff. *J. Hydrol.*, **44**, 135–147.

Allen, R. G., Pereira, L. S., Raes, K. and Smith, M. (1998). Crop evapotranspiration: guidelines for computing crop water requirements. Irrigation and Drainage Paper 56. Rome: Food and Agriculture Organization of the United Nations.

Aller, L., Bennett, T., Lehr, J. H., Petty, R. J. and Hackett, G. (1985). DRASTIC: A standardized system for evaluating ground water pollution potential using hydrogeologic settings. US Environmental Protection Agency, Robert S. Kerr Environmental Research Laboratory, Office of Research and Development. EPA/600/2-85/018.

Alley, W. M. (1984). On the treatment of evapotranspiration, soil moisture accounting, and aquifer recharge in monthly water balance models. *Water Resour. Res.*, **20**, 1137–1149.

Alley, W. M., Healy, R. W., LaBaugh, J. W. and Reilly, T. E. (2002). Flow and storage in groundwater systems. *Science*, **296**, 1985–1990.

Alley, W. M. and Leake, S. A. (2004). The journey from safe yield to sustainability. *Ground Water*, **42**, 12–16.

Allison, G. B. (1987). A review of some of the physical, chemical, and isotopic techniques available for estimating groundwater recharge. In *Estimation of Natural Groundwater Recharge*, ed. I. Simmers. Dordrecht, Holland: D. Reidel, 49–72.

Allison, G. B., Cook, P. G., Barnett, S. R. *et al.* (1990). Land clearance and river salinisation in the western Murray Basin, Australia. *J. Hydrol.*, **119**, 1–20.

Allison, G. B., Gee, G. W. and Tyler, S. W. (1994). Vadose zone techniques for estimating groundwater recharge in arid and semiarid regions. *Soil Sci. Soc. Am. J.*, **58**, 6–14.

Allison, G. B. and Hughes, M. W. (1977). The history of tritium fallout in southern Australia as inferred from rainfall and wine samples. *Earth Planet. Sci. Lett.*, **36**, 334–340.

Allison, G. B. and Hughes, M. W. (1978). The use of environmental chloride and tritium to estimate total recharge to an unconfined aquifer. *Austral. J. Soil Res.*, **16**, 181–195.

Allison, G. B. and Hughes, M. W. (1983). The use of natural tracers as indicators of soil-water movement in a temperate semi-arid region. *J. Hydrol.*, **60**, 157–173.

Anderson, M. P. (2005). Heat as a ground water tracer. *Ground Water*, **43**, 951–968.

Andraski, B. J. and Scanlon, B. R. (2002). Thermocouple psychrometry. In *Methods of Soil Analysis. Part 4: Physical Methods*, ed. J. H. Dane and G. C. Topp. Madison, Wisconsin: Soil Science Society of America, 609–642.

Anton, H. (1984). *Calculus with Analytical Geometry*, 2nd edn. New York: John Wiley and Sons.

Appleyard, S. J. (1995). The impact of urban development on the utilisation of groundwater resources in Perth, Western Australia. *Hydrogeol. J.*, **3**, 65–75.

Appleyard, S. J., Davidson, W. A. and Commander, D. P. (1999). The effects of urban development on the utilisation of groundwater resources in Perth, Western Australia. In *Groundwater in the Urban Environment: Selected City Profiles*, ed. J. Chilton. Rotterdam: A. A. Balkema, 97–104.

Aranyossy, J. F. and Gaye, C. B. (1992). La recherche du pic de tritium thermonucleaire en zone non saturee profonde sous climate semi-aride pour la mesure de la recharge des nappes: premiere application au Sahel. *C.R. Acad. Sci. Paris*, **315**, **Serie II**, 637–643.

Arihood, L. D. and Glatfelter, D. R. (1991). Method for estimating low-flow characteristics of ungaged streams in Indiana. US Geological Survey Water Supply Paper 2372.

Arnold, J. G. and Allen, P. M. (1996). Estimating hydrologic budgets for three Illinois watersheds. *J. Hydrol.*, **176**, 57–77.

Arnold, J. G. and Allen, P. M. (1999). Automated methods for estimating baseflow and ground water recharge from streamflow records. *J. Amer. Water Resour. Assoc.*, **35**, 411–424.

Arnold, J. G., Allen, P. M., Muttiah, R. S. and Bernhardt, G. (1995). Automated baseflow separation and recession analysis techniques. *Ground Water*, **33**, 1010–1018.

Arnold, J. G., Muttiah, R. S., Srinivasan, R. and Allen, P. M. (2000). Regional estimation of base flow and groundwater recharge in the Upper Mississippi River Basin. *J. Hydrol.*, **227**, 21–40.

Arnold, J. G., Srinivasan, R., Muttiah, R. S. and Williams, J. R. (1998). Large area hydrologic modeling and assessment. Part I: Model development. *J. Amer. Water Resour. Assoc.*, **34**, 73–89.

Arya, L. M., Farrell, D. A. and Blake, G. R. (1975). Field study of soil water depletion patterns in presence of growing soybean roots. Part 1: Determination of hydraulic properties of the soil. *Soil Sci. Soc. Am. Proc.*, **39**, 424–430.

Arya, L. M. and Paris, J. (1981). Physicoempirical model to predict the soil moisture characteristic from particle-size distribution and bulk density data. *Soil Sci. Soc. Amer. J.*, **45**, 1023–1030.

ASTM (2008). Standard guide for selection of methods for assessing ground water or aquifer sensitivity and vulnerability. ASTM D6030–96(2008). American Society of Testing Materials.

Athavale, R. N. and Rangarajan, R. (1990). Natural recharge measurements in the hard rock regions of semi-arid India using tritium injection – a review. In *Groundwater Recharge: A Guide to Understanding and Estimating Natural Recharge, International Contributions to Hydrogeology Vol. 8*, ed. D. N. Lerner, A. S. Issar and I. Simmers. Hanover: Verlag Heinz Heise, 235–245.

Aubinet, M., Grelle, A., Ibrom, A. *et al.* (2000). Estimates of the annual net carbon and water exchange of European forests: the EUROFLUX methodology. *Adv. Ecol. Res.*, **30**, 133–175.

Baalousha, H. (2009). Stochastic water balance model for rainfall recharge quantification in Ruataniwha Basin, New Zealand. *Environ. Geol.*, **58**, 85–93.

Baillie, M. N., Hogan, J. F., Ekwurzel, B., Wahi, A. K. and Eastoe, C. J. (2007). Quantifying water sources to a semiarid riparian ecosystem, San Pedro River, Arizona. *J. Geophys. Res. G*, **112**, G03502, doi:10.1029/2006JG000263.

Bakalowicz, M. (2005). Karst groundwater: a challenge for new resources. *Hydrogeol. J.*, **13**, 148–160.

Bakr, M. I. and Butler, A. P. (2004). Worth of head data in well-capture zone design: deterministic and stochastic analysis. *J. Hydrol.*, **290**, 202–216.

Baldocchi, D., Falge, E., Gu, L. *et al.* (2001). FLUXNET: a new tool to study the temporal and spatial variability of ecosystem-scale carbon dioxide, water vapor, and energy flux densities. *Bull. Amer. Met. Soc.*, **82**, 2415–2434.

Bardenhagen, I. (2000). Groundwater reservoir characterisation based on pumping test curve diagnosis in fractured formation. In *Groundwater: Past Achievements and Future Challenges*, ed. O. Sililo. Rotterdam: Balkema, 81–86.

Bardsley, W. E. and Campbell, D. I. (2007). An expression for land surface water storage monitoring using a two-formation geological weighing lysimeter. *J. Hydrol.*, **335**, 240–246.

Barlow, P. M., Desimone, L. A. and Moench, A. F. (2000). Aquifer response to stream-stage and recharge variations, II. Convolution method and applications. *J. Hydrol.*, **230**, 211–229.

Barnes, B. S. (1939). The structure of discharge recession curves. *Trans. Amer. Geophys. Union*, **20**, 721–725.

Barnes, C. J. and Allison, G. B. (1983). The distribution of deuterium and oxygen-18 in dry soils I. Theory. *J. Hydrol.*, **60**, 141–156.

Barnes, C. J. and Allison, G. B. (1988). Tracing of water movement in the unsaturated zone using stable isotopes of hydrogen and oxygen. *J. Hydrol.*, **100**, 143–176.

Barnston, A. G. (1993). *Atlas of Frequency Distribution, Auto-correlation and Cross-correlation of Daily Temperature and Precipitation at Stations in the US, 1948–1991*. Camp Springs, MD: National Oceanic and Atmospheric Administration.

Barr, A. G., van der Kamp, G., Schmidt, R. and Black, T. A. (2000). Monitoring the moisture balance of a boreal aspen forest using a deep groundwater piezometer. *Agric. Forest Met.*, **102**, 13–24.

Barth, S. (1998). Application of boron isotopes for tracing sources of anthropogenic contamination in groundwater. *Water Res.*, **32**, 685–690.

Bartolino, J. R. and Niswonger, R. G. (1999). Numerical simulation of vertical ground-water flux of the Rio Grande from ground-water temperature profiles, central New Mexico. US Geological Survey Water-Resources Investigations Report 99–4212.

Batelaan, O. and De Smedt, F. (2007). GIS-based recharge estimation by coupling surface-subsurface water balances. *J. Hydrol.*, **337**, 337–355.

Batu, V. (1998). *Aquifer Hydraulics: A Comprehensive Guide to Hydrogeological Data Analysis*. Hoboken, NJ: John Wiley and Sons.

Bauer, H. H. and Mastin, M. C. (1997). Recharge from precipitation in three small glacial-till-mantled catchments in the Puget Sound Lowland, Washington. US Geological Survey Water-Resources Investigations Report 96–4219.

Bauer, H. H. and Vaccaro, J. J. (1987). Documentation of a deep percolation model for estimating ground-water recharge. US Geological Survey Open-File Report 86–536.

Bauer, H. H. and Vaccaro, J. J. (1990). Estimates of ground-water recharge to the Columbia Plateau regional aquifer system, Washington, Oregon, and Idaho, for predevelopment and current land-use conditions. US Geological Survey Water-Resources Investigations Report 88–4108.

Bear, J. (1972). *Dynamics of Fluids in Porous Media*. New York: Elsevier.

Becker, M. W. (2006). Potential for satellite remote sensing of ground water. *Ground Water*, **44**, 306–318.

Becker, M. W., Georgian, T., Ambrose, H., Siniscalchi, J. and Fredrick, K. (2004). Estimating flow and flux of ground water discharge using water temperature and velocity. *J. Hydrol.*, **296**, 221–233.

Bekesi, G. and McConchie, J. (1999). Groundwater recharge modelling using the Monte Carlo technique, Manawatu region, New Zealand. *J. Hydrol.*, **224**, 137–148.

Belanger, T. V., Mikutel, D. F. and Churchill, P. A. (1985). Groundwater seepage nutrient loading in a Florida lake. *Water Res.*, **19**, 773–782.

Belanger, T. V. and Montgomery, M. T. (1992). Seepage meter errors. *Limnol. Oceanogr.*, **37**, 1787–1795.

Beltrami, H., Ferguson, G. and Harris, R. N. (2005). Long-term tracking of climate change by underground temperatures. *Geophys. Res. Lett.*, **32**, 1–4.

Bencala, K. E., McKnight, D. M. and Zellweger, G. W. (1987). Evaluation of natural tracers in an acidic and metal-rich stream. *Water Resour. Res.*, **23**, 827–836.

Bendjoudi, H., Cheviron, B., Guérin, R. and Tabbagh, A. (2005). Determination of upward/downward groundwater fluxes using transient variations of soil profile temperature: test of the method with Voyons (Aube, France) experimental data. *Hydrol. Proc.*, **19**, 3735–3745.

Bentley, H. W., Phillips, F. M. and Davis, S. N. (1986). 36Cl in the terrestrial environment. In *Handbook of Environmental Isotope Geochemistry*, ed. P. Fritz and J.-C. Fontes. New York: Elsevier Science, **2b**, 422–475.

Berenbrock, C., Rousseau, J. P. and Twining, B. V. (2007). Hydraulic characteristics of bedrock constrictions and an evaluation of one- and two-dimensional models of flood flow on the Big Lost River at the Idaho National Engineering and Environmental Laboratory, Idaho. US Geological Survey Scientific Investigations Report 2007–5080.

Berendrecht, W. L., Heemink, A. W., van Geer, F. C. and Gehrels, J. C. (2006). A non-linear state space approach to model groundwater fluctuations. *Adv. Water Resour.*, **29**, 959–973.

Besbes, M. (2006). Aquifer recharge by floods in ephemeral streams. *IAHS-AISH Publication*, **305**, 43–72.

Besbes, M. and de Marsily, G. (1984). From infiltration to recharge: use of a parametric transfer function. *J. Hydrol.*, **74**, 271–293.

Bevan, M. J., Endres, A. L., Rudolph, D. L., Parkin, G. (2003). The non-invasive characterization of pumping-induced dewatering using ground penetrating radar. *J. Hydrol.*, **28**, 55–69.

Bevans, H. E. (1986). Estimating stream-aquifer interactions in coal areas of eastern Kansas by using streamflow records. US Geological Survey Water Supply Paper 2290, 51–64.

Bidaux, P. and Drouge, C. (1993). Calculation of low-range flow velocities in fractured carbonate media from borehole hydrochemical logging data comparison with thermometric results. *Ground Water*, **31**, 19–26.

Bierkens, M. F. P. (1998). Modeling water table fluctuations by means of a stochastic differential equation. *Water Resour. Res.*, **34**, 2485–2499.

Binley, A., Winship, P., Middleton, R., Pokar, M. and West, J. (2001). High-resolution characterization of vadose zone dynamics using cross-borehole radar. *Water Resour. Res.*, **37**, 2639–2652.

Black, T. A., Gardner, W. R. and Thurtell, G. W. (1969). The prediction of evaporation, drainage, and soil water storage for a bare soil. *Soil Sci. Soc. Amer. Proc.*, **33**, 655–660.

Blasch, K. W., Constantz, J. and Stonestrom, D. (2007). Thermal methods to investigate ground-water recharge. In *Ground-water Recharge in the Arid and Semiarid Southwestern United States*, ed. D. A. Stonestrom, J. Constantz, T. P. A. Ferre and S. A. Leake. US Geological Survey Professional Paper 1703, Appendix 1.

Blasch, K. W., Ferré, T. P. A. and Hoffmann, J. P. (2004). A statistical technique for interpreting streamflow timing using streambed sediment thermographs. *Vadose Zone J.*, **3**, 936–946.

Blasch, K. W., Ferré, T. P. A., Hoffmann, J. P. and Fleming, J. B. (2006). Relative contributions of transient and steady state infiltration during ephemeral streamflow. *Water Resour. Res.*, **42**, W08405, doi:10.1029/2005WR004049.

Blonquist, J. M. Jr., Jones, S. B. and Robinson, D. A. (2005). Standardizing characterization of electromagnetic water content sensors, Part 2 Evaluation of seven sensing systems. *Vadose Zone J.*, **4**, 1059–1069.

Bogena, H., Kunkel, R., Montzka, C. and Wendland, F. (2005). Uncertainties in the simulation of groundwater recharge at different scales. *Adv. Geosciences*, **5**, 25–30.

Böhlke, J. K. (2002). Groundwater recharge and agricultural contamination. *Hydrogeol. J.*, **10**, 153–179.

Bonan, G. B. and Levis, S. (2006). Evaluating aspects of the Community Land and Atmosphere Models (CLM3 and CAM3) using a dynamic global vegetation model. *J. Climate*, **19**, 2290–2301.

Boonstra, J. and Bhutta, M. N. (1996). Ground water recharge in irrigated agriculture: the theory and practice of inverse modelling. *J. Hydrol.*, **174**, 357–374.

Bossong, C. R., Caine, J. S., Stannard, D. I. *et al.* (2003). Hydrologic conditions and assessment of water resources in the Turkey Creek watershed, Jefferson County, Colorado, 1998–2001. US Geological Survey Water-Resources Investigations Report 2003–4263.

Boulton, N. S. (1963). Analysis of data from non-equilibrium pumping tests allowing for delayed yield from storage. *Proc. Inst. Civil Eng.*, **26**, 469–482.

Bowen, I. S. (1926). The ratio of heat losses by conduction and by evaporation from any water surface. *Physical Rev.*, **27**, 779–787.

Bowman, R. S. and Rice, R. C. (1986). Transport of conservative tracers in the field under intermittent flood irrigation. *Water Resour. Res.*, **22**, 1531–1536.

Bowman, R. S., Schroeder, J., Bulusu, R., Remmenga, M. and Heightman, R. (1997). Plant toxicity and plant uptake of fluorobenzoate and bromide water tracers. *J. Env. Qual.*, **25**, 1292–1299.

Brandes, D., Cavallo, G. J. and Nilson, M. L. (2005). Base flow trends in urbanizing watersheds of the Delaware River Basin. *J. Amer. Water Resour. Assoc.*, **41**, 1377–1391.

Bredehoeft, J. D. (2002). The water budget myth revisited: why hydrogeologists model. *Ground Water*, **40**, 340–345.

Bredehoeft, J. D. (2005). The conceptualization model problem: surprise. *Hydrogeol. J.*, **13**, 37–46.

Bredehoeft, J. D. and Papadopulos, I. S. (1965). Rates of vertical groundwater movement estimated from the Earth's thermal profile. *Water Resour. Res.*, **1**, 325–328.

Bredehoeft, J. D., Papadopulos, S. S. and Cooper, H. H. (1982). The water budget myth. In *Scientific Basis of Water Resource Management, Studies in Geophysics*. Washington, DC: National Academy Press, 51–57.

Briggs, L. J. and Shantz, H. L. (1912). The wilting coefficient for different plants and its indirect determination. US Department of Agriculture, Bureau of Plant Industry Bulletin 230.

Bristow, K. L. (2002). Thermal conductivity. In *Methods of Soil Analysis. Part 4: Physical Methods*, ed. J. H. Dane and G. C. Topp. Madison, Wisconsin: Soil Science Society of America, 1209–1226.

Brooks, R. H. and Corey, A. T. (1964). Hydraulic properties of porous media. Hydrology Paper 3. Colorado State University. Fort Collins, Colorado.

Brunner, P., Hendricks Franssen, H. J., Kgotlhang, L., Bauer-Gottwein, P. and Kinzelbach, W. (2007). How can remote sensing contribute in groundwater modeling? *Hydrogeol. J.*, **15**, 5–18.

Brutsaert, W. (1982). *Evaporation into the Atmosphere*. Dordrecht: D. Reidel.

Brye, K. R., Norman, J. M., Bundy, L. G. and Gower, S. T. (1999). An equilibrium tension lysimeter for measuring drainage through soil. *Soil Sci. Soc. Amer. J.*, **63**, 536–543.

Brye, K. R., Norman, J. M., Bundy, L. G. and Gower, S. T. (2000). Water-budget evaluation of prairie and maize ecosystems. *Soil Sci. Soc. Amer. J.*, **64**, 715–724.

Bu, X. and Warner, M. J. (1995). Solubility of chlorofluorocarbon 113. *Water and Seawater*, **42**, 1151–1161.

Burkhardt, M., Kasteel, R., Vanderborght, J. and Vereecken, H. (2008). Field study on colloid transport using fluorescent microspheres. *Eur. J. Soil Sci.*, **59**, 82–93.

Busenberg, E. and Plummer, L. N. (1992). Use of chlorofluorocarbons (CCl_3F and CCl_2F_2) as hydrologic tracers and age-dating tools: the alluvium and terrace system of central Oklahoma. *Water Resour. Res.*, **28**, 2257–2283.

Busenberg, E. and Plummer, L. N. (2000). Dating young groundwater with sulfur hexafluoride: natural and anthropogenic sources of sulfur hexafluoride. *Water Resour. Res.*, **36**, 3011–3030.

Butler, J. J. Jr. (1997). *The Design, Performance, and Analysis of Slug Tests*. Boca Raton, Florida: Lewis.

Buttle, J. M. and Peters, D. L. (1997). Inferring hydrological processes in a temperate basin using isotopic and geochemical hydrograph separation: a re-evaluation. *Hydrol. Proc.*, **11**, 557–573.

Buytaert, W., Iñiguez, V. and Bièvre, B. D. (2007). The effects of afforestation and cultivation on water yield in the Andean páramo. *Forest Ecol. Man.*, **251**, 22–30.

Callow, J. N. and Smettem, K. R. J. (2007). Channel response to a new hydrological regime in southwestern Australia. *Geomorphology*, **84**, 254–276.

Campbell, G. S. (1985). *Soil Physics with BASIC*. New York: Elsevier.

Campbell, K., Wolfsberg, A., Fabryka-Martin, J. and Sweetkind, D. (2003). Chlorine-36 data at Yucca Mountain: statistical tests of conceptual models for unsaturated-zone flow. *J. Contam. Hydrol.*, **62–63**, 43–61.

Carrera, J., Alcolea, A., Medina, A., Hidalgo, J. and Slooten, L. J. (2005). Inverse problem in hydrogeology. *Hydrogeol. J.*, **13**, 206–222.

Carrera, J. and Neuman, S. P. (1986). Estimation of aquifer parameters under transient and steady state conditions: I. Maximum likelihood method incorporating prior information. *Water Resour. Res.*, **22**, 199–210.

Carsel, R. F. and Parrish, R. S. (1988). Developing joint probability distributions of soil water retention characteristics. *Water Resour. Res.*, **24**, 755–769.

Carslaw, H. S. and Jaeger, J. C. (1959). *Conduction of Heat in Solids*. 2nd edn. New York: Oxford University Press.

Carter, R. W. and Anderson, I. E. (1963). Accuracy of current-meter measurements. *Amer. Soc. Civil Eng. J.*, **89**, 105–115.

Carter, R. W., Anderson, W. L., Isherwood, W. L. *et al.* (1963). Automation of streamflow records. US Geological Survey Circular 474.

Cartwright, K. (1970). Groundwater discharge in the Illinois Basin as suggested by temperature anomalies. *Water Resour. Res.*, **6**, 912–918.

Cartwright, K. (1974). Tracing shallow groundwater systems by soil temperatures. *Water Resour. Res.*, **10**, 847–855.

Cassel, D. K. and Nielsen, D. R. (1986). Field capacity and available water capacity. In *Methods of Soil Analysis. Part 1: Physical and Mineralogical Methods*. 2nd edn., ed. A. Klute. Madison, Wisconsin: Soil Science Society of America, 901–926.

Cedergren, H. R. (1988). *Seepage, Drainage, and Flow Nets*, 3rd edn. New York: John Wiley and Sons, Inc.

Chang, A. T. C., Foster, J. L., Hall, D. K. *et al.* (1997). Snow parameters derived from microwave measurements during the BOREAS winter field campaign. *J. Geophys. Res. D*, **102**, 29663–29671.

Chapman, D. S., Sahm, E. and Gettings, P. (2008). Monitoring aquifer recharge using repeated high-precision gravity measurements: a pilot study in South Weber, Utah. *Geophysics*, **73**, WA83–WA93.

Chapman, T. (1999). A comparison of algorithms for stream flow recession and baseflow separation. *Hydrol. Proc.*, **13**, 701–714.

Charbeneau, R. J. (1984). Kinematic models for soil moisture and solute transport (unsaturated groundwater recharge). *Water Resour. Res.*, **20**, 699–706.

Chen, W. P. and Lee, C. H. (2003). Estimating groundwater recharge from streamflow records. *Environ. Geol.*, **44**, 257–265.

Cherkauer, D. S. (2004). Quantifying ground water recharge at multiple scales using PRMS and GIS. *Ground Water*, **42**, 97–110.

Cherkauer, D. S. and Ansari, S. A. (2005). Estimating ground water recharge from topography, hydrogeology, and land cover. *Ground Water*, **43**, 102–112.

Cheviron, B., Guérin, R., Tabbagh, A. and Bendjoudi, H. (2005). Determining long-term effective groundwater recharge by analyzing vertical soil temperature profiles at meteorological stations. *Water Resour. Res.*, **41**, 1–6.

Childs, E. C. (1960). The nonsteady state of the water table in drained land. *J. Geophys. Res.*, **65**, 780–782.

Childs, E. C. (1969). *An Introduction to the Physical Basis of Soil Water Phenomena*. London: John Wiley and Sons.

Chong, S., Green, R. E. and Ahuja, L. R. (1981). Simple in situ determination of hydraulic conductivity by power function descriptions of drainage. *Water Resour. Res.*, **17**, 1109–1114.

Chow, V. T. (ed.) (1964). *Handbook of Applied Hydrology*. New York: McGraw-Hill.

Chow, V. T., Maidment, D. R. and Mays, L. W. (1988). *Applied Hydrology*. New York: McGraw-Hill.

Clark, W. O. (1917). Groundwater for irrigation in the Morgan Hill area, California. US Geological Survey Water-Supply Paper 400-E.

Clarke, R., Lawrence, A. and Foster, S. (1996). Groundwater: A threatened resource. Nairobi, Kenya: United Nations Environment Programme Environment Library No. 15.

Clarke, W. B., Jenkins, W. J. and Top, Z. (1976). Determination of tritium by mass spectrometric measurements of $3He^+$. *Int. J. Appl. Radiat. Isotopes*, **27**, 512–522.

Clow, D. W. and Fleming, A. C. (2008). Tracer gauge: an automated dye dilution gauging system for ice-affected streams. *Water Resour. Res.*, **44**, W12441, doi:10.1029/2008WR007090.

Clow, G. D. (1992). The extent of temporal smearing in surface-temperature histories derived from borehole temperature measurements. *Palaeogeography, Palaeoclimatology, Palaeoecology*, **98**, 81–86.

Coes, A. L., Spruill, T. B. and Thomasson, M. J. (2007). Multiple-method estimation of recharge rates at diverse locations in the North Carolina Coastal Plain, USA. *Hydrogeol. J.*, **15**, 773–788.

Colbeck, S. C. (1972). A theory of water percolation in snow. *J. Glaciology*, **2**, 369–385.

Conant Jr., B. (2004). Delineating and quantifying ground water discharge zones using streambed temperatures. *Ground Water*, **42**, 243–257.

Constantz, J., Niswonger, R. and Stewart, A. E. (2008). Analysis of temperature gradients to determine stream exchanges with ground water. US Geological Survey Techniques and Methods 4-D2 Chapter 4.

Constantz, J., Stonestrom, D., Stewart, A. E., Niswonger, R. and Smith, T. R. (2001). Analysis of streambed temperatures in ephemeral channels to determine streamflow frequency and duration. *Water Resour. Res.*, **37**, 317–328.

Constantz, J., Tyler, S. W. and Kwicklis, E. M. (2003). Temperature-profile methods for estimating percolation rates in arid environments. *Vadose Zone J.*, **2**, 12–24.

Cook, P. G. and Böhlke, J. K. (2000). Determining timescales for groundwater flow and solute transport. In *Environmental Tracers in Subsurface Hydrology*, ed. P. G. Cook and A. L. Herczeg. Boston: Kluwer Academic Publishers, 1–30.

Cook, P. G. and Herczeg, A. L. (1998). Groundwater chemical methods for recharge studies. In *Part 2: Basics of Recharge and Discharge*, ed. L. Zhang. Victoria, Australia: CSIRO, 1–17.

Cook, P. G. and Herczeg, A. L. (eds.) (2000). *Environmental Tracers in Subsurface Hydrology*. Boston: Kluwer Academic Publishers.

Cook, P. G., Jolly, I. D., Leaney, F. W. and Walker, G. R. (1994). Unsaturated zone tritium and chlorine 36 profiles from southern Australia: their use as tracers of soil water movement. *Water Resour. Res.*, **30**, 1709–1719.

Cook, P. G. and Kilty, S. (1992). A helicopter-borne electromagnetic survey to delineate groundwater recharge rates. *Water Resour. Res.*, **28**, 2953–2961.

Cook, P. G. and Solomon, D. K. (1995). Transport of atmospheric trace gases to the water table: implications for groundwater dating with chlorofluorocarbons and krypton 85. *Water Resour. Res.*, **31**, 263–270.

Cook, P. G. and Solomon, D. K. (1997). Recent advances in dating young groundwater: chlorofluorocarbons, H-3/He-3 and Kr-85. *J. Hydrol.*, **191**, 245–265.

Cook, P. G., Solomon, D. K., Plummer, L. N., Busenberg, E. and Schiff, S. L. (1995). Chlorofluorocarbons as tracers of groundwater transport processes in a shallow, silty sand aquifer. *Water Resour. Res.*, **31**, 425–434.

Cook, P. G., Walker, G. R. and Jolly, I. D. (1989). Spatial variability of groundwater recharge in a semiarid region. *J. Hydrol.*, **111**, 195–212.

Cooley, R. L. (1979). A method of estimating parameters and assessing reliability for models of steady state groundwater flow: 2. Application of statistical analysis. *Water Resour. Res.*, **15**, 603–617.

Cooley, R. L. (1983). Incorporation of prior information on parameters into nonlinear regression groundwater flow models: 2. Applications. *Water Resour. Res.*, **19**, 662–676.

Cooley, R. L. and Naff, R. L. (1990). Regression modeling of ground-water flow. US Geological Survey Techniques in Water-Resources Investigations, Book 3, Chapter B4.

Cooper, J. D., Gardner, C. M. K. and Mackenzie, N. (1990). Soil controls on recharge to aquifers. *J. Soil Sci.*, **41**, 613–630.

Coulibaly, P., Anctil, F., Aravena, R. and Bobée, B. (2001). Artificial neural network modeling of water table depth fluctuations. *Water Resour. Res.*, **37**, 885–896.

Crawford, N. H. and Linsley, R. K. (1962). The synthesis of continuous streamflow hydrographs on a digital computer. Technical Report 12. Civil Engineering Department Stanford University.

Crosbie, R. S., Binning, P. and Kalma, J. D. (2005). A time series approach to inferring groundwater recharge using the water table fluctuation method. *Water Resour. Res.*, **41**, 1–9.

Culler, R. C., Hanson, R. L., Myrick, R. M., Turner, R. M. and Kipple, F. P. (1982). Evapotranspiration before and after clearing phreatophytes, Gila River flood plain, Graham County, Arizona. US Geological Survey Professional Paper 655-P.

Cunnold, D. M., Fraser, P. J., Weiss, R. F. *et al.* (1994). Global trends and annual releases of CCl_3F and CCl_2F_2 estimated from Ale/Gage and other measurements from July 1978 to June 1991. *J. Geophys. Res. D*, **99**, 1107–1126.

Cushing, E. M., Kantrowitz, E. M. and Taylor, K. R. (1973). Water resources of the Delmarva Peninsula. US Geological Survey Professional Paper 822.

Czarnecki, J. B., Gillip, J. A., Jones, P. M. and Yeatts, D. S. (2009). Groundwater-flow model of the Ozark Plateaus aquifer system, northwestern Arkansas, Southeastern Kansas, Southwestern Missouri, and Northeastern Oklahoma. US Geological Survey Scientific Investigations Report 2009–5148.

Dages, C., Voltz, M., Bsaibes, A. *et al.* (2009). Estimating the role of a ditch network in groundwater recharge in a Mediterranean catchment using a water balance approach. *J. Hydrol.*, **375**, 498–512.

Dahan, O., Shani, Y., Enzel, Y., Yechieli, Y. and Yakirevich, A. (2007). Direct measurements of floodwater infiltration into shallow alluvial aquifers. *J. Hydrol.*, **344**, 157–170.

Daly, C., Halbleib, M., Smith, J. I. *et al.* (2008). Physiographically sensitive mapping of temperature and precipitation across the conterminous United States. *Inter. J. Climatology*, doi:10.1002/joc.1688.

Daly, C., Neilson, R. P. and Phillips, D. L. (1994). A statistical-topographic model for mapping climatological precipitation over mountainous terrain. *J. Appl. Met.*, **33**, 140–158.

Dane, J. H. and Topp, G. C. (eds.) (2002). *Methods of Soil Analysis. Part 4: Physical Methods.* Madison, Wisconsin: Soil Science Society of America.

Daniel, C. C. and Harned, D. A. (1998). Ground-water recharge to and storage in the regolith-fractured crystalline rock aquifer system, Guilford County, North Carolina. US Geological Survey Water-Resources Investigations Report 97–4140.

Daniel, J. F. (1976). Estimating groundwater evapotranspiration from streamflow records. *Water Resour. Res.*, **12**, 360–364.

Danielson, T. W. and Hood, J. W. (1984). Infiltration to the Navajo Sandstone in the lower Dirty Devil River Basin, Utah, with emphasis on techniques used in its determination. US Geological Survey Water-Resources Investigations Report 84–4154.

Davidson, W. A. (1995). Hydrogeology and groundwater resources of the Perth region, Western Australia. *West. Austral. Geol. Surv. Bull.*, **142**.

Davis, S. N., Campbell, D. J., Bentley, H. W. and Flynn, T. J. (1985). *Ground Water Tracers.* Dublin, Ohio: National Water Well Association.

Dawes, W. R., Zhang, L., Hatton, T. J. *et al.* (1997). Evaluation of a distributed parameter ecohydrological model (TOPOG-IRM) on a small cropping rotation catchment. *J. Hydrol.*, **191**, 64–86.

de Marsily, G. (1986). *Quantitative Hydrogeology: Groundwater Hydrology for Engineers.* Orlando, Florida: Academic Press.

de Vries, D. A. (1966). Thermal properties of soil. In *Physics of Plant Environment*, ed. W. R. van Wijk. Amsterdam: North-Holland Publishing Company.

Delin, G. N., Healy, R. W., Landon, M. K. and Böhlke, J. K. (2000). Effects of topography and soil properties on recharge at two sites in an agricultural field. *J. Amer. Water Resour. Assoc.*, **36**, 1401–1416.

Delin, G. N., Healy, R. W., Lorenz, D. L. and Nimmo, J. R. (2007). Comparison of local- to regional-scale estimates of ground-water recharge in Minnesota, USA. *J. Hydrol.*, **334**, 231–249.

Delin, G. N. and Herkelrath, W. N. (2005). Use of soil moisture probes to estimate ground water recharge at an oil spill site. *J. Amer. Water Resour. Assoc.*, **41**, 1259–1277.

Desconnets, J. C., Taupin, J. D., Lebel, T. and Leduc, C. (1997). Hydrology of the HAPEX-Sahel Central Super-Site: surface water drainage and aquifer recharge through the pool systems. *J. Hydrol.*, **188–189**, 155–178.

Dettinger, M. D. (1989). Reconnaissance estimates of natural recharge to desert basins in Nevada, USA, by using chloride-balance calculations. *J. Hydrol.*, **106**, 55–78.

Devlin, J. F. and Sophocleous, M. (2005). The persistence of the water budget myth and its relationship to sustainability. *Hydrogeol. J.*, **13**, 549–554.

Dewalle, D. R., Swistock, B. R. and Sharpe, W. E. (1988). Three-component tracer model for stormflow on a small Appalachian forested catchment. *J. Hydrol.*, **104**, 301–310.

Dewandel, B., Lachassagne, P., Bakalowicz, M., Weng, P. and Al-Malki, A. (2003). Evaluation of aquifer thickness by analysing recession hydrographs: application to the Oman ophiolite hard-rock aquifer. *J. Hydrol.*, **274**, 248–269.

Dickinson, J. E., Hanson, R. T., Ferré, T. P. A. and Leake, S. A. (2004). Inferring time-varying recharge from inverse analysis of long-term water levels. *Water Resour. Res.*, **40**, W074031–W0740315.

Dincer, T., Al-Mugrin, A. and Zimmermann, U. (1974). Study of the infiltration and recharge through the sand dunes in arid zones with special reference to stable isotopes and thermonuclear tritium. *J. Hydrol.*, **23**, 79–109.

Dingman, S. L. (1978). Synthesis of flow-duration curves for unregulated streams in New Hampshire. *Water Resour. Bull.*, **14**, 12.

Doerfliger, N., Jeannin, P. Y. and Zwahlen, F. (2000). Water vulnerability assessment in karst environments: a new method of defining protection areas using a multi-attribute approach and GIS tools (EPIK method). *Environ. Geol.*, **39**, 165–176.

Doherty, J. (2004). *PEST: Model-independent Parameter Estimation, User Manual.* 5th edn. Brisbane, QLD, Australia: Watermark Numerical Computing.

Doherty, J. (2005). *PEST Version 9.01.* Corinda, Australia: Watermark Computing.

Domenico, P. A. and Schwartz, F. W. (1998). *Physical and Chemical Hydrogeology.* 2nd edn. New York: John Wiley & Sons, Inc.

Donato, M. M. (1998). Surface-water/ground-water relations in the Lemhi River Basin, east-central Idaho. US Geological Survey Water-Resources Investigations Report 98–4185.

Dooge, J. C. (1959). A general theory of the unit hydrograph. *J. Geophys. Res.*, **64**, 241–256.

Doorenbos, J. and Pruitt, W. O. (1975). Crop water requirements. Irrigation and drainage Paper No. 24. Rome: Food and Agricultural Organization of the United Nations.

dos Santos, A. G. Jr. and Youngs, E. G. (1969). Study of specific yield in land-drainage situations. *J. Hydrol.*, **8**, 59–81.

Downey, J. S. (1984). Geohydrology of the Madison and associated aquifers in parts of Montana, North Dakota, South Dakota, and Wyoming. US Geological Survey Professional Paper 1273-G.

Dreiss, S. J. and Anderson, L. D. (1985). Estimating vertical soil moisture flux at a land treatment site. *Ground Water*, **23**, 503–511.

Dripps, W. R. and Bradbury, K. R. (2007). A simple daily soil-water balance model for estimating the spatial and temporal distribution of groundwater recharge in temperate humid areas. *Hydrogeol. J.*, **15**, 433–444.

Dripps, W. R., Hunt, R. J. and Anderson, M. P. (2006). Estimating recharge rates with analytic element models and parameter estimation. *Ground Water*, **44**, 47–55.

Dugan, J. T. and Peckenpaugh, J. M. (1985). Effects of climate, vegetation, and soils on consumptive water use and ground-water recharge to the central Midwest regional aquifer system, Mid-continent United States. US Geological Survey Water-Resources Investigations Report 85–4236.

Duke, H. R. (1972). Capillary properties of soils: influence upon specific yield. *Trans. Amer. Soc. Agric. Eng.*, **15**, 688–691.

Dumouchelle, D. H. (2001). Evaluation of ground-water/surface-water relations, Chapman Creek, west-central Ohio, by means of multiple methods. US Geological Survey Water-Resources Investigations Report 2001–4202.

Dumouchelle, D. H. and Schiefer, M. C. (2002). Use of streamflow records and basin characteristics to estimate groundwater recharge rates in Ohio. Columbus, Ohio: Ohio Department of Natural Resources Bulletin 46.

Dunkle, S. A., Plummer, L. N., Busenberg, E. *et al.* (1993). Chlorofluorocarbons (CCl_3F and CCl_2F_2) as dating tools and hydrologic tracers in shallow groundwater of the Delmarva Peninsula, Atlantic coastal plain, United States. *Water Resour. Res.*, **29**, 3837–3860.

Dunne, T., Zhang, W. and Aubry, B. F. (1991). Effects of rainfall, vegetation, and microtopography on infiltration and runoff. *Water Resour. Res.*, **27**, 2271–2285.

Eckhardt, K. (2005). How to construct recursive digital filters for baseflow separation. *Hydrol. Proc.*, **19**, 507–515.

Eckhardt, K. (2008). A comparison of baseflow indices, which were calculated with seven different baseflow separation methods. *J. Hydrol.*, **352**, 168–173.

Eckhardt, K. and Ulbrich, U. (2003). Potential impacts of climate change on groundwater recharge and streamflow in a central European low mountain range. *J. Hydrol.*, **284**, 244–252.

Edmunds, W. M., Fellman, E., Goni, I. B. and Prudhomme, C. (2002). Spatial and temporal distribution of groundwater recharge in northern Nigeria. *Hydrogeol. J.*, **10**, 205–215.

Ek, M. B., Mitchell, K. E., Lin, Y. *et al.* (2003). Implementation of Noah land surface model advances in the National Centers for Environmental Prediction operational mesoscale Eta model. *J. Geophys. Res. D*, **108**, GCP121-GCP1216.

Ekwurzel, B., Schlosser, P., Smethie, W. M., Jr. *et al.* (1994). Dating of shallow groundwater: comparison of the transient tracers 3H/3He, chlorofluorocarbons, and 85Kr. *Water Resour. Res.*, **30**, 1693–1708.

El-Kadi, A. I. (2005). Validity of the generalized Richards equation for the analysis of pumping test data for a coarse-material aquifer. *Vadose Zone J.*, **4**, 196–205.

Endres, A. L., Clement, W. P. and Rudolph, D. L. (2000). Ground penetrating radar imaging of an aquifer during a pumping test. *Ground Water*, **38**, 566–576.

Enfield, C. G., Hsieh, J. J. and Warrick, A. W. (1973). Evaluation of water flux above a deep water table using thermocouple psychrometers. *Soil Sci. Soc. Amer. Proc.*, **37**, 968–970.

Engesgaard, P., Jensen, K. H., Molson, J., Frind, E. O. and Olsen, H. (1996). Large-scale dispersion in a sandy aquifer: simulation of subsurface transport of environmental tritium. *Water Resour. Res.*, **32**, 3253–3266.

Englund, E., Aldahan, A. and Possnert, G. (2008). Tracing anthropogenic nuclear activity with 129I in lake sediment. *J. Environ. Radioact.*, **99**, 219–229.

Engott, J. A. and Vana, T. T. (2007). Effects of agricultural land-use changes and rainfall on groundwater recharge in central and west Maui, Hawaii, 1926–2004. US Geological Survey Scientific Investigations Report 2007-5103.

Entekhabi, D. and Moghaddam, M. (2007). Mapping recharge from space: roadmap to meeting the grand challenge. *Hydrogeol. J.*, **15**, 105–116.

Eriksson, E. and Khunakasem, V. (1969). Chloride concentration in groundwater, recharge rate and rate of deposition of chloride in the Israel Coastal Plain. *J. Hydrol.*, **7**, 178–197.

Essaid, H. I., Zamora, C. M., McCarthy, K. A., Vogel, J. R. and Wilson, J. T. (2008). Using heat to characterize streambed water flux variability in four stream reaches. *J. Environ. Qual.*, **37**, 1010–1023.

Evett, S. R., Warrick, A. W. and Matthias, A. D. (1995). Wall material and capping effects on microlysimeter temperatures and evaporation. *Soil Sci. Soc. Amer. J.*, **59**, 329–336.

Fan, Y., Van den Dool, H. M., Lohmann, D. and Mitchell, K. (2006). 1948–98 US hydrological reanalysis by the Noah land data assimilation system. *J. Climate*, **19**, 1214–1237.

Fassnacht, S. R., Dressler, K. A. and Bales, R. C. (2003). Snow water equivalent interpolation for the Colorado River Basin from snow telemetry (SNOTEL) data. *Water Resour. Res.*, **39**, SWC31–WC310.

Faunt, C. C. (ed.) (2009). Groundwater availability of the Central Valley aquifer, California. US Geological Survey Professional Paper 1766.

Favreau, G., Cappelaere, B., Massuel, S. *et al.* (2009). Land clearing, climate variability, and water resources increase in semiarid southwest Niger: a review. *Water Resour. Res.*, **45**, W00A16, doi:10.1029/2007WR006785.

Faye, R. E. and Mayer, G. C. (1997). Simulation of ground-water flow in southeastern Coastal Plain clastic aquifers in Georgia and adjacent parts of Alabama and South Carolina. US Geological Survey Professional Paper 1410-F.

Fayer, M. J. (2000). UNSAT-H version 3.0: Unsaturated soil water and heat flow model, theory, user manual, and examples. Pacific Northwest National Laboratory PNNL 13249. Richland, WA.

Fayer, M. J. and Gee, G. W. (2006). Multiple-year water balance of soil covers in a semiarid setting. *J. Environ. Qual.*, **35**, 366–377.

Fayer, M. J., Gee, G. W., Rockhold, M. L., Freshley, M. D. and Walters, T. B. (1996). Estimating recharge rates for a groundwater model using a GIS. *J. Environ. Qual.*, **25**, 510–518.

Fernando, L. G. and Gerardo, O. F. (1999). Feasibility study for the attenuation of groundwater exploitation impacts in the urban area of Aquascalientes, Mexico. In *Groundwater in the Urban Environment: Selected City Profiles*, ed. J. Chilton. Rotterdam: A. A. Balkema, 181–187.

Ferré, T. P. A., Binley, A. M., Blasch, K. W. *et al.* (2007). Geophysical methods for investigating ground-water recharge. In *Ground-water recharge in the arid and semiarid southwestern United States*, ed. D. A. Stonestrom, J. Constantz, T. P. A. Ferré, and S. A. Leake, US Geological Survey Professional Paper 1703, 377–414.

Ferré, T. P. A., von Glinski, G. and Ferré, L. A. (2003). Monitoring the maximum depth of drainage in response to pumping using borehole ground penetrating radar. *Vadose Zone J.*, **2**, 511–518.

Feth, J. H. (1964). Hidden recharge. *Ground Water*, **2**, 14–17.

Feth, J. H., Barker, D. A., Moore, L. G., Brown, R. J. and Veirs, C. E. (1966). Lake Bonneville: geology and hydrology of the Weber Delta district, including Ogden, Utah. US Geological Survey Professional Paper 518.

Finch, J. W. (1998). Estimating direct groundwater recharge using a simple water balance model: sensitivity to land surface parameters. *J. Hydrol.*, **211**, 112–125.

Fischer, B., Goldberg, V. and Bernhofer, C. (2008). Effect of a coupled soil water-plant gas exchange on forest energy fluxes: simulations with the coupled vegetation-boundary layer model HIRVAC. *Ecological Modelling*, **214**, 75–82.

Fisher, L. H. and Healy, R. W. (2008). Water movement within the unsaturated zone in four agricultural areas of the United States. *J. Environ. Qual.*, **37**, 1051–1063.

Flint, A. L. and Flint, L. E. (2007). Application of the Basin Characterization Model to estimate in-place recharge and runoff potential in the Basin and Range Carbonate-Rock aquifer system, White Pine County, Nevada, and adjacent areas in Nevada and Utah. US Geological Survey Scientific Investigations Report 2007–5099.

Flint, A. L., Flint, L. E., Bodvarsson, G. S., Kwicklis, E. M. and Fabryka-Martin, J. (2001a). Evolution of the conceptual model of unsaturated zone hydrology at Yucca Mountain, Nevada. *J. Hydrol.*, **247**, 1–30.

Flint, A. L., Flint, L. E., Hevesi, J. A. and Blainey, J. M. (2004). Fundamental concepts of recharge in the Desert Southwest: a regional modeling perspective. In *Groundwater Recharge in a Desert Environment, the Southwestern United States*, ed. J. F. Hogan, F. M. Phillips and B. R. Scanlon. Washington, DC: American Geophysical Union Water Science and Application Series, **9**, 159–184.

Flint, A. L., Flint, L. E., Kwicklis, E. M., Bodvarsson, G. S. and Fabryka-Martin, J. (2001b). Hydrology of Yucca Mountain, Nevada. *Rev. Geophys.*, **39**, 447–470.

Flint, A. L., Flint, L. E., Kwicklis, E. M., Fabryka-Martin, J. T. and Bodvarsson, G. S. (2002). Estimating recharge at Yucca Mountain, Nevada, USA: comparison of methods. *Hydrogeol. J.*, **10**, 180–204.

Flury, M. and Papritz, A. (1993). Bromide in the natural environment: occurrence and toxicity. *J. Environ. Qual.*, **22**, 747–758.

Flynn, R. H. and Tasker, G. D. (2004). Generalized estimates from streamflow data of annual and seasonal ground-water-recharge rates for drainage basins in New Hampshire. US Geological Survey Scientific Investigations Report 2004–5019.

Forchheimer, P. (1930). *Hydraulik*, 3rd edn. Berlin: B. G. Teubner.

Franke, O. L., Reilly, T. E., Pollock, D. W. and LaBaugh, J. W. (1998). Estimating areas contributing recharge to wells, lessons from previous studies. US Geological Survey Circular 1174.

Frankenberger, W. T. Jr., Tabatabai, M. A., Adriano, D. C. and Doner, H. E. (1996). Bromine, chlorine, and fluorine. In *Methods of Soil Analysis. Part 3: Chemical Methods*, ed. J. M. Bartels. Madison, Wisconsin. Soil Science of America, 833–867.

Freeman, L. A., Carpenter, M. C., Rosenberry, D. O. et al. (2004). Use of submersible pressure transducers in water-resources investigations. US Geological Survey Techniques of Water-Resources Investigation Report 08-A3.

Freethey, G. W. and Cordy, G. E. (1991). Geohydrology of Mesozoic rocks in the upper Colorado River Basin in Arizona, Colorado, New Mexico, Utah, and Wyoming, excluding the San Juan Basin. US Geological Survey Professional Paper 1411-C.

Freeze, R. A. and Cherry, J. A. (1979). *Groundwater*. Englewood Cliffs, NJ: Prentice-Hall Inc.

Gardner, W. R. (1964). Water movement below the root zone. 8th International Congress on Soil Science, Bucharest: Rompresfilatelia.

Gardner, W. R. (1967). Water uptake and salt distribution patterns in saline soils. In *Proceedings of the Symposium on Isotope and Radiation Techniques in Soil Physics and irrigation studies*. June 12–16, 1967. Istanbul. Vienna: International Atomic Energy Agency, 335–340.

Garen, D. C. and Moore, D. S. (2005a). Reply to discussion by M. Todd Walter and Stephen B. Shaw on curve number hydrology in water quality modeling: uses, abuses, and future directions. *J. Amer. Water Resour. Assoc.*, **41**, 1493–1494.

Garen, D. C. and Moore, D. S. (2005b). Curve number hydrology in water quality modeling: uses, abuses, and future directions. *J. Amer. Water Resour. Assoc.*, **41**, 377–388.

Gat, J. R. (1996). Oxygen and hydrogen isotopes in the hydrologic cycle. *Ann. Rev. Earth Planet. Sci.*, **24**, 225–262.

Gates, J. B., Edmunds, W. M., Ma, J. and Scanlon, B. R. (2008). Estimating groundwater recharge in a cold desert environment in northern China using chloride. *Hydrogeol. J.*, **16**, 893–910.

Gburek, W. J. and Folmar, G. J. (1999). A ground water recharge field study: site characterization and initial results. *Hydrol. Proc.*, **13**, 2813–2831.

Gburek, W. J., Folmar, G. J. and Urban, J. B. (1999). Field data and ground water modeling in a layered fractured aquifer. *Ground Water*, **37**, 175–184.

Gebert, W. A., Radloff, M. J., Considine, E. J. and Kennedy, J. L. (2007). Use of streamflow data to estimate base flow ground-water recharge for Wisconsin. *J. Amer. Water Resour. Assoc.*, **43**, 220–236.

Gee, G. W., Fayer, M. J., Rockhold, M. L. and Campbell, M. D. (1992). Variations in recharge at the Hanford Site. *Northwest Science*, **66**, 237–250.

Gee, G. W. and Hillel, D. (1988). Groundwater recharge in arid regions: review and critique of estimation methods. *Hydrol. Proc.*, **2**, 255–266.

Gee, G. W., Newman, B. D., Green, S. R. *et al.* (2009). Passive wick fluxmeters: design considerations and field applications. *Water Resour. Res.*, **45**, W04420, doi:10.1029/2008/WR007088.

Gee, G. W., Ward, A. L., Caldwell, T. G. and Ritter, J. C. (2002). A vadose zone water fluxmeter with divergence control. *Water Resour. Res.*, **38**(8) 1141, doi:10.1029/2001WR000816.

Gee, G. W., Wierenga, P. J., Andraski, B. J. *et al.* (1994). Variations in water balance and recharge potential at three western desert sites. *Soil Sci. Soc. Amer. J.*, **58**, 63–72.

Genereux, D. P. (1998). Quantifying uncertainty in tracer-based hydrograph separations. *Water Resour. Res.*, **34**, 915–919.

Genereux, D. P., Leahy, S., Mitasova, H., Kennedy, C. D. and Corbett, D. R. (2008). Spatial and temporal variability of streambed hydraulic conductivity in West Bear Creek, North Carolina, USA. *J. Hydrol.*, **358**, 332–353.

Gerhart, J. M. (1986). Ground-water recharge and its effects on nitrate concentration beneath a manured field site in Pennsylvania. *Ground Water*, **24**, 483–489.

Gerla, P. J. (1999). Estimating the ground-water contribution in wetlands using modeling and digital terrain analysis. *Wetlands*, **19**, 394–402.

Ghodrati, M. and Jury, W. A. (1990). A field study using dyes to characterize preferential flow of water. *Soil Sci. Soc. Am. J.*, **54**, 1558–1563.

Gillham, R. W. (1984). The capillary fringe and its effect on water-table response. *J. Hydrol.*, **67**, 307–324.

Glover, R. E. (1964). Ground-water movement. US Bureau of Reclamation Engineering Monograph **13**.

Glynn, J. E., Carroll, T. R., Holman, P. B. and Grasty, R. L. (1988). An airborne gamma ray snow survey of a forest covered area with a deep snowpack. *Remote Sens. Environ.*, **26**, 149–160.

Glynn, P. and Busenberg, E. (1996). Unsaturated zone investigations and chlorofluorocarbon dating of ground waters in the Pinal Creek Basin, Arizona. In *Proceedings of the US Geological Survey Toxic Substances Hydrology Program Technical Meeting*. Colorado Springs, CO, September 20–24, 1993.

ed. G. L. Mallard and D.A. Aronson. US Geological Survey Water-Resources Investigations Report 93–4015.

Gogu, R. C. and Dassargues, A. (2000). Current trends and future challenges in groundwater vulnerability assessment using overlay and index methods. *Environ. Geol.*, **39**, 549–559.

Goldscheider, N. (2005). Karst groundwater vulnerability mapping: application of a new method in the Swabian Alb, Germany. *Hydrogeol. J.*, **13**, 555–564.

Graham, D. N., Butts, M. B. and Frevert, D. K. (2006). Flexible integrated watershed modeling with MIKE SHE. In *Watershed Models*, ed. V. P. Singh and D. K. Frevert. Boca Raton, Florida: CRC Press.

Greene, E. A. (1997). Tracing recharge from sinking streams over spatial dimensions of kilometers in a karst aquifer. *Ground Water*, **35**, 898–904.

Gu, A., Gray, F., Eastoe, C. J. *et al.* (2008). Tracing ground water input to base flow using sulfate (S, O) isotopes. *Ground Water*, **46**, 502–509.

Gunderson, L. C. S. and Wanty, R. B. (1991). Field studies of radon in rocks, soils, and water. US Geological Survey Bulletin 1971.

Gurdak, J. J. and Roe, C. D. (2009). Recharge rates and chemistry beneath playas of the High Plains aquifer: a literature review and synthesis. US Geological Survey Circular 1333.

Gvirtzman, H. and Magaritz, M. (1986). Investigation of water movement in the unsaturated zone under an irrigated area using environmental tritium. *Water Resour. Res.*, **22**, 635–642.

Gvirtzman, H., Ronen, D. and Magaritz, M. (1986). Anion exclusion during transport through the unsaturated zone. *J. Hydrol.*, **87**, 267–283.

Ha, K., Koh, D. C., Yum, B. W. and Lee, K. K. (2008). Estimation of river stage effect on groundwater level, discharge, and bank storage and its field application. *Geosciences J.*, **12**, 191–204.

Haitjema, H. M. (1995). *Analytic Element Modeling of Groundwater Flow*. San Diego: Academic Press.

Halford, K. J. (1997). Effects of unsaturated zone on aquifer test analysis in a shallow-aquifer system. *Ground Water*, **35**, 512–522.

Halford, K. J. and Mayer, G. C. (2000). Problems associated with estimating ground water discharge and recharge from stream-discharge records. *Ground Water*, **38**, 331–342.

Hall, D. W. and Risser, D. W. (1993). Effects of agricultural nutrient management on nitrogen fate and transport in Lancaster County, Pennsylvania. *Water Resour. Bull.*, **29**, 55–76.

Hall, F. R. (1968). Base-flow recessions: a review. *Water Resour. Res.*, **4**, 973–983.

Hall, F. R. and Moench, A. F. (1972). Application of the convolution equation to stream-aquifer relationships. *Water Resour. Res.*, **8**, 487–493.

Hamon, W. R. (1963). Computation of direct run-off amounts from storm rainfall. *Inter. Assoc. Sci. Hydrol. Pub.*, **63**, 52–62.

Hanson, R. T., McLean, J. S. and Miller, R. S. (1994). Hydrogeologic framework and preliminary simulation of ground-water flow in the Mimbres Basin, southwestern New Mexico. US Geological Survey Water-Resources Investigations Report 94–4011.

Hantush, M. S. (1956). Analysis of data from pumping tests in leaky aquifers. *Trans. Amer. Geophys. Union*, **37**, 702–714.

Harbaugh, A. W. (2005). MODFLOW-2005: the US Geological Survey modular ground-water model, the ground-water flow process. US Geological Survey Techniques and Methods Report 6-A16.

Hardman, G. (1936). Nevada precipitation and acreages of land by rainfall zones. University of Nevada, Reno Agricultural Experiment Station Report.

Hart, G. L. and Lowery, B. (1998). Measuring instantaneous solute flux and loading with time domain reflectometry. *Soil Sci. Soc. Amer. J.*, **62**, 23–35.

Harvey, J. W., Wagner, B. J. and Bencala, K. E. (1996). Evaluating the reliability of the stream tracer approach to characterize stream-subsurface water exchange. *Water Resour. Res.*, **32**, 2441–2451.

Hatch, C. E., Fisher, A. T., Revenaugh, J. S., Constantz, J. and Ruehl, C. (2006). Quantifying surface water-groundwater interactions using time series analysis of streambed thermal records: method development. *Water Resour. Res.*, **42**, W10410, doi:10.1029/2005WR004787.

Healy, R. W. (1989). Seepage through a hazardous-waste trench cover. *J. Hydrol.*, **108**, 213–234.

Healy, R. W. and Cook, P. G. (2002). Using groundwater levels to estimate recharge. *Hydrogeol. J.*, **10**, 91–109.

Healy, R. W., Gray, J. R., de Vries, M. P. and Mills, P. C. (1989). Water balance at a low-level radioactive-waste disposal site. *Water Resour. Bull.*, **25**, 381–390.

Healy, R. W. and Mills, P. C. (1991). Variability of an unsaturated sand unit underlying a radioactive-waste trench. *Soil Sci. Soc. Amer. J.*, **55**, 899–907.

Healy, R. W., Rice, C. A., Bartos, T. T. and McKinley, M. P. (2008). Infiltration from an impoundment for coal-bed natural gas, Powder River Basin, Wyoming: evolution of water and sediment chemistry. *Water Resour. Res.*, **44**, W06424, doi:10.1029/2007WR006396.

Healy, R. W. and Ronan, A. D. (1996). Documentation of computer program VS2DH for simulation of energy transport in variably saturated porous media. US Geological Survey Water-Resources Investigations Report 96–4230.

Healy, R. W., Winter, T. C., LaBaugh, J. W. and Franke, O. L. (2007). Water budgets: foundations for effective water-resources and environmental management. US Geological Survey Circular 1308.

Heath, R. C. (1983). Basic ground-water hydrology. US Geological Survey Water-Supply Paper 2220.

Heath, R. C. (1984). Ground-water regions of the United States. US Geological Survey Water-Supply Paper 2242.

Heaton, T. H. E. and Vogel., J. C. (1981). Excess air in groundwater. *J. Hydrol.*, **50**, 201–216.

Heilweil, V. M. and Freethey, G. W. (1992). Simulation of ground-water flow and water-level declines that could be caused by proposed withdrawals, Navajo Sandstone, southwestern Utah and northwestern Arizona. US Geological Survey Water-Resources Investigations Report 90–4105.

Heilweil, V. M., Solomon, D. K. and Gardner, P. M. (2006). Borehole environmental tracers for evaluating net infiltration and recharge through desert bedrock. *Vadose Zone J.*, **5**, 98–120.

Heilweil, V. M., Solomon, D. K. and Gardner, P. M. (2007). Infiltration and recharge at Sand Hollow, an upland bedrock basin in southwestern Utah. In *Ground-water Recharge in the Arid and Semiarid Southwestern United States*, ed. D. A. Stonestrom, J. Constantz, T. P. A. Ferre, and S. A. Leake. US Geological Survey Professional Paper 1703, Chapter I, 221–252.

Helsel, D. R. and Hirsch, R. M. (2002). Statistical Methods in Water Resources. US Geological Survey Techniques of Water Resources Investigations, Book 4, Chapter A3.

Hendricks Franssen, H. J., Brunner, P., Kgothlang, L. and Kinzelbach, W. (2006). Inclusion of remote sensing information to improve groundwater flow modelling in the Chobe region (Botswana). In *Calibration and Reliability in Groundwater Modeling: From Uncertainty to Decision Making. IAHS-AISH Publication 304*, ed. M. F. P. Bierkens, J. C. Gehrels and K. Kovar: International Association of Hydrological Sciences, 31–37.

Hendry, M. J. (1983). Groundwater recharge through heavy-textured soil. *J. Hydrol.*, **63**, 201–209.

Heppner, C. S. and Nimmo, J. R. (2005). A computer program for predicting recharge with a master recession curve. US Geological Survey Scientific Investigations Report 2005–5172.

Heppner, C. S., Nimmo, J. R., Folmar, G. J., Gburek, W. J. and Risser, D. W. (2007). Multiple-methods investigation of recharge at a humid-region fractured rock site, Pennsylvania, USA. *Hydrogeol. J.*, **15**, 915–927.

Herczeg, A. L. and Edmunds, W. M. (2000). Inorganic ions as tracers. In *Environmental Tracers in Subsurface Hydrology*, ed. P. G. Cook and A. L. Herczeg. Boston: Kluwer Academic Publishers, 31–77.

Hevesi, J. A., Flint, A. L. and Istok, J. D. (1992a). Precipitation estimation in mountainous terrain using multivariate geostatistics. Part II: Isohyetal maps. *J. Appl. Met.*, **31**, 677–688.

Hevesi, J. A., Istok, J. D. and Flint, A. L. (1992b). Precipitation estimation in mountainous terrain using multivariate geostatistics. Part I: Structural analysis. *J. Appl. Met.*, **31**, 661–676.

Hignett, C. and Evett, S. R. (2002). Neutron thermalization. In *Methods of Soil Analysis. Part 4: Physical Methods*, ed. J. H. Dane and G. C. Topp. Madison, Wisconsin: Soil Science Society of America.

Hill, M. C. (1992). A computer program (MODFLOWP) for estimating parameters of a transient, three-dimensional ground-water flow model using non-linear regression. US Geological Survey Open-File Report 91–484.

Hill, M. C. (1998). Methods and guidelines for effective model calibration. US Geological Survey Water-Resources Investigations Report 98–4005.

Hill, M. C. and Tiedeman, C. R. (2007). *Effective Groundwater Model Calibration*. Hoboken, NJ: John Wiley and Sons, Inc.

Hillel, D. (1980). *Fundamentals of Soil Physics*. New York: Academic Press.

Ho, D. T. and Schlosser, P. (2000). Atmospheric SF_6 near a large urban area. *Geophys. Res. Lett.*, **27**, 1679–1682.

Hodnett, M. G. and Bell, J. P. (1990). Processes of water movement through a chalk Coombe deposit in southeast England. *Hydrol. Proc.*, **4**, 361–372.

Hoffmann, J. P., Blasch, K. W. and Ferre, T. P. (2003). Combined use of heat and soil-water content to determine stream/ground-water exchanges, Rillito Creek, Tucson, Arizona. In *Heat as a Tool for Studying the Movement of Ground Water Near Streams*, ed. D. A. Stonestrom and J. Constantz. US Geological Survey Circular 1260, 48–56.

Hoffmann, J. P., Blasch, K. W., Pool, D. R., Bailey, M. A. and Callegary, J. B. (2007). Estimated infiltration, percolation, and recharge rates at the Rillito Creek focused recharge investigation site, Pima County, Arizona. In *Ground-water Recharge in the Arid and Semiarid Southwestern United States*, ed. D. A. Stonestrom, J. Constantz, T. P. A. Ferre and S. A. Leake. US Geological Survey Professional Paper 1703, Chapter H, 185–220.

Hogan, J. F., Phillips, F. M. and Scanlon, B. R. (eds.) (2004). *Groundwater Recharge in a Desert Environment. The Southwestern United States*. Washington, DC: American Geophysical Union.

Holder, M., Brown, K. W., Thomas, J. C., Zabcik, D. and Murray, H. E. (1991). Capillary-wick unsaturated zone soil pore water sampler. *Soil Sci. Soc. Amer. J.*, **55**, 1195–1202.

Holser, W. T. (1979). Mineralogy of evaporites. In *Marine Minerals*, ed. R. G. Burns. Mineralogical Society of America, **6**, 211–294.

Holtschlag, D. J. (1997). A generalized estimate of ground-water-recharge rates in the lower peninsula of Michigan. US Geological Survey Water-Supply Paper 2437.

Hooper, R. P. and Shoemaker, C. A. (1986). Comparison of chemical and isotopic hydrograph separation. *Water Resour. Res.*, **22**, 1444–1454.

Hoos, A. B. (1990). Recharge rates and aquifer hydraulic characteristics for selected drainage basins in middle and east Tennessee. US Geological Survey Water-Resources Investigations Report 90–4015.

Hsieh, P. A., Wingle, W. and Healy, R. W. (1999). VS2DI: a graphical software package for simulating fluid flow and solute or energy transport in variably saturated porous media. US Geological Survey Water-Resources Investigations Report 99–4130.

Huang, S., Pollack, H. N. and Shen, P. Y. (2000). Temperature trends over the past five centuries reconstructed from borehole temperatures. *Nature*, **403**, 756–758.

Hubbell, J. M. and Sisson, J. B. (1998). Advanced tensiometer for shallow or deep soil water potential. *Soil Sci.*, **163**, 271–277.

Hughes, A. G., Mansour, M. M. and Robins, N. S. (2008). Evaluation of distributed recharge in an upland semi-arid karst system: the West Bank Mountain Aquifer, Middle East. *Hydrogeol. J.*, **16**, 845–854.

Huisman, J. A., Snepvangers, J. J., Bouten, W. and Heuvelink, G. B. (2003). Monitoring temporal development of spatial soil water content variation: comparison of ground penetrating radar and time domain reflectometry. *Vadose Zone J.*, **2**, 519–529.

Hunt, R. J., Anderson, M. P. and Kelson, V. A. (1998). Improving a complex finite-difference ground water flow model through the use of an analytic element screening model. *Ground Water*, **36**, 1011–1017.

Hunt, R. J., Doherty, J. and Tonkin, M. J. (2007). Are models too simple? Arguments for increased parameterization. *Ground Water*, **45**, 254–262.

Hunt, R. J., Krabbenhoft, D. P. and Anderson, M. P. (1996). Groundwater inflow measurements in wetland systems. *Water Resour. Res.*, **32**, 495–507.

Hunt, R. J. and Steuer, J. J. (2001). Evaluating the effects of urbanization and land-use planning using ground-water and surface-water models. US Geological Survey Fact Sheet 102-01.

Hunt, R. J., Steuer, J. J., Mansor, M. T. C. and Bullen, T. D. (2001). Delineating a recharge area for a spring using numerical modeling, Monte Carlo techniques, and geochemical investigation. *Ground Water*, **39**, 702–712.

Hurr, T. R. and Litke, D. W. (1989). Estimating pumping time and ground-water withdrawals using energy-consumption data. US Geological Survey Water-Resources Investigations Report 89–4107.

Hutson, J. L. and Wagenet, R. J. (1992). LEACHM: Leaching estimation and chemistry model: a process-based model of water and solute movement, transformations, plant uptake and chemical reactions in the unsaturated zone continuum. Version 3. Ithaca, NY. Water Resources Institute, Cornell University.

Imes, J. L. and Emmett, L. F. (1994). Geohydrology of the Ozark Plateaus aquifer system in parts of Missouri, Arkansas, Oklahoma, and Kansas. US Geological Survey Professional Paper 1414-D.

Ingebritsen, S. E., Sanford, W. E. and Neuzil, C. E. (2006). *Groundwater in Geologic Processes*, 2nd edn. New York: Cambridge University Press.

Ingraham, N. L. and Shadel, C. (1992). A comparison of the toluene distillation and vacuum/heat methods for extracting soil water for stable isotopic analysis. *J. Hydrol.*, **140**, 371–387.

Institute of Hydrology (1980). Low flow studies: Research Report 1. Institute of Hydrology. Wallingford, UK.

Izuka, S. K., Oki, D. S. and Chen, C. -H. (2005). Effects of irrigation and rainfall reduction on groundwater recharge in the Lihue Basin, Kauai, Hawaii. US Geological Survey Scientific Investigations Report 2005–5146.

Jamison, V. C. and Kroth, E. M. (1958). Available moisture storage capacity in relation to textural composition and organic matter content of several Missouri soils. *Soil Sci. Soc. Amer. Proc.*, **22**, 189–192.

Jan, C. D., Chen, T. H. and Lo, W. C. (2007). Effect of rainfall intensity and distribution on groundwater level fluctuations. *J. Hydrol.*, **332**, 348–360.

Jarvis, N. (2002). The MACRO model (Version 4.3) technical description. Uppsala. Department of Soil Sciences, Swedish University of Agricultural Sciences.

Jensen, M. E., Burman, R. D. and Allen, R. G. (1990). *Evapotranspiration and Irrigation Water Requirements*. New York: American Society of Civil Engineers.

Jensen, M. E. and Haise, H. R. (1963). Estimating evapotranspiration from solar radiation. *J. Irrig. Drain. Div. Amer. Soc. Civ. Engin.*, **89**, 15–41.

Johnson, A. I. (1967). Specific yield: compilation of specific yields for various materials. US Geological Survey Water-Supply Paper 1662-D.

Johnson, A. I., Prill, R. C. and Morris, D. A. (1963). Specific yield: column drainage and centrifuge moisture content. US Geological Survey Water-Supply Paper 1662-A.

Johnson, J. B. and Schaefer, G. L. (2002). The influence of thermal, hydrologic, and snow deformation mechanisms on snow water equivalent pressure sensor accuracy. *Hydrol. Proc.*, **16**, 3529–3542.

Jolly, I. D., Cook, P. G., Allison, G. B. and Hughes, M. W. (1989). Simultaneous water and solute movement through an unsaturated soil following an increase in recharge. *J. Hydrol.*, **111**, 391–396.

Jones, J. P., Sudicky, E. A., Brookfield, A. E. and Park, Y. J. (2006). An assessment of the tracer-based approach to quantifying groundwater contributions to streamflow. *Water Resour. Res.*, **42**, W02407, doi:10.1029/2005WR004130.

Journel, A. G. and Huijbregts, C. J. (1978). *Mining Geostatistics*. New York: Academic Press.

Juckem, P. F. and Hunt, R. J. (2007). Simulation of the shallow ground-water flow system near Grindstone Creek and the community of New Post, Sawyer County, Wisconsin. US Geological Survey Scientific Investigations Report 2007-5014.

Jury, W. A. (1982). Simulation of solute transport using a transfer function model. *Water Resour. Res.*, **18**, 363–368.

Jyrkama, M. I., Sykes, J. F. and Normani, S. D. (2002). Recharge estimation for transient ground water modeling. *Ground Water*, **40**, 638–648.

Kachanoski, G., Pringle, E. and Ward, A. (1992). Field measurement of solute travel times using time domain reflectometry. *Soil Sci. Soc. Amer. J.*, **56**, 47–52.

Kalin, R. M. (2000). Radiocarbon dating of groundwater systems. In *Environmental Tracers in Subsurface Hydrology*, ed. P. G. Cook and A. L. Herczeg. Boston: Kluwer Academic Publishers, 111–144.

Kalnay, E., Kanamitsu, M. and Baker, W. E. (1990). Global numerical weather prediction at the National Meteorological Center. *Bull. Amer. Met. Soc.*, **71**, 1410–1428.

Karl, T. R. and Knight, R. W. (1985). *Atlas of Monthly Palmer Moisture Anomaly Indices (1931–1983) for the Contiguous United States*. Ashville, NC: National Climatic Data Center.

Katz, B. G., Coplen, T. B., Bullen, T. D. and Hal Davis, J. (1997). Use of chemical and isotopic tracers to characterize the interactions between ground water and surface water in mantled karst. *Ground Water*, **35**, 1014–1028.

Kaufman, S. and Libby, W. F. (1954). The natural distribution of tritium. *Phys. Rev.*, **93**, 1337–1344.

Keery, J., Binley, A., Crook, N. and Smith, J. W. N. (2007). Temporal and spatial variability of groundwater-surface water fluxes: development and application of an analytical method using temperature time series. *J. Hydrol.*, **336**, 1–16.

Keese, K. E., Scanlon, B. R. and Reedy, R. C. (2005). Assessing controls on diffuse groundwater recharge using unsaturated flow modeling. *Water Resour. Res.*, **41**, 1–12.

Kendall, C. and McDonnell, J. J. (eds.) (1998). *Isotope Tracers in Catchment Hydrology*. Amsterdam: Elsevier Science Publishing.

Kengni, L., Vachaud, G., Thony, J. L. *et al.* (1994). Field measurements of water and nitrogen losses under irrigated maize. *J. Hydrol.*, **162**, 23–46.

Kennedy, C. D., Genereux, D. P., Mitasova, H., Corbett, D. R. and Leahy, S. (2008). Effect of sampling density and design on estimation of streambed attributes. *J. Hydrol.*, **355**, 164–180.

Kennedy, E. J. (1984). Discharge ratings at gaging stations. US Geological Survey Techniques of Water-Resources Investigations 03-A10.

Kernodle, J. M., McAda, D. P. and Thorn, C. R. (1995). Simulation of ground-water flow in the Albuquerque Basin, central New Mexico, 1901–1994, with projections to 2020. US Geological Survey Water-Resources Investigations Report 94-4251.

Ketchum, J. N. Jr., Donovan, J. J. and Avery, W. H. (2000). Recharge characteristics of a phreatic aquifer as determined by storage accumulation. *Hydrogeol. J.*, **8**, 579–593.

Kilpatrick, F. A. and Cobb, E. D. (1985). Measurement of discharge using tracers. US Geological Survey Techniques of Water-Resource Investigation, Chapter 03-A16.

Kimball, B. A., Broshears, R. E., Bencala, K. E. and McKnight, D. M. (1994). Coupling of hydrologic transport and chemical reactions in a stream affected by acid mine drainage. *Environ. Sci. Technol.*, **28**, 2065–2073.

Kimball, B. A., Runkel, R. L., Cleasby, T. E. and Nimick, D. A. (2004). Quantification of metal loading by tracer injection and synoptic sampling, 1997–98. In *Integrated Investigations of Environmental Effects of Historical Mining in the Basin and Boulder Mining Districts, Boulder River Watershed, Jefferson County, Montana*, ed. D. A. Nimick, S. E. Church and S. E. Finger. US Geological Survey Professional Paper 1652, 191–262.

Kimball, B. A., Walton-Day, K. and Runkel, R. L. (2007). Quantification of metal loading by tracer injection and synoptic sampling, 1996–2000. US Geological Survey Professional Paper 1651-E9.

King, F. H. (1899). Principles and conditions of the movements of groundwater. US Geological Survey Nineteenth Annual Report, 86–91.

Kinzelbach, W., Aeschbach, W., Alberich, C. *et al.* (2002). A survey of methods for groundwater recharge in arid and semi-arid regions. UNEP/DEWA/RS.02.2. Nairobi, Kenya: United Nations Environment Programme.

Kipp, K. L. Jr. (1997). Guide to the revised heat and solute transport simulator, HST3D. US Geological Survey Water-Resources Investigations Report 97-4157.

Kitching, R. and Shearer, T. R. (1982). Construction and operation of a large undisturbed lysimeter to measure recharge to the Chalk aquifer, England. *J. Hydrol.*, **58**, 267–277.

Knott, J. F. and Olimpio, J. C. (1986). Estimation of recharge rates to the sand and gravel aquifer using environmental tritium, Nantucket Island, Massachusetts. US Geological Survey Water Supply Paper 2297.

Kontis, A. L., Randall, A. D. and Mazzaferro, D. L. (2004). Regional hydrology and simulation of flow of stratified-drift aquifers in the glaciated northeastern United States. US Geological Survey Professional Paper 1415-C.

Krajewski, W. F., Anderson, M. C., Eichinger, W. E. *et al.* (2006). A remote sensing observatory for hydrologic sciences: a genesis for scaling to continental hydrology. *Water Resour. Res.*, **42**, W07301, doi:10.1029/2005WR004435.

Krul, W. F. and Liefrinck, F. A. (1946). *Recent Groundwater Investigations in the Netherlands: Monograph on the Progress of Research in Holland*. New York: Elsevier.

Kung, K. J. S. (1990a). Influence of plant uptake on the performance of bromide tracer. *Soil Sci. Soc. Am. J.*, **54**, 975–979.

Kung, K. J. S. (1990b). Preferential flow in a sandy vadose zone: 1. Field observation. *Geoderma*, **46**, 51–58.

Kung, K. J. S., Kladivko, E. J., Gish, T. J. *et al.* (2000). Quantifying preferential flow by breakthrough of sequentially applied tracers: silt loam soil. *Soil Sci. Soc. Am. J.*, **64**, 1296–1304.

Kuniansky, E. L. (1989). Geohydrology and simulation of groundwater flow in the "400-foot," "600-foot," and adjacent aquifers, Baton Rouge area, Louisiana. Louisiana Department of Transportation and Development Technical Report 49.

Kwicklis, E. M. (1999). Analysis of percolation flux based on heat flow estimated in boreholes. In *Hydrogeology of the Unsaturated Zone, North Ramp Area of the Exploratory Studies Facility, Yucca Mountain, Nevada*, ed. J. P. Rousseau, E. M. Kwicklis and D. C. Gillies. US Geological Survey Water-Resources Investigations Report 98–4050, 184–208.

Kwicklis, E. M., Flint, A. L. and Healy, R. W. (1993). Estimation of unsaturated zone liquid water flux at borehole UZ#4, UZ#5, UZ#7, and UZ#13, Yucca Mountain, Nevada, from saturation and water potential profiles. In *Proceedings Topical Meeting on Site Characterization and Model Validation, Focus 93*. LaGrange Park, Illinois: American Nuclear Society, 39–57.

Labaugh, J. W. and Rosenberry, D. O. (2008). Introduction and characteristics of flow. In *Field Techniques for Estimating Water Fluxes Between Surface Water and Ground Water*, ed. D. I. Rosenberry and J. W. Labaugh. US Geological Survey Techniques and Methods 4-D2, 1–38.

Labaugh, J. W., Rosenberry, D. O. and Winter, T. C. (1995). Groundwater contribution to the water and chemical budgets of Williams Lake, Minnesota, 1980-1991. *Can. J. Fish. Aquatic Sci.*, **52**, 754–767.

Laczniak, R. J., Flint, A. L., Moreo, M. T. *et al.* (2008). Ground-water budgets. In *Water Resources of the Basin and Range Carbonate-rock Aquifer System, White Pine County, Nevada, and Adjacent Areas in Nevada and Utah*, ed. A. H. Welch, D. J. Bright and L. A. Knochenmus. US Geological Survey Scientific Investigations Report 2007–5261.

Laenen, A. and Risley, J. C. (1997). Precipitation-runoff and streamflow-routing models for the Willamette River Basin, Oregon. US Geological Survey Water-Resources Investigations Report 05–4284.

Langsholt, E. (1992). A water balance study in lateritic terrain. *Hydrol. Proc.*, **6**, 11–27.

Lapham, W. W. (1989). Use of temperature profiles beneath streams to determine rates of vertical ground-water flow and vertical hydraulic conductivity. US Geological Survey Water-Supply Paper 2337.

Leake, S. A. (1984). A method for estimating ground-water return flow to the Colorado River in the Parker area, Arizona and California. US Geological Survey Water-Resources Investigations Report 84–4299.

Leavesley, G. H., Lichty, R. W., Troutman, B. M. and Saindon, L. G. (1983). Precipitation-runoff modeling system: user's manual. US Geological Survey Water-Resources Investigations Report 83–4238.

Leavesley, G. H., Markstrom, S. L., Brewer, M. S. and Viger, R. J. (1996). The Modular Modeling System (MMS): the physical process modeling component of a database-centered decision support system for water and power management. *Water, Air, and Soil Pollution*, **90**, 303–311.

Leblanc, M. J., Favreau, G., Massuel, S. *et al.* (2008). Land clearance and hydrological change in the Sahel: SW Niger. *Global and Planetary Change*, **61**, 135–150.

Leduc, C., Favreau, G. and Schroeter, P. (2001). Long-term rise in a Sahelian water-table: the Continental Terminal in South-West Niger. *J. Hydrol.*, **243**, 43–54.

Lee, C. H., Chen, W. P. and Lee, R. H. (2006). Estimation of groundwater recharge using water balance coupled with base-flow-record estimation and stable-base-flow analysis. *Environ. Geol.*, **51**, 73–82.

Lee, D. R. (1977). A device for measuring seepage flux in lakes and estuaries. *Limnol. Oceanogr.*, **22**, 140–147.

Lee, D. R. and Cherry, J. A. (1978). A field exercise on ground-water flow using seepage meters and mini-piezometers. *J. Geol. Educ.*, **27**, 6–20.

Lee, E. S. and Krothe, N. C. (2001). A four-component mixing model for water in a karst terrain in south-central Indiana, USA: using solute concentration and stable isotopes as tracers. *Chem. Geol.*, **179**, 129–143.

Lee, J. Y., Yi, M. J. and Hwang, D. (2005). Dependency of hydrologic responses and recharge estimates on water-level monitoring locations within a small catchment. *Geosciences J.*, **9**, 277–286.

Lee, X., Massman, W. and Law, B. (eds.) (2004). *Handbook of Micrometeorology: A Guide for Surface Flux Measurement and Analysis*. New York: Springer.

Lehmann, B. E., Davis, S. N. and Fabryka-Martin, J. T. (1993). Atmospheric and subsurface sources of stable and radioactive nuclides used for groundwater dating. *Water Resour. Res.*, **29**, 2027–2040.

Lerner, D. N. (1986). Leaking pipes recharge groundwater. *Ground Water*, **24**, 654–662.

Lerner, D. N. (2002). Identifying and quantifying urban recharge: a review. *Hydrogeol. J.*, **10**, 143–152.

Lerner, D. N., Isaar, A. S. and Simmers, I. (eds.) (1990). *Groundwater Recharge: A Guide to Understanding and Estimating Natural Recharge, International Contributions to Hydrogeology Vol. 8*. Hanover: Verlag Heinz Heise.

Lim, K. J., Engel, B. A., Tang, A. *et al.* (2005). Automated WEB GIS based hydrograph analysis tool, WHAT. *J. Amer. Water Resour. Assoc.*, **41**, 10.

Lin, R. F. and Wei, K. Q. (2006). Tritium profiles of pore water in the Chinese loess unsaturated zone: implications for estimation of groundwater recharge. *J. Hydrol.*, **328**, 192–199.

Lin, Y. F. and Anderson, M. P. (2003). A digital procedure for ground water recharge and discharge pattern recognition and rate estimation. *Ground Water*, **41**, 306–315.

Lin, Y. F., Wang, J. and Valocchi, A. J. (2009). PROGRADE: GIS toolkits for ground water recharge and discharge estimation. *Ground Water*, **47**, 122–128.

Linsley, R. K., Kohler, M. A. and Paulhus, J. L. H. (1982). *Hydrology for Engineers*, 3rd edn. New York: McGraw-Hill.

Liu, B., Phillips, F., Hoines, S., Campbell, A. R. and Sharma, P. (1995). Water movement in desert soil traced by hydrogen and oxygen isotopes, chloride, and chlorine-36, southern Arizona. *J. Hydrol.*, **168**, 91–110.

Liu, J., Chen, J. M. and Cihlar, J. (2003). Mapping evapotranspiration based on remote sensing: an application to Canada's landmass. *Water Resour. Res.*, **39**, SWC41-SWC415.

Loeltz, O. J., and Leake, S. A. (1983). A method for estimating ground-water return flow to the lower Colorado River in the Yuma area, Arizona and California. US Geological Survey Water-Resources Investigations Report 83–4220.

Loheide, S. P. Jr., Butler, J. J. Jr. and Gorelick, S. M. (2005). Estimation of groundwater consumption by phreatophytes using diurnal water table fluctuations: a saturated-unsaturated flow assessment. *Water Resour. Res.*, **41**, 1–14.

Loheide, S. P. Jr. and Gorelick, S. M. (2006). Quantifying stream-aquifer interactions through the analysis of remotely sensed thermographic profiles and in situ temperature histories. *Environ. Sci. Technol.*, **40**, 3336–3341.

Lorenz, D. L. and Delin, G. N. (2007). A regression model to estimate regional ground water recharge. *Ground Water*, **45**, 196–208.

Louie, M. J., Shelby, P. M., Smesrud, J. S. *et al.* (2000). Field evaluation of passive capillary samplers for estimating groundwater recharge. *Water Resour. Res.*, **36**, 2407–2416.

Lowry, C. S., Walker, J. F., Hunt, R. J. and Anderson, M. P. (2007). Identifying spatial variability of groundwater discharge in a wetland stream using a distributed temperature sensor. *Water Resour. Res.*, **43**, W10408, doi:10.1029/2007WR006145.

Lu, N. and Ge, S. (1996). Effect of horizontal heat and fluid flow on the vertical temperature distribution in a semiconfining layer. *Water Resour. Res.*, **32**, 1449–1453.

Lucas, L. L. and Unterweger, M. P. (2000). Comprehensive review and critical evaluation of the half-life of tritium. *J. Res. Nat. Inst. Stand. Tech.*, **105**, 541–549.

Luckey, R. R. and Becker, M. F. (1999). Hydrogeology, water use, and simulation of flow in the High Plains aquifer in southwestern Oklahoma, southeastern Colorado, southwestern Kansas, northeastern New Mexico, and northwestern Texas. US Geological Survey Water-Resources Report 99–4104.

Luckey, R. R., Gutentag, E. D., Heimes, F. J. and Weeks, J. B. (1986). Digital simulation of groundwater flow in the High Plains aquifer in parts of Colorado, Kansas, Nebraska, New Mexico, Oklahoma, South Dakota, Texas, and Wyoming. US Geological Survey Professional Paper 1400-D.

Maidment, D. R. (ed.) (1993). *Handbook in Hydrology*. New York: McGraw-Hill.

Makkeasorn, A., Chang, N. B., Beaman, M., Wyatt, C. and Slater, C. (2006). Soil moisture estimation in a semiarid watershed using RADARSAT-1 satellite imagery and genetic programming. *Water Resour. Res.*, **42**, W09401, doi:10.1029/2005WR004033.

Mallants, D., Vanclooster, M., Toride, N. *et al.* (1996). Comparison of three methods to calibrate TDR for monitoring solute movement in undisturbed soil. *Soil Sci. Soc. Amer. J.*, **60**, 747–754.

Maloszewski, P. and Zuber, A. (1982). Determining the turnover time of groundwater systems with the aid of environmental tracers. 1. Models and their applicability. *J. Hydrol.*, **57**, 207–231.

Mandle, R. J. and Kontis, A. L. (1992). Simulation of regional ground-water flow in the Cambrian-Ordovician aquifer system in the northern Midwest, United States. US Geological Survey Professional Paper 1405-C.

Manning, A. H. and Solomon, D. K. (2004). Constraining mountain-block recharge to the eastern Salt lake valley, Utah with dissolved noble gas and tritium data. In *Groundwater Recharge in a Desert Environment, the Southwestern United States*, ed. J. F. Hogan, F. M. Phillips and B. R. Scanlon. Washington, DC: American Geophysical Union, 139–148.

Markstrom, S. L., Niswonger, R. G., Regan, R. S., Prudic, D. E. and Barlow, P. M. (2008). GSFLOW: coupled ground-water and surface-water flow model based on the integration of the Precipitation-Runoff Modeling System (PRMS) and the Modular Ground-Water Flow Model (MODFLOW-2005). US Geological Survey Techniques and Methods Report 6-D1.

Masarik, K. C., Norman, J. M., Brye, K. R. and Baker, J. M. (2004). Improvements to measuring water flux in the vadose zone. *J. Environ. Qual.*, **33**, 1152–1158.

Mau, D. P. and Winter, T. C. (1997). Estimating ground-water recharge from streamflow hydrographs for a small mountain watershed in a temperate humid climate, New Hampshire, United States. *Ground Water*, **35**, 291–304.

Maxey, G. B. and Eakin, T. E. (1949). Ground water in White River Valley, White Pine, Nye, and Lincoln Counties, Nevada. Nevada State Engineer Water Resources Bulletin 8.

McCabe, G. J. and Markstrom, S. L. (2007). A monthly water-balance model driven by a graphical user interface, US Geological Survey Open-File Report 2007–1088.

McCarthy, R. L., Bower, F. A. and Jesson, J. P. (1977). Fluorocarbon-ozone theory: 1. Production and release – world production and release of CCl_3F and CCl_2F_2 (Fluorocarbons 11 and 12) through 1975. *Atmos. Environ.*, **11**, 491–497.

McConville, C., Kalin, R. M., Johnston, H. and McNeill, G. W. (2001). Evaluation of recharge in a small temperate catchment using natural and applied 18O profiles in the unsaturated zone. *Ground Water*, **39**, 616–624.

McDonald, M. G. and Harbaugh, A. W. (1988). A modular three-dimensional finite-difference ground-water flow model. US Geological Survey Techniques of Water-Resources Investigations, Volume 6, Chapter A1.

McDonnell, J. J. (1990). A rationale for old water discharge through macropores in a steep, humid catchment. *Water Resour. Res.*, **26**, 2821–2832.

McGuire, V. L., Johnson, M. R., Schieffer, R. L. et al. (2003). Water in storage and approaches to ground-water management, High Plains aquifer, 2000. US Geological Survey Circular 1243.

McMahon, P. B., Böhlke, J. K. and Christenson, S. C. (2004). Geochemistry, radiocarbon ages, and paleorecharge conditions along a transect in the central High Plains aquifer, southwestern Kansas, USA. *Applied Geochem.*, **19**, 1655–1686.

McMahon, P. B., Böhlke, J. K., Kauffman, L. J. et al. (2008). Source and transport controls on the movement of nitrate to public supply wells in selected principal aquifers of the United States. *Water Resour. Res.*, **44**, W04401, doi:10.1029/2007WR006252.

McMahon, P. B., Dennehy, K. F., Bruce, B. W. et al. (2006). Storage and transit time of chemicals in thick unsaturated zones under rangeland and irrigated cropland, High Plains, United States. *Water Resour. Res.*, **42**, W03413.

McMahon, P. B., Dennehy, K. F., Michel, R. L. et al. (2003). Water movement through thick unsaturated zones overlying the central High Plains aquifer, southwestern Kansas, 2000–2001. US Geological Survey Water-Resources Investigations Report 2003-4171.

Meinzer, O. E. (1923). The occurrence of ground water in the United States with a discussion of principles. US Geological Survey Water-Supply Paper 489.

Meinzer, O. E. and Stearns, N. D. (1929). A study of groundwater in the Pomperaug Basin, Conn, with special reference to intake and discharge. US Geological Survey Water-Supply Paper 597-B.

Mertes, L. A. K. (2002). Remote sensing of riverine landscapes. *Freshwater Biology*, **47**, 799–816.

Merz, Z. and Bloschl, G. (2004). Regionalisation of catchment model parameters. *J. Hydrol.*, 95–123.

Meyboom, P. (1961). Estimating ground-water recharge from stream hydrographs. *J. Geophys. Res.*, **66**, 1203–1214.

Meyboom, P. (1967). Groundwater studies in the Assiniboine River Drainage Basin. Part II:

hydrologic characteristics of phreatophytic vegetation in south-central Saskatchewan. Geological Survey of Canada Bulletin 139.

Meyer, S. C. (2005). Analysis of base flow trends in urban streams, Northeastern Illinois, USA. *Hydrogeol. J.*, **13**, 871–885.

Meyer, W. R. (1962). Use of a neutron moisture probe to determine the storage coefficient of an unconfined aquifer. US Geological Survey Professional Paper 450-E.

Micovic, Z. and Quick, M. C. (1999). A rainfall and snowmelt runoff modeling approach to flow estimation at ungauged sites in British Columbia. *J. Hydrol.*, **226**, 101–120.

Minasny, B. and McBratney, A. B. (2007). Estimating the water retention shape parameter from sand and clay content. *Soil Sci. Soc. Amer. J.*, **71**, 1105–1110.

Minasny, B., McBratney, A. B. and Bristow, K. L. (1999). Comparison of different approaches to the development of pedotransfer functions for water-retention curves. *Geoderma*, **93**, 225–253.

Minor, T. B., Russell, C. E. and Mizell, S. A. (2007). Development of a GIS-based model for extrapolating mesoscale groundwater recharge estimates using integrated geospatial data sets. *Hydrogeol. J.*, **15**, 183–195.

Modica, E., Buxton, H. T. and Plummer, L. N. (1998). Evaluating the source and residence times of groundwater seepage to streams, New Jersey Coastal Plain. *Water Resour. Res.*, **34**, 2797–2810.

Modica, E., Reilly, T. E. and Pollock, D. W. (1997). Patterns and age distribution of ground-water flow to streams. *Ground Water*, **35**, 523–537.

Moench, A. F. (1994). Specific yield as determined by type-curve analysis of aquifer-test data. *Ground Water*, **32**, 949–957.

Moench, A. F. (1995). Combining the Neuman and Boulton models for flow to a well in an unconfined aquifer. *Ground Water*, **33**, 378–384.

Moench, A. F. (1996). Flow to a well in a water-table aquifer: an improved Laplace transform solution. *Ground Water*, **34**, 593–604.

Moench, A. F. (2003). Estimation of hectare-scale soil-moisture characteristics from aquifer-test data. *J. Hydrol.*, **281**, 82–95.

Moench, A. F. and Barlow, P. M. (2000). Aquifer response to stream-stage and recharge variations: I. Analytical step-response functions. *J. Hydrol.*, **230**, 192–210.

Molini, A., Lanza, L. G. and La Barbera, P. (2005). Improving the accuracy of tipping-bucket rain

records using disaggregation techniques. *Atmos. Res.*, **77**, 203–217.

Monteith, J. L. (1963). Gas exchange in plant communities. In *Environmental Control of Plant Growth*, ed. L. T. Evans. New York: Academic Press, 95–112.

Moore, G. K. (1992). Hydrograph analysis in a fractured rock terrane. *Ground Water*, **30**, 390–395.

Morel-Seytoux, H. J. (1984). From excess infiltration to aquifer recharge: a derivation based on the theory of flow of water in unsaturated soils (unit hydrograph). *Water Resour. Res.*, **20**, 1230–1240.

Moreo, M. T., Laczniak, R. J. and Stannard, D. I. (2007). Evapotranspiration rate measurements of vegetation typical of ground-water discharge areas in the Basin and Range carbonate-rock aquifer system, White Pine County, Nevada, and adjacent areas in Nevada and Utah, September 2005–August 2006. US Geological Survey Scientific Investigations Report 2007-5078.

Morgan, C. P. and Stolt, M. H. (2004). A comparison of several approaches to monitor water-table fluctuations. *Soil Sci. Soc. Amer. J.*, **68**, 562–566.

Morton, F. I. (1978). Estimating evapotranspiration from potential evaporation: practicality of an iconoclastic approach. *J. Hydrol.*, **38**, 1–32.

Moutsopoulos, K. N., Gemitzi, A. and Tsihrintzis, V. A. (2008). Delineation of groundwater protection zones by the backward particle tracking method: theoretical background and GIS-based stochastic analysis. *Environ. Geol.*, **54**, 1081–1090.

Moysey, S., Davis, S. N., Zreda, M. and Cecil, L. D. (2003). The distribution of meteoric Cl-36/Cl in the United States: a comparison of models. *Hydrogeol. J.*, **11**, 615–627.

Murdoch, L. C. and Kelly, S. E. (2003). Factors affecting the performance of conventional seepage meters. *Water Resour. Res.*, **39**, SWC21-SWC210.

Nace, R. L. (1967). Are we running out of water? US Geological Survey Circular 536.

Nachabe, M. H. (2002). Analytical expressions for transient specific yield and shallow water table drainage. *Water Resour. Res.*, **38**, 1193, doi:10.1029/2001WR001071

Nathan, R. J. and McMahon, T. A. (1990). Evaluation of automated techniques for base flow and recession analyses. *Water Resour. Res.*, **26**, 1465–1473.

Nativ, R., Adar, E., Dahan, O. and Geyh, M. (1995). Water recharge and solute transport through the vadose zone of fractured chalk under desert conditions. *Water Resour. Res.*, **31**, 253–261.

Nativ, R., Günay, G., Hötzl, H. *et al.* (1999). Separation of groundwater-flow components in a karstified aquifer using environmental tracers. *Applied Geochem.*, **14**, 1001–1014.

Natural Resources Conservation Service (2004). Chapter 10: Estimation of direct runoff from storm rainfall, *Part 630, National Engineering Handbook*. Washington, DC: US Department of Agriculture.

Neff, B. P., Day, S. M., Piggott, A. R. and Fuller, L. M. (2005). Base flow in the Great Lakes Basin. US Geological Survey Scientific Investigations Report 2005–5217.

Nelms, D. L., Harlow, G. E. Jr. and Hayes, D. C. (1997). Base-flow characteristics of streams in the Valley and Ridge, the Blue Ridge, and the Piedmont physiographic provinces of Virginia. US Geological Survey Water-Supply Paper 2457.

Neukum, C., Hötzl, H. and Himmelsbach, T. (2008). Validation of vulnerability mapping methods by field investigations and numerical modelling. *Hydrogeol. J.*, **16**, 641–658.

Neuman, S. P. (1972). Theory of flow in unconfined aquifers considering delayed response of the water table. *Water Resour. Res.*, **8**, 1031–1045.

Neuman, S. P. and Witherspoon, P. A. (1972). Field determination of the hydraulic properties of leaky multiple aquifer systems. *Water Resour. Res.*, **8**, 1284–1298.

Neuzil, C. E. (1986). Groundwater flow in low-permeability environments. *Water Resour. Res.*, **22**, 1163–1195.

Nichols, D. S. and Verry, E. S. (2001). Stream flow and ground water recharge from small forested watersheds in north central Minnesota. *J. Hydrol.*, **245**, 89–103.

Nichols, W. D. (2000). Regional ground-water evapotranspiration and ground-water budgets, Great Basin, Nevada. US Geological Survey Professional Paper 1628.

Nielsen, T. H., Well, R. and Myrold, D. D. (1997). Combination probe for nitrogen-15 soil labeling and sampling of soil atmosphere to measure subsurface denitrification activity. *Soil Sci. Soc. Amer. J.*, **61**, 802–811.

Nimmo, J. R. and Perkins, K. S. (2008). Effect of soil disturbance on recharging fluxes: case study on the Snake River Plain, Idaho National Laboratory, USA. *Hydrogeol. J.*, **16**, 829–844.

Nimmo, J. R., Perkins, K. S. and Lewis, A. M. (2002a). Steady-state centrifuge. In *Methods of Soil Analysis. Part 4: Physical Methods*, ed. J. H. Dane and G. C. Topp. Madison, Wisconsin: Soil Science Society of America.

Nimmo, J. R., Perkins, K. S., Rose, P. E. *et al.* (2002b). Kilometer-scale rapid transport of naphthalene sulfonate tracer in the unsaturated zone at the Idaho National Engineering and Environmental Laboratory. *Vadose Zone J.*, **1**, 89–101.

Nimmo, J. R., Stonestrom, D. A. and Akstin, K. C. (1994). The feasibility of recharge rate determinations using the steady-state centrifuge method. *Soil Sci. Soc. Amer. J.*, **58**, 49–56.

Niswonger, R. G. and Prudic, D. E. (2003). Modeling heat as a tracer to estimate streambed seepage and hydraulic conductivity. In *Heat as a Tool for Studying the Movement of Ground Water Near Streams*, ed. D. A. Stonestrom and J. Constantz. US Geological Survey Circular 1260, 81–89.

Niswonger, R. G., Prudic, D. E., Fogg, G. E., Stonestrom, D. A. and Buckland, E. M. (2008). Method for estimating spatially variable seepage loss and hydraulic conductivity in intermittent and ephemeral streams. *Water Resour. Res.*, **44**, W05418, doi:10.1029/2007WR006626.

Niswonger, R. G., Prudic, D. E. and Regan, R. S. (2006). Documentation of the unsaturated-zone flow (UZF1) package for modeling unsaturated flow between the land surface and the water table with MODFLOW-2005. US Geological Survey Techniques and Methods 6-A19.

Nœtinger, B., Artus, V. and Zargar, G. (2005). The future of stochastic and upscaling methods in hydrogeology. *Hydrogeol. J.*, **13**, 184–201.

Noilhan, J. and Planton, S. (1989). A simple parameterization of land surface processes for meteorological models. *Monthly Weather Rev.*, **117**, 536–549.

Nolan, B. T., Healy, R. W., Taber, P. E. *et al.* (2007). Factors influencing ground-water recharge in the eastern United States. *J. Hydrol.*, **332**, 187–205.

Norin, M., Hulten, A. -M. and Svensson, C. (1999). Groundwater studies conducted in Göteborg, Sweden. In *Groundwater in the Urban Environment: Selected City Profiles*, ed. J. Chilton. Rotterdam: A. A. Balkema, 209–216.

Norman, J. M., Kustas, W. P. and Humes, K. S. (1995). Source approach for estimating soil and vegetation energy fluxes in observations of directional radiometric surface temperature. *Agric. Forest Met.*, **77**, 263–293.

Normand, B., Recous, S., Vachaud, G., Kengni, L. and Garino, B. (1997). Nitrogen-15 tracers combined with tensio-neutronic method to estimate the nitrogen balance of irrigated maize. *Soil Sci. Soc. Amer. J.*, **61**, 1508–1518.

Norris, A. E., Wolfsberg, K., Gifford, S. K., Bentley, H. W. and Elmore, D. (1987). Infiltration at Yucca Mountain, Nevada, traced by 36Cl. *Nucl. Inst. Meth. Phys. Res.*, **B29**, 376–379.

Nwankwor, G. I., Cherry, J. A. and Gillham, R. W. (1984). A comparative study of specific yield determinations for a shallow sand aquifer. *Ground Water*, **22**, 764–772.

Oberg, K. A., Morlock, S. E. and Caldwell, W. S. (2005). Quality-assurance plan for discharge measurements using acoustic Doppler current profilers. US Geological Survey Scientific Investigations Report 2005–5183.

Ockerman, D. J. (2002). Simulation of runoff and recharge and estimation of constituent loads in runoff: Edwards aquifer recharge zone (outcrop) and catchment area, Bexar County, Texas, 1997–2000. US Geological Survey Water-Resources Investigations Report 02–4241.

Ockerman, D. J. (2005). Simulation of streamflow and estimation of recharge to the Edwards aquifer in the Hondo Creek, Verde Creek, and San Geronimo Creek watersheds, south-central Texas, 1951–2003. US Geological Survey Scientific Investigations Report 2005–5252.

Ockerman, D. J. (2007). Simulation of streamflow and estimation of ground-water recharge in the upper Cibolo Creek watershed, south-central Texas, 1992–2004. US Geological Survey Scientific Investigations Report 2007–5202.

Olson, S. A. and Norris, J. M. (2007). US Geological Survey Streamgaging from the National Streamflow Information Program, US Geological Survey Fact Sheet 2005–3131.

O'Reilly, A. M. (2004). A method for simulating transient ground-water recharge in deep water-table settings in central Florida by using a simple water-balance/transfer-function model. US Geological Survey Scientific Investigations Report 2004–5195.

Ostendorf, D. W., Rees, P. L. S., Kelley, S. P. and Lutenegger, A. J. (2004). Steady, annual, and monthly recharge implied by deep unconfined aquifer flow. *J. Hydrol.*, **290**, 259–274.

Ostlund, H. G. and Dorsey, H. G. (1977). Rapid electrolytic enrichment and hydrogen gas proportional counting of tritium. In *Proceedings of the International Conference on Low-Radioactivity Measurements and Application*, 6–10 October 1975: The High Tatras, Czechoslovakia, Slovenske Pedagogicke Nakladatelstvo, Bratislava.

Padilla, A., Pulido-Bosch, A. and Mangin, A. (1994). Relative importance of baseflow and quickflow from hydrographs of Karst spring. *Ground Water*, **32**, 267–277.

Parizek, R. R. and Lane, B. E. (1970). Soil-water sampling using pan and deep pressure-vacuum lysimeters. *J. Hydrol.*, **11**, 1–21.

Parsons, M. L. (1970). Groundwater thermal regime in a glacial complex. *Water Resour. Res.*, **6**, 1701–1720.

Payne, D. F., Rumman, M. A. and Clarke, J. S. (2005). Simulation of ground-water flow in coastal Georgia and adjacent parts of South Carolina and Florida predevelopment, 1980, and 2000. US Geological Survey Scientific Investigations Report 2005–5089.

Penman, H. L. (1948). Natural evapotranspiration from open water, bare soil, and grass. *Proc. Roy. Soc. London*, **A193**, 120–145.

Pérez, E. S. (1997). Estimation of basin-wide recharge rates using spring flow, precipitation, and temperature data. *Ground Water*, **35**, 1058–1065.

Pettyjohn, W. A. and Henning, R. (1979). Preliminary estimate of ground-water recharge rates, related streamflow, and water quality in Ohio. Ohio State University Water Resources Center Project Completion Report 552.

Philip, J. R. and de Vries, D. A. (1957). Moisture movement in porous materials under temperature gradients. *Trans. Amer. Geophys. Union*, **38**, 222–232.

Phillips, F. M. (1994). Environmental tracers for water movement in desert soils of the American southwest. *Soil Sci. Soc. Amer. J.*, **58**, 15–24.

Phillips, F. M. (2000). Chlorine-36. In *Environmental Tracers in Subsurface Hydrology*, ed. P. G. Cook and A. L. Herczeg. Boston: Kluwer Academic Publishers.

Phillips, F. M., Mattick, J. L., Duval, T. A., Elmore, D. and Kubik, P. W. (1988). Chlorine 36 and tritium from nuclear weapons fallout as tracers for long-term liquid and vapor movement in desert soils. *Water Resour. Res.*, **24**, 1877–1891.

Piggott, A. R., Moin, S. and Southam, C. (2005). A revised approach to the UKIH method for the calculation of baseflow. *Hydrol. Sci. J.*, **50**, 911–920.

Plummer, L. N. (2005). Dating of young groundwater. In *Isotopes in the Water Cycle: Past, Present, and Future of a Developing Science*, ed. P. K. Aggarwal, J. R. Gat and K. F. O. Froehlich. Dordrecht, The Netherlands: Springer, 193–220.

Plummer, L. N., Bexfield, L. M., Anderholm, S. K., Sanford, W. E. and Busenberg, E. (2004). Hydrochemical tracers in the Middle Rio Grande

Basin, USA: 1. Conceptualization of groundwater flow. *Hydrogeol. J.*, **12**, 359–388.

Plummer, L. N. and Busenberg, E. (2000). Chlorofluorocarbons. In *Environmental Tracers in Subsurface Hydrology*, ed. P. G. Cook and A. L. Herczeg. Boston: Kluwer Academic Publishers, 441–478.

Plummer, L. N., Busenberg, E., Böhlke, J. K. *et al.* (2001). Groundwater residence times in Shenandoah National Park, Blue Ridge Mountains, Virginia, USA: a multi-tracer approach. *Chem. Geol.*, **179**, 93–111.

Plummer, L. N., Busenberg, E., McConnell, J. B. *et al.* (1998). Flow of river water into a karstic limestone aquifer. 1. Tracing the young fraction in groundwater mixtures in the Upper Floridan Aquifer near Valdosta, Georgia. *Applied Geochem.*, **13**, 995–1015.

Plummer, L. N., Prestemon, E. C. and Parkhurst, D. L. (1994). An interactive code (NETPATH) for modeling NET geochemical reactions along a flow PATH, version 2.0. US Geological Survey Water-Resources Investigations Report 94–4169.

Poeter, E. P. and Hill, M. C. (1997). Inverse models: a necessary next step in groundwater modeling. *Ground Water*, **35**, 250–260.

Poeter, E. P. and Hill, M. C. (1998). Documentation of UCODE: a computer code for universal inverse modeling. US Geological Survey Water-Resources Investigations Report 98–4080.

Poeter, E. P., Hill, M. C., Banta, E. R. and Mehl, S. W. (2005). UCODE_2005 and three post-processors: computer codes for universal sensitivity analysis, inverse modeling, and uncertainty evaluation. US Geological Survey Techniques and Methods Report TM 6-A11.

Pollock, D. W. (1994). User's guide for MODPATH/MODPATH-PLOT, Version 3: a particle tracking post-processing package for MODFLOW, the US Geological Survey finite-difference ground-water flow model. US Geological Survey Open-File Report 94–464.

Pool, D. R. (2005). Variations in climate and ephemeral channel recharge in southeastern Arizona, United States. *Water Resour. Res.*, **41**, 1–24.

Pool, D. R. (2008). The utility of gravity and water-level monitoring at alluvial aquifer wells in southern Arizona. *Geophysics*, **73**, WA49–WA59.

Pool, D. R. and Eychaner, J. H. (1995). Measurements of aquifer-storage change and specific yield using gravity surveys. *Ground Water*, **33**, 425–432.

Pool, D. R. and Schmidt, W. (1997). Measurement of ground-water storage change and specific yield using the temporal-gravity method near Rillito Creek, Tucson, Arizona, US Geological Survey Water-Resources Investigations Report 97–4125.

Portniaguine, O. and Solomon, D. K. (1998). Parameter estimation using groundwater age and head data, Cape Cod, Massachusetts. *Water Resour. Res.*, **34**, 637–645.

Price, M., Low, R. G. and McCann, C. (2000). Mechanisms of water storage and flow in the unsaturated zone of the Chalk aquifer. *J. Hydrol.*, **233**, 54–71.

Prickett, T. A. (1965). Type-curve solution to aquifer tests under water-table conditions. *Ground Water*, **3**, 5–14.

Priestley, C. H. B. and Taylor, R. J. (1972). On the assessment of surface heat flux and evaporation using large-scale parameters. *Monthly Weather Rev.*, **100**, 81–92.

Prill, R. C., Johnson, A. I. and Morris, D. A. (1965). Specific yield-laboratory experiments showing the effect of time on column drainage. US Geological Survey Water-Supply Paper 1662-B.

Prince, K. R. (1981). Use of flow-duration curves to evaluate effects of urbanization on streamflow patterns on Long Island, New York. US Geological Survey Water-Resources Investigations Report 80–114.

Prudic, D. E., Niswonger, R., Harrill, J. R. and Wood, J. L. (2007). Streambed infiltration and ground-water flow from the Trout Creek drainage, an intermittent tributary to the Humboldt River, north-central Nevada. In *Ground-water Recharge in the Arid and Semiarid Southwestern United States*, ed. D. A. Stonestrom, J. Constantz, T. P. A. Ferre and S. A. Leake. US Geological Survey Professional Paper 1703, Chapter K, 313–351.

Prudic, D. E., Niswonger, R. G., Wood, J. L. and Henkelman, K. K. (2003). Trout Creek: estimating flow duration and seepage losses along an intermittent stream tributary to the Humboldt River, Lander and Humboldt Counties, Nevada. In *Heat as a Tool for Studying the Movement of Ground Water Near Streams*, ed. D. A. Stonestrom and J. Constantz. US Geological Survey Circular 1260, 58–71.

Pruess, K., Oldenburg, C. and Moridis, G. (1999). TOUGH2 user's guide. Version 2.0. Lawrence Berkeley National Laboratory Report LBNL-43134. Berkeley, CA.

Pruitt, W. O. and Angus, D. E. (1960). Large weighing lysimeter for measuring evapotranspiration. *Trans. Amer. Soc. Agric. Eng.*, **3**, 13–18.

Prych, E. A. (1998). Using chloride and chlorine-36 as soil-water tracers to estimate deep percolation at selected locations on the US Department of Energy Hanford Site, Washington. US Geological Survey Water-Supply Paper 2481.

Puente, C. (1978). Method of estimating natural recharge to the Edwards aquifer in the San Antonio area, Texas. US Geological Survey Water-Resources Investigations Report 78–10.

Qi, J., Chehbouni, A., Huete, A. R., Kerr, Y. H. and Sorooshian, S. (1994). A modified soil adjusted vegetation index. *Remote Sens. Environ.*, **48**, 119–126.

Ragab, R., Finch, J. and Harding, R. (1997). Estimation of groundwater recharge to chalk and sandstone aquifers using simple soil models. *J. Hydrol.*, **190**, 19–41.

Rangarajan, R. and Athavale, R. N. (2000). Annual replenishable ground water potential of India: an estimate based on injected tritium studies. *J. Hydrol.*, **234**, 38–53.

Rantz, S. E. *et al.* (1982). Measurement and computation of streamflow. US Geological Survey Water Supply Paper 2175.

Rasmussen, T. C. and Crawford, L. A. (1997). Identifying and removing barometric pressure effects in confined and unconfined aquifers. *Ground Water*, **35**, 502–511.

Rasmussen, T. C. and Mote, T. L. (2007). Monitoring surface and subsurface water storage using confined aquifer water levels at the Savannah River Site, USA. *Vadose Zone J.*, **6**, 327–335.

Rasmussen, W. C. and Andreasen, G. E. (1959). Hydrologic budget of the Beaverdam Creek Basin, Maryland. US Geological Survey Water-Supply Paper 1472.

Rawls, W. J., Brakensick, D. L. and Saxton, K. E. (1982). Estimation of soil water properties. *Trans. Amer. Soc. Agric. Eng.*, **25**, 1316–1320.

Rehm, B. W., Moran, S. R. and Groenewold, G. H. (1982). Natural groundwater recharge in an upland area of central North Dakota, USA. *J. Hydrol.*, **59**, 293–314.

Reilly, T. E., Plummer, L. N., Phillips, P. J. and Busenberg, E. (1994). The use of simulation and multiple environmental tracers to quantify groundwater flow in a shallow aquifer. *Water Resour. Res.*, **30**, 421–433.

Reiter, M. (2003). Hydrogeothermal studies in the Albuquerque Basin: a geophysical investigation of ground water flow characteristics. Circular 211. New Mexico Bureau of Geology and Mineral Resources.

Remson, I. and Lang, S. M. (1955). A pumping test method for the determination of specific yield. *Trans. Amer. Geophys. Union*, **36**, 321–325.

Reynolds Electrical and Engineering Company (1994). Site characterization and monitoring data from Area 5 pilot wells, Nevada Test Site, Nye County, Nevada. Contract Report DOE/NB/11432-74. US Department of Energy, Las Vegas, Nevada.

Reynolds, R. J. (1982). Base flow of streams on Long Island, New York. US Geological Survey Water-Resources Investigations Report 81–48.

Reynolds, W. D., Elrick, D. E., Youngs, E. G. *et al.* (2002). Saturated and field-saturated water flow parameters. In *Methods of Soil Analysis. Part 4: Physical Methods*, ed. J. H. Dane and G. C. Topp. Madison, Wisconsin: Soil Science Society of America.

Richards, L. A., Gardner, W. R. and Ogata, G. (1956). Physical process determining water loss from soil. *Soil Sci. Soc. Amer. Proc.*, **20**, 310–314.

Rimmer, A. and Salingar, Y. (2006). Modelling precipitation-streamflow processes in karst basin: the case of the Jordan River sources, Israel. *J. Hydrol.*, **331**, 524–542.

Risser, D. W. (2008). Spatial distribution of ground-water recharge estimated with a water-budget method for the Jordan Creek watershed, Lehigh County, Pennsylvania. US Geological Survey Scientific Investigations Reports 2008–5041.

Risser, D. W., Conger, R. W., Ulrich, J. E. and Asmussen, M. P. (2005a). Estimates of ground-water recharge based on streamflow-hydrograph methods: Pennsylvania. US Geological Survey Open-File Report 2005–1333.

Risser, D. W., Gburek, W. J. and Folmar, G. J. (2005b). Comparison of methods for estimating ground-water recharge and base flow at a small watershed underlain by fractured bedrock in the Eastern United States. US Geological Survey Scientific Investigations Report 2005–5038.

Risser, D. W., Gburek, W. J. and Folmar, G. J. (2009). Comparison of recharge estimates at a small watershed in east-central Pennsylvania, USA. *Hydrogeol. J.*, **17**, 287–298.

Roark, D. M. and Healy, D. F. (1998). Quantification of deep percolation from two flood-irrigated alfalfa fields, Roswell Basin, New Mexico. US Geological Survey Water-Resources Investigations Report 98-4096.

Robertson, W. D. and Cherry, J. A. (1989). Tritium as an indicator of recharge and dispersion in a groundwater system in central Ontario. *Water Resour. Res.*, **25**, 1097–1109.

Robinson, D. A., Binley, A., Crook, N. *et al.* (2008a). Advancing process-based watershed hydrological research using near-surface geophysics: a vision for, and review of, electrical and magnetic geophysical methods. *Hydrol. Proc.*, **22**, 3604–3635.

Robinson, D. A., Campbell, C. S., Hopmans, J. W. *et al.* (2008b). Soil moisture measurement for ecological and hydrological watershed-scale observatories: a review. *Vadose Zone J.*, **7**, 358–389.

Robock, A. and Li, H. (2006). Solar dimming and CO_2 effects on soil moisture trends. *Geophys. Res. Lett.*, **33**, L20708, doi:10.1029/2006GL027585.

Robock, A., Vinnikov, K. Y., Srinivasan, G. *et al.* (2000). The global soil moisture data bank. *Bull. Amer. Met. Soc.*, **81**, 1281–1299.

Rodell, M., Chen, J., Kato, H. *et al.* (2007). Estimating groundwater storage changes in the Mississippi River Basin (USA) using GRACE. *Hydrogeol. J.*, **15**, 159–166.

Rodell, M., Famiglietti, J. S., Chen, J. *et al.* (2004). Basin scale estimates of evapotranspiration using GRACE and other observations. *Geophys. Res. Lett.*, **31**, L20504, doi:10.1029/2004GL020873.

Rogowski, A. S. (1996). GIS modeling of recharge on a watershed. *J. Environ. Qual.*, **25**, 463–474.

Rojstaczer, S. (1988). Determination of fluid flow properties from the response of water levels in wells to atmospheric loading. *Water Resour. Res.*, **24**, 1927–1938.

Roman, R., Caballero, R., Bustos, A. *et al.* (1996). Water and solute movement under conventional corn in Central Spain: I. Water balance. *Soil Sci. Soc. Amer. J.*, **60**, 1530–1536.

Romano, N. and Santini, A. (2002). Water retention and storage: field. In *Methods of Soil Analysis. Part 4: Physical Methods*, ed. J. H. Dane and G. C. Topp. Madison, Wisconsin: Soil Science Society of America, 721–738.

Ronan, A. D., Prudic, D. E., Thodal, C. E. and Constantz, J. (1998). Field study and simulation of diurnal temperature effects on infiltration and variably saturated flow beneath an ephemeral stream. *Water Resour. Res.*, **34**, 2137–2153.

Rorabaugh, M. I. (1960). Use of water levels in estimating aquifer constants. *Inter. Assoc. Sci. Hydrol. Pub.*, **52**, 314–323.

Rorabaugh, M. I. (1964). Estimating changes in bank storage and groundwater contribution to streamflow. *Inter. Assoc. Sci. Hydrol. Pub.*, **63**, 432–441.

Rosenberg, N. J., Blad, B. L. and Verma, S. V. (1983). *The Biological Environment.* New York: John Wiley and Sons, Inc.

Rosenberry, D. O. (2008). A seepage meter designed for use in flowing water. *J. Hydrol.*, **359**, 118–130.

Rosenberry, D. O., Labaugh, J. W. and Hunt, R. J. (2008). Use of monitoring wells, portable piezometers, and seepage meters to quantify flow between surface water and ground water. In *Field Techniques for Estimating Water Fluxes Between Surface Water and Ground Water*, ed. D. I. Rosenberry and J. W. Labaugh. US Geological Survey Techniques and Methods 4-D2, 39–70.

Rosenberry, D. O. and Menheer, M. A. (2006). A system for calibrating seepage meters used to measure flow between ground water and surface water. US Geological Survey Scientific Investigations Report 2006-5053.

Rosenberry, D. O. and Morin, R. H. (2004). Use of an electromagnetic seepage meter to investigate temporal variability in lake seepage. *Ground Water*, **42**, 68–77.

Rosqvist, H. and Destouni, G. (2000). Solute transport through preferential pathways in municipal solid waste. *J. Contam. Hydrol.*, **46**, 39–60.

Royer, J. M. and Vachaud, G. (1974). Determination directe de l'evapotranspiration et de l'infiltration par mesures des teneurs en eau et des succions. *Hydrol. Sci. Bull.*, **19**, 319–336.

Rugh, D. F. and Burbey, T. J. (2008). Using saline tracers to evaluate preferential recharge in fractured rocks, Floyd County, Virginia, USA. *Hydrogeol. J.*, **16**, 251–262.

Rutledge, A. T. (1998). Computer programs for describing the recession of ground-water discharge and for estimating mean groundwater recharge and discharge from streamflow records: update. US Geological Survey Water-Resources Investigations Report 98-4148.

Rutledge, A. T. (2000). Considerations for use of the RORA program to estimate ground-water recharge from streamflow records. US Geological Survey Open-File Report 2000-156.

Rutledge, A. T. (2007). Update on the use of the RORA program for recharge estimation. *Ground Water*, **45**, 374–382.

Rutledge, A. T. and Daniel, C. C. (1994). Testing an automated method to estimate ground-water recharge from streamflow records. *Ground Water*, **32**, 180–189.

Rutledge, A. T. and Mesko, T. O. (1996). Estimated hydrologic characteristics of shallow aquifer systems in the Valley and Ridge, the Blue Ridge, and the Piedmont physiographic provinces based on analysis of streamflow recession and base flow. US Geological Survey Professional Paper 1422-B.

Saha, D. and Agrawal, A. K. (2006). Determination of specific yield using a water balance approach: case study of Torla Odha watershed in the Deccan Trap province, Maharastra State, India. *Hydrogeol. J.*, **14**, 625–635.

Salama, R. B., Bartle, G. A. and Farrington, P. (1994a). Water use of plantation *Eucalyptus camaldulensis* estimated by groundwater hydrograph separation techniques and heat pulse method. *J. Hydrol.*, **156**, 163–180.

Salama, R. B., Tapley, I., Ishii, T. and Hawkes, G. (1994b). Identification of areas of recharge and discharge using Landsat-TM satellite imagery and aerial photography mapping techniques. *J. Hydrol.*, **162**, 119–141.

Sami, K. and Hughes, D. A. (1996). A comparison of recharge estimates to a fractured sedimentary aquifer in South Africa from a chloride mass balance and an integrated surface-subsurface model. *J. Hydrol.*, **179**, 111–136.

Sammis, T. W., Evans, D. D. and Warrick, A. W. (1982). Comparison of methods to estimate deep percolation rate. *Water Resour. Bull.*, **18**, 465–470.

Sanford, W. E. (2002). Recharge and groundwater models: an overview. *Hydrogeol. J.*, **10**, 110–120.

Sanford, W. E., Plummer, L. N., McAda, D. P., Bexfield, L. M. and Anderholm, S. K. (2004). Hydrochemical tracers in the Middle Rio Grande Basin, USA: 2. Calibration of a ground-water flow model. *Hydrogeol. J.*, **12**, 389–407.

Santhi, C., Allen, P. M., Muttiah, R. S., Arnold, J. G. and Tuppad, P. (2008). Regional estimation of base flow for the conterminous United States by hydrologic landscape regions. *J. Hydrol.*, **351**, 139–153.

Sauer, T. J. (2002). Heat flux density. In *Methods of Soil Analysis. Part 4: Physical Methods*, ed. J. H. Dane and G. C. Topp. Madison, Wisconsin: Soil Science Society of America, 1233–1252.

Scanlon, B. R. (1991). Evaluation of moisture flux from chloride data in desert soils. *J. Hydrol.*, **128**, 137–156.

Scanlon, B. R. (1992). Evaluation of liquid and vapor water flow in desert soils based on chlorine 36 and tritium tracers and nonisothermal flow simulations. *Water Resour. Res.*, **28**, 285–297.

Scanlon, B. R. (2000). Uncertainties in estimating water fluxes and residence times using environmental tracers in an arid unsaturated zone. *Water Resour. Res.*, **36**, 395–409.

Scanlon, B. R., Andraski, B. J. and Bilskie, J. (2002a). Miscellaneous methods for measuring matric or water potential. In *Methods of Soil Analysis. Part 4: Physical Methods*, ed. J. H. Dane and G. C. Topp. Madison, Wisconsin: Soil Science Society of America.

Scanlon, B. R. and Goldsmith, R. S. (1997). Field study of spatial variability in unsaturated flow beneath and adjacent to playas. *Water Resour. Res.*, **33**, 2239–2252.

Scanlon, B. R., Healy, R. W. and Cook, P. G. (2002b). Choosing appropriate techniques for quantifying groundwater recharge. *Hydrogeol. J.*, **10**, 18–39.

Scanlon, B. R., Keese, K., Reedy, R. C., Simunek, J. and Andraski, B. J. (2003). Variations in flow and transport in thick desert vadose zones in response to paleoclimatic forcing (0–90 kyr): field measurements, modeling, and uncertainties. *Water Resour. Res.*, **39**, Art. No. 1179.

Scanlon, B. R. and Milly, P. C. D. (1994). Water and heat fluxes in desert soils: 2. Numerical simulations. *Water Resour. Res.*, **30**, 709–719.

Scanlon, B. R., Paine, J. G. and Goldsmith, R. S. (1999) Evaluation of electromagnetic induction as a reconnaissance technique to characterize unsaturated flow in an arid setting. *Ground Water*, **37**, 296–304.

Scanlon, B. R., Reedy, R. C., Stonestrom, D. A., Prudic, D. E. and Dennehy, K. F. (2005). Impact of land use and land cover change on groundwater recharge and quality in the southwestern US. *Global Change Biology*, **11**, 1577–1593.

Scanlon, B. R., Reedy, R. C. and Tachovsky, J. A. (2007). Semiarid unsaturated zone chloride profiles: archives of past land use change impacts on water resources in the southern High Plains, United States. *Water Resour. Res.*, **43**, W06423, doi:10.1029/2006WR005769.

Schaap, M. G., Leij, F. J. and van Genuchten, M. T. (2001). Rosetta: a computer program for estimating soil hydraulic parameters with hierarchical pedotransfer functions. *J. Hydrol.*, **251**, 163–176.

Schemenauer, R. S. and Cereceda, P. (1994). A proposed standard fog collector for use in high-elevation regions. *J. Appl. Met.*, **33**, 1313–1322.

Schicht, R. J. and Walton, W. C. (1961). Hydrologic budgets for three small watersheds in Illinois. Illinois State Water Survey Report of Investigation 40.

Schilling, K. E. and Kiniry, J. R. (2007). Estimation of evapotranspiration by reed canarygrass using field observations and model simulations. *J. Hydrol.*, **337**, 356–363.

Schilling, K. E. and Wolter, C. F. (2005). Estimation of streamflow, base flow, and nitrate-nitrogen loads in Iowa using multiple linear regression models. *J. Amer. Water Resour. Assoc.*, **41**, 1333–1346.

Schlosser, P., Stute, M., Dörr, H., Sonntag, C. and Münich, K. O. (1988). Tritium/3He dating of shallow groundwater. *Earth Planet. Sci. Lett.*, **89**, 353–362.

Schlosser, P., Stute, M., Sonntag, C. and Münich, K. O. (1989). Tritiogenic 3He in shallow groundwater. *Earth Planet. Sci. Lett.*, **94**, 245–256.

Schmidt, C., Conant, B. Jr., Bayer-Raich, M. and Schirmer, M. (2007). Evaluation and field-scale application of an analytical method to quantify groundwater discharge using mapped streambed temperatures. *J. Hydrol.*, **347**, 292–307.

Schmugge, T. J., Kustas, W. P., Ritchie, J. C., Jackson, T. J. and Rango, A. (2002). Remote sensing in hydrology. *Adv. Water Resour.*, **25**, 1367–1385.

Scholl, M., Christenson, S., Cozzarelli, I., Ferree, D. and Jaeshke, J. (2005). Recharge processes in an alluvial aquifer riparian zone, Norman Landfill, Norman, Oklahoma, 1998–2000. US Geological Survey Scientific Investigations Report 2004–5238.

Schroeder, P. R., Dozier, P. R., Zappi, P. A. *et al.* (1994). The Hydrologic Evaluation of Landfill Performance (HELP) model, engineering documentation for Version 3. EPA/600/9-94/1686. US Environmental Protection Agency Risk Reduction Engineering Laboratory. Washington, DC.

Schultz, T. R., Randall, J. H., Wilson, L. G. and Davis, S. V. (1976). Tracing sewage effluent recharge, Tucson, Arizona. *Ground Water*, **14**, 463–470.

Schwartz, M. O. (2006). Numerical modelling of groundwater vulnerability: the example Namibia. *Environ. Geol.*, **50**, 237–249.

Schwartz, R. C., Baumhardt, R. L. and Howell, T. A. (2008). Estimation of soil water balance components using an iterative procedure. *Vadose Zone J.*, **7**, 115–123.

Seiler, K. P. and Alvarado Rivas, J. (1999). Recharge and discharge of the Caracas aquifer, Venezuela. In *Groundwater in the Urban Environment, Selected City Profiles*, ed. J. Chilton. Rotterdam: A. A. Balkema, 233–238.

Selker, J., van de Giesen, N. C., Westhoff, M., Luxemburg, W. and Parlange, M. B. (2006a). Fiber optics opens window on stream dynamics. *Geophys. Res. Lett.*, **33**.

Selker, J. S., Thévenaz, L., Huwald, H. *et al.* (2006b). Distributed fiber-optic temperature sensing for hydrologic systems. *Water Resour. Res.*, **42**, W12202, doi: 10.1029/2006WR005326.

Seo, D. J. (1998). Real-time estimation of rainfall fields using radar rainfall and rain gage data. *J. Hydrol.*, **208**, 37–52.

Seo, D. J., Breidenbach, J., Fulton, R., Miller, D. and O'Bannon, T. (2000). Real-time adjustment of range-dependent biases in WSR-88D rainfall estimates due to nonuniform vertical profile of reflectivity. *J. Hydromet.*, **1**, 222–240.

Seo, D. J., Breidenbach, J. P. and Johnson, E. R. (1999). Real-time estimation of mean field bias in radar rainfall data. *J. Hydrol.*, **223**, 131–147.

Shan, C. and Bodvarsson, G. (2004). An analytical solution for estimating percolation rate by fitting temperature profiles in the vadose zone. *J. Contam. Hydrol.*, **68**, 83–95.

Sharma, M. L., Bari, M. and Byrne, J. (1991). Dynamics of seasonal recharge beneath a semi-arid vegetation on the Gnangara mound, western Australia. *Hydrol. Proc.*, **5**, 383–398.

Sharma, M. L. and Hughes, M. W. (1985). Groundwater recharge estimation using chloride, deuterium and oxygen-18 profiles in the deep coastal sands of western Australia. *J. Hydrol.*, **81**, 93–109.

Sheffield, J., Ferguson, C. R., Troy, T. J., Wood, E. F. and McCabe, M. F. (2009). Closing the terrestrial water budget from satellite remote sensing. *Geophys. Res. Lett.*, **36**, L07403, doi:10.1029/2009GL037338.

Shevenell, L. (1996). Analysis of well hydrographs in a karst aquifer: estimates of specific yields and continuum transmissivities. *J. Hydrol.*, **174**, 331–355.

Shi, J. and Dozier, J. (1997). Mapping seasonal snow with SIR-C/X-SAR in mountainous areas. *Remote Sens. Environ.*, **59**, 294–307.

Shiklomanov, I. A. and Rodda, J. C. (2003). *World Water Resources at the Beginning of the Twenty-first Century*. Cambridge: Cambridge University Press.

Silliman, S. E. and Booth, D. F. (1993). Analysis of time-series measurements of sediment temperature for identification of gaining vs. losing portions of Juday Creek, Indiana. *J. Hydrol.*, **146**, 131–148.

Silliman, S. E., Ramirez, J. and McCabe, R. L. (1995). Quantifying downflow through creek sediments using temperature time series: one-dimensional solution incorporating measured surface temperature. *J. Hydrol.*, **167**, 99–119.

Simmers, I. (ed.) (1988). *Estimation of Natural Groundwater Recharge*. Dordrecht, Holland: D. Reidel.

Simmers, I. (1990). Aridity, groundwater recharge and water resources management. In *Groundwater Recharge, A Guide to Understanding and Estimating Natural Recharge. International Contributions to Hydrogeology Vol. 8*, ed. D. N. Lerner, A. S. Isaar and I. Simmers. Hanover: Verlag Heinz Heise, 3–22.

Simmers, I. (ed.) (1997). *Recharge of Phreatic Aquifers in (Semi-) Arid Areas*. Rotterdam: A. A. Balkema.

Simmons, C. T., Hong, H., Wye, D., Cook, P. G. and Love, A. J. (1999). Signal propagation and periodic response in aquifers: the effect of fractures and signal measurement methods. *Water 99 Joint Congress*, 727–732.

Simonds, F. W., Longpre, C. I. and Justin, G. B. (2004). Ground-water system in the Chimacum Creek Basin and surface water/ground water interaction in Chimacum and Tarboo Creeks and the Big and Little Quilcene Rivers, Eastern Jefferson County, Washington. US Geological Survey Scientific Investigations Report 2004–5058.

Simunek, J., Sejna, M. and van Genuchten, M. T. (1999). The HYDRUS-2D software package for simulating the two-dimensional movement of water, heat, and multiple solutes in variably saturated media. Version 2.0. IGWMC-TPS 53. International Ground Water Modeling Center, Colorado School of Mines.

Singh, K. P. (1971). Model flow duration and stream-flow variability. *Water Resour. Res.*, **7**, 1031–1036.

Singh, V. P. (ed.) (1995). *Computer Models of Watershed Hydrology*. Highlands Ranch, Colorado: Water Resources Publications.

Singh, V. P. and Frevert, D. K. (eds.) (2006). *Watershed Models*. Boca Raton, Florida: CRC Press.

Sisson, J. B. (1987). Drainage from layered field soils: fixed gradient models. *Water Resour. Res.*, **23**, 2071–2075.

Sloto, R. A. and Crouse, M. Y. (1996). HYSEP: A computer program for streamflow hydrograph separation and analysis. US Geological Survey Water-Resources Investigations Report 96–4040.

Smakhtin, V. U. (2001). Low flow hydrology: a review. *J. Hydrol.*, **240**, 147–186.

Smerdon, B. D., Mendoza, C. A. and Devito, K. J. (2008). Influence of subhumid climate and water table depth on groundwater recharge in shallow outwash aquifers. *Water Resour. Res.*, **44**, W08427, doi:10.1029/2007WR0059550.

Smith, J. L., Laczniak, R. J., Moreo, M. T. and Welborn, T. L. (2007). Mapping evapotranspiration units in the Basin and Range carbonate-rock aquifer system, White Pine County, Nevada, and adjacent parts of Nevada and Utah, US Geological Survey Scientific Investigations Report 2007–5087.

Smith, R. E. (1983). Approximate soil water movement by kinematic characteristics. *Soil Sci. Soc. Amer. J.*, **47**, 3–8.

Solomon, D. K. and Cook, P. G. (2000). ^{3}H and ^{3}He. In *Environmental Tracers in Subsurface Hydrology*, ed. P. G. Cook and A. L. Herczeg. Boston: Kluwer Academic Publishers, 397–424.

Solomon, D. K., Poreda, R. J., Cook, P. G. and Hunt, A. (1995). Site characterization using 3H/3He ground-water ages, Cape Cod, MA. *Ground Water*, **33**, 988–996.

Solomon, D. K., Poreda, R. J., Schiff, S. L. and Cherry, J. A. (1992). Tritium and helium 3 as groundwater age tracers in the Borden aquifer. *Water Resour. Res.*, **28**, 741–755.

Solomon, D. K., Schiff, S. L., Poreda, R. J. and Clark, W. B. (1993). A validation of the 3H/3He method for determining groundwater recharge. *Water Resour. Res.*, **29**, 2951–2962.

Sophocleous, M. (1992). Groundwater recharge estimation and regionalization: the Great Bend Prairie of central Kansas and its recharge statistics. *J. Hydrol.*, **137**, 113–140.

Sophocleous, M. (2000). From safe yield to sustainable development of water resources: the Kansas experience. *J. Hydrol.*, **235**, 27–43.

Sophocleous, M., Bardsley, E. and Healey, J. (2006). A rainfall loading response recorded at 300 meters depth: implications for geological weighing lysimeters. *J. Hydrol.*, **319**, 237–244.

Sophocleous, M., Devlin, J. F. and Bredehoeft, J. D. (2004). Discussion of "the water budget myth revisited: why hydrogeologists model," by John D. Bredehoeft. July–August 2002 issue, v. 40, no. 4: 340–345. *Ground Water*, **42**, 618–619.

Sophocleous, M. and Perkins, S. P. (2000). Methodology and application of combined watershed and ground-water models in Kansas. *J. Hydrol.*, **236**, 185–201.

Sophocleous, M. A. (1991). Combining the soilwater balance and water-level fluctuation methods to estimate natural groundwater recharge: practical aspects. *J. Hydrol.*, **124**, 229–241.

Sophocleous, M. A., Kluitenberg, G. and Healey, J. (2002). Southwestern Kansas High Plains unsaturated zone pilot study to estimate Darcian-based groundwater recharge at three instrumented sites. Kansas Geological Survey Open-File Report 2001-11.

Sorey, M. L. (1971). Measurement of vertical groundwater velocity from temperature profiles in wells. *Water Resour. Res.*, **7**, 963–970.

Soulsby, C., Rodgers, P. J., Petry, J., Hannah, D. M., Malcolm, I. A. and Dunn, S. M. (2004). Using tracers to upscale flow path understanding in mesoscale mountainous catchments: two examples from Scotland. *J. Hydrol.*, **291**, 174–196.

Spinello, A. G. and Simmons, D. L. (1992). Base flow of 10 south-shore streams, Long Island, New York, 1976–85, and the effects of urbanization on base flow and flow duration. US Geological Survey Water-Resources Investigations Report 90–4205.

Stallman, R. W. (1963). Computation of ground-water velocity from temperature data. In Methods of Collecting and Interpreting Ground-Water Data, US Geological Survey Water-Supply Paper 1544-H, 36–47.

Stallman, R. W. (1965). Steady one-dimensional fluid flow in a semi-infinite porous medium with sinusoidal surface temperature. *J. Geophys. Res.*, **70**, 2821–2827.

Stallman, R. W. (1967). Flow in the zone of aeration. In *Advances in Hydrosciences, Volume 4*, ed. V. T. Chow. New York: Academic Press, 151–197.

Stallman, R. W. (1971). Aquifer-test design, observation and data analysis. Techniques of Water Resource Investigations of the US Geological Survey, Chapter B1.

Stannard, D. I. and Weltz, M. A. (2006). Partitioning evapotranspiration in sparsely vegetated rangeland using a portable chamber. *Water Resour. Res.*, **42**, W02413, doi:10.1029/2005WR004251.

Steenhuis, T. S., Jackson, C. D., Kung, S. K. J. and Brutsaert, W. (1985). Measurement of groundwater recharge on eastern Long Island, New York, USA. *J. Hydrol.*, **79**, 145–169.

Steenhuis, T. S. and van der Molen, W. H. (1986). The Thornthwaite-Mather procedure as a simple engineering method to predict recharge. *J. Hydrol.*, **84**, 221–229.

Stephens, D. B. and Knowlton, R. Jr. (1986). Soil water movement and recharge through sand at a semiarid site in New Mexico. *Water Resour. Res.*, **22**, 881–889.

Steuer, J. J. and Hunt, R. J. (2001). Use of a watershed-modeling approach to assess hydrologic effects of urbanization, North Fork Pheasant Branch Basin near Middleton, Wisconsin. US Geological Survey Water-Resources Investigations Report 2001-4113.

Stewart, M., Cimino, J. and Ross, M. (2007). Calibration of base flow separation methods with streamflow conductivity. *Ground Water*, **45**, 17–27.

Stoertz, M. W. and Bradbury, K. R. (1989). Mapping recharge areas using a ground-water flow model: a case study. *Ground Water*, **27**, 220–228.

Stonestrom, D. and Blasch, K. W. (2003). Determining temperature and thermal properties for heat-based studies of surface-water groundwater interactions. In *Heat as a Tool for Studying the Movement of Ground Water Near Streams*, ed. D. A. Stonestrom and J. Constantz. US Geological Survey Circular 1260, 73–80.

Stonestrom, D. A. and Constantz, J. (eds.) (2003). *Heat as a Tool for Studying the Movement of Ground Water near Streams*. US Geological Survey Circular 1260.

Stonestrom, D. A., Constantz, J., Ferré, T. P. A. and Leake, S. A. (eds.) (2007). Ground-water recharge in the arid and semiarid southwestern United States. US Geological Survey Professional Paper 1703.

Stonestrom, D. A. and Harrill, J. R. (2007). Ground-water recharge in the arid and semiarid southwestern United States: climatic and geologic framework. In *Ground-water Recharge in the Arid and Semiarid Southwestern United States*, ed. D. A. Stonestrom, J. Constantz, T. P. A. Ferre and S. A. Leake. US Geological Survey Professional Paper 1703, Chapter A, 1–28.

Stonestrom, D. A., Prudic, D. E., Laczniak, R. J. and Akstin, K. C. (2004). Tectonic, climatic, and land-use controls of groundwater recharge in an arid alluvial basin: Amargosa Desert, USA. In *Groundwater Recharge in a Desert Environment. The southwestern United States*, ed. J. F. Hogan, F. M. Phillips and B. R. Scanlon. Washington, DC: American Geophysical Union, 29–48.

Strangeways, I. (2004). Improving precipitation measurement. *Inter. J. Climatology*, **24**, 1443–1460.

Strassberg, G., Scanlon, B. R. and Rodell, M. (2007). Comparison of seasonal terrestrial water storage variations from GRACE with groundwater-level measurements from the High Plains Aquifer (USA). *Geophys. Res. Lett.*, **34**, L14402, doi:10.1029/2007GL030139.

Stricker, V. (1983). Base flow of streams in the outcrop area of southeastern sand aquifer, South Carolina, Georgia, Alabama, and Mississippi. US Geological Survey Water-Resources Investigations Report 83-4106.

Su, N. (1994). A formula for computation of time-varying recharge of groundwater. *J. Hydrol.*, **160**, 123-135.

Sumner, D. M. (1996). Evapotranspiration from successional vegetation in a deforested area of the Lake Wales Ridge, Florida. US Geological Survey Water-Resources Investigations Report 96-4244.

Suzuki, S. (1960). Percolation measurements based on heat flow through soil with special reference to paddy fields. *J. Geophys. Res.*, **65**, 2883-2885.

Swenson, F. A. (1968). New theory of recharge in the artesian basin of the Dakotas. *Geol. Soc. Amer. Bull.*, **79**, 163-182.

Szabo, Z., Rice, D. E., Plummer, L. N., Busenberg, E., Drenkard, S. and Schlosser, P. (1996). Age dating of shallow groundwater with chlorofluorocarbons, tritium/helium 3, and flow path analysis, southern New Jersey coastal plain. *Water Resour. Res.*, **32**, 1023-1038.

Szilagyi, J., Harvey, F. E. and Ayers, J. F. (2003). Regional estimation of base recharge to ground water using water balance and a base-flow index. *Ground Water*, **41**, 504-513.

Szilagyi, J., Harvey, F. E. and Ayers, J. F. (2005). Regional estimation of total recharge to ground water in Nebraska. *Ground Water*, **43**, 63-69.

Tabbagh, A., Bendjoudi, H. and Benderitter, Y. (1999). Determination of recharge in unsaturated soils using temperature monitoring. *Water Resour. Res.*, **35**, 2439-2446.

Tamura, Y., Sato, T., Ooe, M. and Ishiguro, M. (1991). A procedure for tidal analysis with a Bayesian information criterion. *Geophys. J. Inter.*, **104**, 507-516.

Tan, S. B. K., Shuy, E. B. and Chua, L. H. C. (2007). Regression method for estimating rainfall recharge at unconfined sandy aquifers with an equatorial climate. *Hydrol. Proc.*, **21**, 3514-3526.

Taniguchi, M. and Fukuo, Y. (1993). Continuous measurements of ground-water seepage using an automatic seepage meter. *Ground Water*, **31**, 675-679.

Tankersley, C. D., Graham, W. D. and Hatfield, K. (1993). Comparison of univariate and transfer function models of groundwater fluctuations. *Water Resour. Res.*, **29**, 3517-3533.

Tarantola, A. (2005). *Inverse Problem Theory*. Philadelphia: Society for Industrial and Applied Mathematics.

Taylor, C. J. and Alley, W. M. (2001). Ground-water-level monitoring and the importance of long-term water-level data. US Geological Survey Circular 1217.

Theis, C. V. (1937). Amount of ground-water recharge in the Southern High Plains. *Trans. Amer. Geophys. Union*, **18**, 564-568.

Thom, A. S. and Oliver, H. R. (1977). On Penman's equation for estimating regional evaporation. *Quart. J. Roy. Met. Soc.*, **103**, 345-357.

Thomas, H. E. (1952). Ground-water regions of the United States: their storage facilities, v. 3 of The physical and economic foundation of natural resources: US 83rd Congress, House Committee on Interior and Insular Affairs, 3-78.

Thompson, G. M. and Hayes, J. M. (1979). Trichlorofluoromethane in groundwater: a possible tracer and indicator of groundwater age. *Water Resour. Res.*, **15**, 546-554.

Thompson, G. M., Hayes, J. M. and Davis, S. V. (1974). Fluorocarbon tracers in hydrology. *Geophys. Res. Lett.*, **1**, 177-180.

Thornthwaite, C. W. (1948). An approach toward a rational classification of climate. *Geograph. Rev.*, **38**, 55-94.

Thornthwaite, C. W. and Mather, J. R. (1955). The water balance. *Publications in Climatology*, **8**, 1-104.

Thornthwaite, C. W. and Mather, J. R. (1957). Instructions and tables for computing potential evapotranspiration and the water balance. *Publications in Climatology*, **10**(3), 185-311.

Thornton, P. E., Running, S. W. and White, M. A. (1997). Generating surfaces of daily meteorology variables over large regions of complex terrain. *J. Hydrol.*, **190**, 214-251.

Thorstenson, D. C., Weeks, E. P., Haas, H. *et al.* (1998). Chemistry of unsaturated zone gases sampled in open boreholes at the crest of Yucca Mountain, Nevada: data and basic concepts of chemical and physical processes in the mountain. *Water Resour. Res.*, **34**, 1507-1529.

Tiedeman, C. R., Goode, D. J. and Hsieh, P. A. (1997). Numerical simulation of ground-water flow through glacial deposits and crystalline bedrock in the Mirror Lake area, Grafton County, New Hampshire. US Geological Survey Professional Paper 1572.

Tiedeman, C. R., Kernodle, J. M. and McAda, D. P. (1998). Application of nonlinear-regression methods to a ground-water flow model of the Albuquerque Basin, New Mexico. US Geological Survey Water-Resources Investigations Report 98-4172.

Timmermans, W. J., Kustas, W. P., Anderson, M. C. and French, A. N. (2007). An intercomparison of the Surface Energy Balance Algorithm for Land (SEBAL) and the Two-Source Energy Balance (TSEB) modeling schemes. *Remote Sens. Environ.*, **108**, 369–384.

Todd, D. K. (1980). *Groundwater Hydrology*, 2nd edn. New York: John Wiley and Sons.

Todd, D. K. and Mays, L. W. (2005). *Groundwater Hydrology*, 3rd edn. New York: John Wiley & Sons, Inc.

Toll, N. J. and Rasmussen, T. C. (2007). Removal of barometric pressure effects and earth tides from observed water levels. *Ground Water*, **45**, 101–105.

Topp, G. C., Davis, J. L. and Annan, A. P. (1980). Electromagnetic determination of soil water content: measurements in coaxial transmission lines. *Water Resour. Res.*, **16**, 574–582.

Topp, G. C. and Ferre, T. P. A. (2002). Thermogravimetric method using convective oven-drying. In *Methods of Soil Analysis. Part 4: Physical Methods*, ed. J. H. Dane and G. C. Topp. Madison, Wisconsin: Soil Science Society of America.

Trommer, J. T., Sacks, L. A. and Kuniansky, E. L. (2007). Hydrology, water quality, and surface- and ground-water interactions in the upper Hillsborough River watershed, west-central Florida. US Geological Survey Scientific Investigations Report 2007-5121.

Troxell, H. C. (1936). The diurnal fluctuation in the groundwater and flow of the Santa Ana River and its meaning. *Trans. Amer. Geophys. Union*, **17**, 496–504.

Turco, M. J., East, J. W. and Milburn, M. S. (2007). Base flow (1966–2005) and streamflow gain and loss (2006) of the Brazos River, McLennan County to Fort Bend County, Texas. US Geological Survey Scientific Investigations Report 2007-5286.

Tuteja, N. K., Vaze, J., Teng, J. and Mutendeudzi, M. (2007). Partitioning the effects of pine plantations and climate variability on runoff from a large catchment in southeastern Australia. *Water Resour. Res.*, **43**, W08415, doi:10.1029/2006/WR005016.

Tweed, S. O., Leblanc, M., Webb, J. A. and Lubczynski, M. W. (2007). Remote sensing and GIS for mapping groundwater recharge and discharge areas in salinity prone catchments, southeastern Australia. *Hydrogeol. J.*, **15**, 75–96.

Twine, T. E., Kustas, W. P., Norman, J. M. *et al.* (2000). Correcting eddy-covariance flux underestimates over a grassland. *Agric. Forest Met.*, **103**, 279–300.

Tyler, S. W., Chapman, J. B., Conrad, S. H. *et al.* (1996). Soil-water flux in the southern Great Basin, United States: temporal and spatial variations over the last 120 000 years. *Water Resour. Res.*, **32**, 1481–1499.

Tyler, S. W., McKay, W. A. and Mihevc, T. M. (1992). Assessment of soil moisture movement in nuclear subsidence craters. *J. Hydrol.*, **139**, 159–181.

Tyler, S. W. and Walker, G. R. (1994). Root zone effects on tracer migration in arid zones. *Soil Sci. Soc. Am. J.*, **58**, 26–31.

Uhlenbrook, S., Frey, M., Leibundgut, C. and Maloszewski, P. (2002). Hydrograph separations in a mesoscale mountainous basin at event and seasonal timescales. *Water Resour. Res.*, **38** (6), 1096, doi:10.1029/2001WR000938.

University of Idaho (2003). Ref-ET: Reference evapotranspiration calculation software for FAO and ASCE standardized equations.

US National Research Council (1993). *Ground Water Vulnerability Assessment. Predicting Relative Contamination Potential Under Conditions of Uncertainty*. Washington, DC: National Academy Press.

US Nuclear Regulatory Commission (1993). 10 CFR Part 61 Licensing requirements for land disposal of radioactive waste.

Vaccaro, J. J. (1992). Sensitivity of groundwater recharge estimates to climate variability and change, Columbia Plateau, Washington. *J. Geophys. Res.*, **97**, 2821–2833.

Vaccaro, J. J. (2007). A deep percolation model for estimating ground-water recharge: documentation of modules for the Modular Modeling System of the US Geological Survey. US Geological Survey Scientific Investigations Report 2006-5318.

Vaccaro, J. J. and Olsen, T. D. (2007). Estimates of ground-water recharge to the Yakima River Basin aquifer system, Washington, for predevelopment and current land-use and land-cover conditions. US Geological Survey Scientific Investigations Report 2007-5007.

Vachaud, G. and Dane, J. H. (2002). Instantaneous profile. In *Methods of Soil Analysis. Part 4: Physical Methods*, ed. J. H. Dane and G. C. Topp. Madison, Wisconsin: Soil Science Society of America.

van Bavel, C. H. M. (1961). Lysimetric measurements of evapotranspiration rates in the eastern United States. *Soil Sci. Soc. Amer. Proc.*, **25**, 138–141.

van der Kamp, G. and Schmidt, R. (1997). Monitoring the total soil moisture on a scale of hectares using groundwater piezometers. *Geophys. Res. Lett.*, **24**, 719–722.

van Genuchten, M. T. (1980). A closed-form equation for predicting the hydraulic conductivity of unsaturated soils. *Soil Sci. Soc. Amer. J.*, **44**, 892–898.

van Stempvoort, D., Evert, L. and Wassenaar, L. (1993). Aquifer vulnerability index: a GIS-compatible method for groundwater vulnerability mapping. *Can. Water Resour. J.*, **18**, 25–37.

VanderKwaak, J. E. and Loague, K. (2001). Hydrologic-response simulations for the R-5 catchment with a comprehensive physics-based model. *Water Resour. Res.*, **37**, 999–1013.

Vecchia, A. V. and Cooley, R. L. (1987). Simultaneous confidence and prediction intervals for nonlinear regression models with application to a groundwater flow model. *Water Resour. Res.*, **22**, 95–108.

Viger, R. J. and Leavesley, G. H. (2007). The GIS Weasel user's manual. US Geological Survey Techniques and Methods Report 6-B4 Section 4.

Viswanathan, M. N. (1984). Recharge characteristics of an unconfined aquifer from the rainfall-water table relationship. *J. Hydrol.*, **70**, 233–250.

Vogel, J. C. (1967). Investigation of groundwater flow with radiocarbon. In *Isotopes in Hydrology*. Vienna: International Atomic Energy Agency, **SM-83/24**, 355–369.

Vogel, R. M. and Fennessey, N. M. (1995). Flow duration curves II: a review of applications in water resources planning. *Water Resour. Bull.*, **31**, 1029–1039.

von Asmuth, J. R., Mass, K., Bakker, M. and Petersen, J. (2008). Modeling time series of ground water head fluctuations subjected to multiple stresses. *Ground Water*, **46**, 30–40.

Voronin, L. M. (2004). Documentation of revisions to the regional aquifer system analysis model of the New Jersey coastal plain. US Geological Survey Water-Resources Investigations Report 2003-4268.

Voss, C. I. and Provost, A. M. (2002). SUTRA, a model for saturated-unsaturated variable-density ground-water flow with solute or energy transport. US Geological Survey Water-Resources Investigations Report 02-4231.

Wagner, B. J. (1995). Sampling design methods for groundwater modeling under uncertainty. *Water Resour. Res.*, **31**, 2581–2591.

Wahl, K. L. and Wahl, T. L. (1988). Effects of regional ground water level declines on streamflow in the Oklahoma panhandle. In *Proceedings of the Symposium on Water-Use Data for Water Resources Management*, American Water Resources Association.

Walker, G. R. (1998). Using soil water tracers to estimate recharge. In *The Basics of Recharge and Discharge*, ed. L. Zhang and G. R. Walker: CSIRO Publication, **7**.

Walker, G. R., Jolly, I. D. and Cook, P. G. (1991). A new chloride leaching approach to the estimation of diffuse recharge following a change in land use. *J. Hydrol.*, **128**, 49–67.

Walton, W. C. (1970). *Groundwater Resource Evaluation*. New York: McGraw-Hill.

Walton, W. C. (2007). *Aquifer Test Modeling*. Boca Raton, Florida: CRC Press.

Walton-Day, K., Flynn, J. L., Kimball, B. A. and Runkel, R. L. (2005). Mass loading of selected major and trace elements in Lake Fork Creek near Leadville, Colorado, September–October 2001. US Geological Survey Scientific Investigations Report 2005-5151.

Walvoord, M. A. and Phillips, F. M. (2004). Identifying areas of basin-floor recharge in the Trans-Pecos region and the link to vegetation. *J. Hydrol.*, **292**, 59–74.

Walvoord, M. A., Phillips, F. M., Tyler, S. W. and Hartsough, P. C. (2002a). Deep arid system hydrodynamics. 2: Application to paleohydrologic reconstruction using vadose zone profiles from the northern Mojave Desert. *Water Resour. Res.*, **38**, 1291, doi: 10.1029/ 2001WR000825.

Walvoord, M. A., Plummer, M. A., Phillips, F. M. and Wolfsberg, A. V. (2002b). Deep arid system hydrodynamics: 1, Equilibrium states and response times in thick desert vadose zones. *Water Resour. Res.*, **38**, 1308, doi: 10.1029/ 2001WR000824.

Wang, B., Jin, M., Nimmo, J. R., Yang, L. and Wang, W. (2008). Estimating groundwater recharge in Hebei Plain, China, under varying land use practices using tritium and bromide tracers. *J. Hydrol.*, **356**, 209–222.

Warner, M. J. and Weiss, R. F. (1985). Solubilities of chlorofluorocarbons 11 and 12 in water and seawater. *Deep-Sea Res.*, **32**, 1485–1497.

Watson, P., Sinclair, P. and Waggoner, R. (1976). Quantitative evaluation of a method for estimating recharge to the desert basins of Nevada. *J. Hydrol.*, **31**, 335–357.

Webb, R. H. and Leake, S. A. (2006). Ground-water surface-water interactions and long-term change in riverine riparian vegetation in the southwestern United States. *J. Hydrol.*, **320**, 302–323.

Webb, R. M. T., Wieczorek, M. E., Nolan, B. T. *et al.* (2008). Variations in pesticide leaching related to land use, pesticide properties, and unsaturated zone thickness. *J. Environ. Qual.*, **37**, 1145–1157.

Weeks, E. P. (1979). Barometric fluctuations in wells tapping deep unconfined aquifers. *Water Resour. Res.*, **15**, 1167–1176.

Weeks, E. P. (2002). The Lisse effect revisited. *Ground Water*, **40**, 652–656.

Weeks, E. P., Earp, D. E. and Thompson, G. M. (1982). Use of atmospheric fluorocarbons F-11 and F-12 to determine the diffusion parameters of the unsaturated zone in the southern High Plains of Texas. *Water Resour. Res.*, **18**, 1365–1378.

Weeks, E. P. and Sorey, M. L. (1973). Use of finite-difference arrays of observation wells to estimate evapotranspiration from groundwater in the Arkansas River Valley, Colorado. US Geological Survey Water-Supply Paper 2029-C.

Weeks, J. B., Gutentag, E. D., Heimes, F. J. and Luckey, R. R. (1988). Summary of the High Plains regional aquifer-system analysis in parts of Colorado, Kansas, Nebraska, New Mexico, Oklahoma, South Dakota, Texas, and Wyoming. US Geological Survey Professional Paper 1400-A.

Weeks, J. B., Leavesley, G. H., Welder, F. A. and Saulnier, G. J. Jr. (1974). Simulated effects of oil-shale development on the hydrology of Piceance Basin, Colorado. US Geological Survey Professional Paper 908.

Weiss, M. and Gvirtzman, H. (2007). Estimating ground water recharge using flow models of perched karstic aquifers. *Ground Water*, **45**, 761–773.

Weissmann, G. S., Zhang, Y., LaBolle, E. M. and Fogg, G. E. (2002). Dispersion of groundwater age in an alluvial aquifer system. *Water Resour. Res.*, **38**(10), 1198–1211.

Wellings, S. R. (1984). Recharge of the Upper Chalk aquifer at a site in Hampshire, England. 1: Water balance and unsaturated flow. *J. Hydrol.*, **69**, 259–273.

Wentz, D. A., Rose, W. J. and Webster, K. E. (1995). Long-term hydrologic and biogeochemical responses of a soft water seepage lake in north central Wisconsin. *Water Resour. Res.*, **31**, 199–212.

Wenzel, L. K. (1942). Methods of determining permeability of water-bearing materials with special reference to discharging well methods. US Geological Survey Water-Supply Paper 887.

Westhoff, M. C., Savenjie, H. H. G., Luxemburg, W. M. J. *et al.* (2007). A distributed stream temperature model using high resolution temperature observations. *Hydrol. Earth Sys. Sci.*, **11**, 1469–1480.

White, W. N. (1930). Preliminary report on the ground-water supply of Mimbres Valley, New Mexico. New Mexico State Engineer Office, Ninth Biennial Report, 1928–30, 133–151.

White, W. N. (1932). A method of estimating ground-water supplies based on discharge by plants and evaporation from soil: results of investigations in Escalante Valley, Utah. US Geological Survey Water-Supply Paper 659-A.

Wierenga, P. J., Hendrickx, J. M. H., Nash, M. H., Ludwig, J. and Daugherty, L. A. (1987). Variation of soil and vegetation with distance along a transect in the Chihuahuan Desert. *J. Arid Environ.*, **13**, 53–63.

Wigley, T. M. L., Plummer, L. N. and Pearson, F. J. Jr. (1978). Mass transfer and carbon isotope evolution in natural water systems. *Geochimica et Cosmochimica Acta*, **42**, 1117–1139.

Wilkison, D. H., Blevins, D. W., Kelly, B. P. and Wallace, W. C. (1994). Hydrology and water quality in claypan soil and glacial till at the Missouri Management Systems Evaluation Area near Centralia, Missouri, May 1991 to September 1993. US Geological Survey Open-File Report 94–705.

Wilson, G. B. and McNeill, G. W. (1997). Noble gas temperatures and excess air component. *Applied Geochem.*, **12**, 747–762.

Wilson, J. L. and Guan, H. (2004). Mountain-block hydrology and mountain-front recharge. In *Groundwater Recharge in a Desert Environment. The Southwestern United States*, ed. J. F. Hogan, F. M. Phillips and B. R. Scanlon. Washington, DC: American Geophysical Union, 113–138.

Wilson, K. B., Hanson, P. J., Mulholland, P. J., Baldocchi, D. D. and Wullschleger, S. D. (2001). A comparison of methods for determining forest evapotranspiration and its components: sap-flow, soil water budget, eddy covariance and catchment water balance. *Agric. Forest Met.*, **106**, 153–168.

Wilson, L. G. (1980). Regional recharge research for southwest alluvial basins. Tucson, Arizona: Water Resources Research Center.

Wilson, R. D. and Mackay, D. M. (1996). SF_6 as a conservative tracer in saturated media with high intragranular porosity or high organic carbon content. *Ground Water*, **34**, 241–249.

Winter, T. C. (2001). The concept of hydrologic landscapes. *J. Amer. Water Resour. Assoc.*, **37**, 335–349.

Winter, T. C., Harvey, J. W., Franke, O. L. and Alley, W. M. (1998). Ground water and surface water; a single resource. US Geological Survey Circular 1139.

Winter, T. C., Labaugh, J. W. and Rosenberry, P. O. (1988). The design and use of a hydraulic potentiomanometer for direct measurement of differences in hydraulic head between groundwater and surface water. *Limnol. Oceanogr.*, **33**, 1209–1214.

Wolf, A., Saliendra, N., Akshalov, K., Johnson, D. A. and Laca, E. (2008). Effects of different eddy covariance correction schemes on energy balance closure and comparisons with the modified Bowen ratio system. *Agric. Forest Met.*, **148**, 942–952.

Wolock, D. M. (2003). Estimated mean annual natural ground-water recharge in the conterminous United States. US Geological Survey Open-File Report 2003-311.

Wood, W. W. (1999). Use and misuse of the chloride-mass balance method in estimating ground water recharge. *Ground Water*, **37**, 2–3.

Wood, W. W. and Sanford, W. E. (1995). Chemical and isotopic methods for quantifying ground-water recharge in a regional, semiarid environment. *Ground Water*, **33**, 458–468.

World Meteorological Organization (1983). *Guide to Meteorological Instruments and Methods of Observation*. Geneva: World Meteorological Organization.

Xiang, Y., Sykes, J. F. and Thomson, N. R. (1993). A composite L1 parameter estimator for model fitting in groundwater flow and solute transport simulation. *Water Resour. Res.*, **29**, 1661–1673.

Yager, R. M. (1996). Simulated three-dimensional ground-water flow in the Lockport group, a fractured-dolomite aquifer near Niagara Falls, New York. US Geological Survey Water-Supply Paper 2487.

Yang, Y., Lerner, D. N., Barrett, M. H. and Tellam, J. H. (1999). Quantification of groundwater recharge in the city of Nottingham, UK. *Environ. Geol.*, **38**, 183–198.

Young, M. B., Gonneea, M. E., Fong, D. A., Moore, W. S., Herrera-Silveira, J. and Paytan, A. (2008). Characterizing sources of groundwater to a tropical coastal lagoon in a karstic area using radium isotopes and water chemistry. *Marine Chemistry*, **109**, 377–394.

Zerle, L., Faestermann, T., Knie, K. *et al.* (1997). The Ca-41 bomb pulse and atmospheric transport of radionuclides. *J. Geophys. Res. D*, **102**, 19517–19527.

Zhang, L., Dawes, W. R., Hatton, T. J., Reece, P. H., Beale, G. T. H. and Packer, I. (1999). Estimation of soil moisture and groundwater recharge using the TOPOG IRM model. *Water Resour. Res.*, **35**, 149–161.

Zheng, C. and Bennett, G. D. (2002). *Applied Contaminant Transport Modeling*, 2nd edn. New York: John Wiley and Sons, Inc.

Zheng, C., Lin, J. and Maidment, D. R. (2006). Internet data sources for ground water modeling. *Ground Water*, **44**, 136–138.

Zheng, C. and Wang, P. P. (1996). Parameter structure identification using tabu search and simulated annealing. *Adv. Water Resour.*, **19**, 215–224.

Zhu, C. (2000). Estimate of recharge from radiocarbon dating of groundwater and numerical flow and transport modeling. *Water Resour. Res.*, **36**, 2607–2620.

Zhu, C., Winterle, J. R. and Love, E. I. (2003). Late Pleistocene and Holocene groundwater recharge from the chloride mass balance method and chlorine-36 data. *Water Resour. Res.*, **39**, SBH41-SBH415.

Zyvoloski, G., Dash, Z. and Kellar, S. (1997). FEHM 1.0: Finite element heat and mass transfer code. Los Alamos National Laboratory Report LA-12062. Los Alamos, NM.

Index

Printed in the United States
By Bookmasters